A Course on

Abstract

Second Edition

A Course on

Abstract

Algebra

Second Edition

Minking Eie
Shou-Te Chang

National Chung Cheng University, Taiwan

W World Scientific

NEW JERSEY · LONDON · SINGAPORE · BEIJING · SHANGHAI · HONG KONG · TAIPEI · CHENNAI · TOKYO

Published by

World Scientific Publishing Co. Pte. Ltd.

5 Toh Tuck Link, Singapore 596224

USA office: 27 Warren Street, Suite 401-402, Hackensack, NJ 07601

UK office: 57 Shelton Street, Covent Garden, London WC2H 9HE

Library of Congress Cataloging-in-Publication Data

Names: Eie, Minking, 1952–   author. | Chang, Shou-Te, author.

Title: A course on abstract algebra / by Minking Eie (National Chung Cheng University, Taiwan),
    Shou-Te Chang (National Chung Cheng University, Taiwan).

Description: Second edition. | New Jersey : World Scientific, 2017. | Includes index.

Identifiers: LCCN 2017038817 | ISBN 9789813229624 (hardcover : alk. paper)

Subjects: LCSH: Algebra, Abstract. | Algebra, Abstract--Textbooks.

Classification: LCC QA162 .E375 2017 | DDC 512/.02--dc23

LC record available at https://lccn.loc.gov/2017038817

British Library Cataloguing-in-Publication Data

A catalogue record for this book is available from the British Library.

Printed in Singapore

Preface

During our many years of teaching the undergraduate course in abstract algebra, we have encountered many excellent textbooks at the introductory level. Why write another one?

The most important reason is that we want to provide a textbook for those students whose mother tongue is not English. Then why don't we just write one in Chinese for our own students? We hope to inspire more students into pursuing a mathematical (or other academic) career, and this will eventually involve reading books and writing papers in English. We feel it is maybe easier to start doing it in elementary courses when the mathematics is not overwhelmingly difficult. We aim to use simple and straightforward English to explain basic notions and terminologies in algebra. The process of abstraction and generalization in abstract algebra already seem foreign enough to most beginners. We don't want the language to be another barrier. Even though we write this book with our own students in mind, we believe that this book will also be beneficial to native speakers of English or other non-Chinese speaking readers who want to learn some abstract algebra.

The existing textbooks tend to be too long, often easily exceeding five or six hundred pages. Some authors put in a lot of information, either in explaining in great detail, or adding historical notes, or even providing additional chapters on applied algebra. As instructors, we greatly appreciate and have benefited from these efforts. However, we seldom find time during the semesters to convey to our students all this extra material. Most of our students, being not fluent in English, are intimidated by the long explaining sections. They never really seem to notice that how much precious knowledge is near at hand. This is why we want to write one that is not so densely packed with extra information, but do contain the most essential reading for any beginning student of algebra. We hope that the size of this book will not be so daunting to our readers. Mastering this book will more than adequately prepare anyone who wants to continue into more advanced courses in any field of algebra.

In our experience of studying mathematics, there is nothing more important than doing the exercises. Hardworking students often find reading explanations and proofs easy. However, one cannot really learn mathematics without doing the exercises. (This does not include reading other people's solutions of the exercises.) Students are encouraged to try the exercises immediately after finishing reading the text for a full learning experience. Exercises will help test if one really has acquired enough understanding of the material. Exercises will also help students discover new problems for themselves, which is an essential process in higher learning. Indeed, we cannot stress enough how important doing exercises is! Especially in this book, some details are left as exercises, so that students will have a feeling of participation in "doing" this book. You do while you learn, and you learn while you do. Some of the exercises are designed to practise the concepts just discussed in the text, while some exercises are designed to introduce new concepts which we may or may not discuss in the coming chapters. Whichever the case these materials will still be important and interesting in their own right. Most of the exercises are designed to help students develop a deeper understanding of the content of the book. Finally, there are a small amount of exercises which are meant to be challenging! We are sure students will feel greatly rewarded after they have solved these more difficult problems.

In this book there are a fair amount of exercises which we hope will not be intimidating and repetitive to our readers. Well designed exercises are extremely helpful to students in becoming familiar with the material.

They also stimulate students' interests in wanting to learn more. We hope that our problem sets are well chosen and not excessive so that they fill the purpose of well designed exercises.

This textbook is designed for a fast-paced two-semester (one-year) course or a leisurely three-semester course. Depending on the pace of the course and certainly on the length of the school terms, the instructor might choose to skip some of the materials. If so, §10.3, Chapter 18, Chapter 19, §20.2 and §22.3 can be skipped without affecting the understanding of the rest of the book. Chapters 10–12, 18, 19, 23, 24 cover more advanced materials which are better suited for a third semester. Under time constraints, the proofs of some of the major theorems, for example, Sylow's Theorems, Structure Theorem of Finite Abelian Groups and Fundamental Theorem of Galois Pairing, can definitely be skipped.

The authors of this book work in analytic number theory and in commutative algebra, respectively. We hope we have contributed in this book a little bit of flavor in each discipline. We also hope our readers will have a great experience in using this book and enjoy learning abstract algebra as we did when we were students ourselves.

For this second edition, three new chapters are added. There are some changes and corrections made to the old chapters and the problem sets. We are sure this book still contains typos and small errors no matter how hard we try to avoid them. Please send us emails if you have any questions or comments. Any reader's input will be greatly appreciated. We can be reached at `minking@math.ccu.edu.tw` and `shoute.chang@gmail.com`.

At last, we would like to thank our editor at World Scientific, Ms Kwong Lai Fun, for being kind, helpful and tolerant. We would also like to dedicate this book to our families for their love and support.

Minking Eie
Shou-Te Chang
June 6, 2017

About the Authors

Mingking Eie is currently a Professor of the Department of Mathematics, National Chung Cheng University in Chiayi, Taiwan. He received his Ph.D. in Mathematics in 1982 from the University of Chicago and began his career as an associate research fellow at the Institute of Mathematics, Academia Sinica of Taiwan. He was promoted to a research fellow in 1986 and became an Outstanding Researcher of National Science Council of Taiwan at the same time. In 1991, he left Academia Sinica and took up the current position. More than 50 research papers have been published about theory of modular forms of several variables, theory of Jacobi forms over Cayley numbers, Bernoulli identities and multiple zeta values. He has also published several sets of mathematical textbooks for senior high school students in Taiwan and a Calculus textbook in English for Business undergraduates.

Shou-Te Chang is currently an Associate Professor at the Department of Mathematics, National Chung Cheng University in Chiayi, Taiwan. She received her Ph.D. in Mathematics in 1993 from the University of Michigan, Ann Arbor. Her research interest is in commutative algebra and homological algebra. She has published papers on Horrocks' question, generalized Hilbert-Kunz functions and local cohomology.

Contents

CHAPTER 1

Preliminaries

At the turn of the 20th century, David Hilbert (1862–1943) aimed to build a sound foundation for mathematics using *axiomization*. On August 8, 1900, Hilbert gave a historical speech at the Paris Conference of the international congress of mathematicians. He proposed 23 problems as the goal of the mathematical research of the 20th century. He aimed to answer the following questions: Is mathematics consistent? Is mathematics complete? Is mathematics decidable? Although this great mission eventually failed, Bertrand Russell (1872–1970) and Alfred North Whitehead (1861–1947) did pioneer work on set theory and logics in their great book *Principia Mathematica*.

Basic notions and terminologies such as sets, functions and relations are discussed without delving into great details. These are basic tools for the whole book.

1.1   Basic Ideas of Set Theory

We will not plunge into a formal definition of sets and its subsequent philosophical discussions. We will simply say that a **set** is a collection of "objects", which we will call **elements**. When given a set, there will be no ambiguity on what elements are contained in it. There are usually two methods to explicitly express a set. One is to recount every element contained in it, and the other is to describe the elements' properties using a logical and verifiable expression. For example,

$$S = \{\, 1, 2, 3, \ldots, 99 \,\}$$

and

$$S = \{\, x : x \text{ is a positive integer and } 1 \leq x \leq 99 \,\}.$$

both describe the same set.

Below we summarize some basic notions and terminologies concerning sets.

(1) **Elements and sets**. We write $x \in S$ to denote that x is an element in a set S. In this case we also that x **belongs to** S. For example, let $S = \{\, 1, 2, 3, a, b \,\}$, then we have

$$1 \in S, \quad 2 \in S, \quad 3 \in S, \quad a \in S \quad \text{and} \quad b \in S.$$

However, 4 is not an element in S and we write $4 \notin S$.

A set without any element is called an **empty set** and is denoted by $\varnothing$.

(2) **Subsets**. We say that a set A is a **subset** of another set B if every element in A is also an element in B. In other words,

$$x \in A \quad \Longrightarrow \quad x \in B.$$

In this case, we write $A \subseteq B$ and say that A is **contained in** B. For example, we have $\mathbb{N} \subseteq \mathbb{Z} \subseteq \mathbb{Q} \subseteq \mathbb{R} \subseteq \mathbb{C}$, where

- $\mathbb{N} =$ the set of natural numbers;
- $\mathbb{Z} =$ the set of integers;

- $\mathbb{Q}$ = the set of rational numbers;
- $\mathbb{R}$ = the set of real numbers;
- $\mathbb{C}$ = the set of complex numbers.

Here is another example. If

$$A = \{1, 3, 5, 7\} \quad \text{and} \quad B = \{1, 2, 3, 4, 5, 6, 7\},$$

then we have the relations $A \subseteq A$, $B \subseteq B$ and $A \subseteq B$. Note that B is not a subset of A, denoted $B \not\subseteq A$. In other words, there is an element x in B which is not in A.

(3) **The equality of sets**. Two sets A and B are **equal** if they contain the same elements. To show that $A = B$ is equivalent to showing both

$$A \subseteq B \quad \text{and} \quad B \subseteq A.$$

If $A \subseteq B$ but $A \neq B$, we will write that $A \subsetneq B$ and say that A is a **proper** subset of B.

Example 1.1.1. Let

$$A = \{2a + 3b : a, b \in \mathbb{Z}\}.$$

Show that $A = \mathbb{Z}$, the set of integers.

Proof. Obviously we have $A \subseteq \mathbb{Z}$ since $2a + 3b$ is an integer when a and b are integers. It suffices to show that $\mathbb{Z} \subseteq A$. For each integer n in $\mathbb{Z}$, we have $n = 2(-n) + 3n \in A$. This shows that $\mathbb{Z} \subseteq A$. We conclude that $A = \mathbb{Z}$. $\qquad\qquad\square$

Next we introduce three important operations among sets. In the following discussion, let A, B and C be subsets of a set S.

(1) The **intersection** of A and B is defined as

$$A \cap B = \{x \in S : x \in A \text{ and } x \in B\}.$$

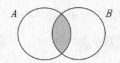

The intersection $A \cap B$

For example, if $A = \{a, b, c, d\}$ and $B = \{b, c, e, f\}$, then $A \cap B = \{b, c\}$. Note that $A \cap B$ contains elements from both A and B, and we have

$$(A \cap B) \subseteq A \quad \text{and} \quad (A \cap B) \subseteq B.$$

If A and B have no common elements, then $A \cap B = \varnothing$. In this case, we say A and B are **disjoint**.

(2) The **union** of A and B is defined as

$$A \cup B = \{x \in S : x \in A \text{ or } x \in B\}.$$

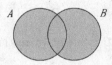

The union $A \cup B$

For example, if $A = \{1, 3, 4\}$ and $B = \{2, 4, 5, 6\}$, then $A \cup B = \{1, 2, 3, 4, 5, 6\}$. Note that $A \cup B$ contains elements from either A or B, and we have

$$A \subseteq (A \cup B) \quad \text{and} \quad B \subseteq (A \cup B).$$

In case when the two sets A and B are disjoint, we also call the union $A \cup B$ the **disjoint union** of A and B, and we denote it by $A \mathbin{\dot\cup} B$.

(3) The **difference** of A and B is defined as

$$A \setminus B = \{x \in S : x \in A \text{ and } x \notin B\}.$$

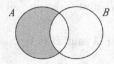

The difference $A \setminus B$

For example, if $A = \{-2, -1, 3, 4\}$ and $B = \{-1, 4, 5\}$, then $A \setminus B = \{-2, 3\}$, and $B \setminus A = \{5\}$. Note that

$$A \setminus B \subseteq A \quad \text{and} \quad (A \setminus B) \cap B = \varnothing.$$

Obviously, the operation union (or intersection, respectively) can be generalized naturally to be performed on three or even more sets. A set of sets is often called a **class** or a **family** of sets. A family of sets is often indexed by another set Λ such as $\mathbb{N}$ or $\mathbb{Z}$:

$$(1.1.1) \qquad \mathscr{F} = \{A_i : i \in \Lambda\} = \{A_i\}_{i \in \Lambda}.$$

We call Λ the **index set** of the family $\mathscr{F}$.

Let S be a set. We may extend the definitions of the union and intersection of sets to a (possibly infinite) family $\mathscr{F}$ of subsets of S as

$$\bigcup \mathscr{F} = \{x \in S : x \in A \text{ for some } A \in \mathscr{F}\};$$
$$\bigcap \mathscr{F} = \{x \in S : x \in A \text{ for all } A \in \mathscr{F}\}.$$

For the family in (1.1.1), it is customary to write

$$\bigcup \mathscr{F} = \bigcup_{i \in \Lambda} A_i \quad \text{and} \quad \bigcap \mathscr{F} = \bigcap_{i \in \Lambda} A_i.$$

Exercises 1.1

1. Suppose that the two sets A and B satisfy

$$A \setminus B = \{-2, 1, 3\}, \quad B \setminus A = \{5, 6\} \quad \text{and} \quad A \cap B = \{0, 7\}.$$

Find A, B and $A \cup B$.

2. Prove or disprove the following statements:

   (a) $A \cup B = A \cup C \implies B = C$;

   (b) $A \cap B = A \cap C \implies B = C$;

   (c) $A \subseteq B, B \subseteq C \implies A \subseteq C$;

(d) $A \subseteq B \implies (A \cup C) \subseteq (B \cup C)$ for any set C.

3. Let A, B and C be subsets of a set S. Prove the following properties on union and intersection:

    (a) $A \cup \varnothing = A$ and $A \cup S = S$;

    (b) $A \cap \varnothing = \varnothing$ and $A \cap S = A$;

    (c) *Associativity:* $(A \cup B) \cup C = A \cup (B \cup C)$ and $(A \cap B) \cap C = A \cap (B \cap C)$.

4. Let A be a subset and let $\{S_i\}_{i \in \Lambda}$ be a (finite or infinite) family of subsets of a larger set S. Show that the following statements are true.

    (a) *Distributivity:* $A \cup (\bigcap_{i \in \Lambda} S_i) = \bigcap_{i \in \Lambda}(A \cup S_i)$ and $A \cap (\bigcup_{i \in \Lambda} S_i) = \bigcup_{i \in \Lambda}(A \cap S_i)$.

    (b) *De Morgan laws:* $A \setminus (\bigcup_{i \in \Lambda} S_i) = \bigcap_{i \in \Lambda}(A \setminus S_i)$ and $A \setminus (\bigcap_{i \in \Lambda} S_i) = \bigcup_{i \in \Lambda}(A \setminus S_i)$.

5. Define 2^S to be the set of all subsets of S, called the **power set** of S. For two sets A and B in 2^S, we define the **symmetric difference** or the **Boolean sum** of A and B to be

$$A + B = (A \setminus B) \cup (B \setminus A).$$

Prove the following assertions:

    (a) $A + (B + C) = (A + B) + C$;

    (b) $A + \varnothing = \varnothing + A = A$;

    (c) $A + A = \varnothing$.

6. Let $A = \{4m + 6n \in \mathbb{Z} : m, n \in \mathbb{Z}\}$ and $B = \{6m + 8n \in \mathbb{Z} : m, n \in \mathbb{Z}\}$. Show that $A = B$.

7. The **Cartesian product** of two sets A and B is defined as

$$A \times B = \{(a,b) : a \in A,\ b \in B\}.$$

Show that the following statements are true:

    (a) $A \times (B \cup C) = (A \times B) \cup (A \times C)$;

    (b) $(A \setminus B) \times C = (A \times C) \setminus (B \times C)$.

1.2   Functions

Let A and B be sets. A **function** $f\colon A \to B$ consists of three parts:

(i) the **domain** A,

(ii) the **codomain** B, and

(iii) a rule assigning to each element $x \in A$ a unique element $f(x) \in B$.

A function is also called a **map** or a **mapping**. Note that even when two functions appear to operate under the same rule, they are not considered to be equal if they don't have the same domain or codomain.

If $f(a) = b$, the element b is called the **image** of a under f and a is called a **preimage** of b under f. In mathematics, when we use the word "a", it means "at least one", and when we use the word "the", it means the unique one. Each element in the domain of a function f has one and only one image under f, while an element in the codomain of f may have no or one or more than one preimages. The **image** of f is defined to be

$$f(A) = \{\, f(x) \in B : x \in A \,\} \subseteq B.$$

Suppose that $f\colon A \to B$ and $g\colon B \to C$ are two functions, the composition of f and g is a function from A to C defined by

$$\begin{aligned} g \circ f\colon \quad A &\longrightarrow \quad C \\ x &\longmapsto \quad g(f(x)). \end{aligned}$$

For example if $f\colon \mathbb{R} \to \mathbb{R}$ and $g\colon \mathbb{R} \to \mathbb{R}$ are given by

$$f(x) = 2x + 1 \quad \text{and} \quad g(x) = x^2,$$

then

$$(g \circ f)(x) = g(2x + 1) = (2x + 1)^2 = 4x^2 + 4x + 1,$$

and similarly,

$$(f \circ g)(x) = f(g(x)) = 2x^2 + 1.$$

From this example we can see that $g \circ f$ and $f \circ g$ are not equal in general. However, composition of functions is indeed associative. Suppose further given a function $h\colon C \to D$. Then for all $x \in A$, we have

$$(h \circ (g \circ f))(x) = h((g \circ f)(x)) = h(g(f(x)))$$

and

$$((h \circ g) \circ f)(x) = (h \circ g)(f(x)) = h(g(f(x))).$$

This shows that

$$(1.2.1) \qquad\qquad h \circ (g \circ f) = (h \circ g) \circ f.$$

A function $f \colon A \to B$ is **one-to-one** or **injective** if it satisfies the condition

$$x_1 \neq x_2 \implies f(x_1) \neq f(x_2),$$

or the equivalent condition

$$f(x_1) = f(x_2) \implies x_1 = x_2.$$

Proposition 1.2.1. *The composition of two injective functions is injective.*

Proof. Suppose that the functions $f \colon A \to B$ and $g \colon B \to C$ are one-to-one. Let $(g \circ f)(x_1) = (g \circ f)(x_2)$. We want to show that $x_1 = x_2$. By definition, we have

$$g(f(x_1)) = g(f(x_2)).$$

As g is one-to-one, it follows that $f(x_1) = f(x_2)$. Since f is also one-to-one, it implies that $x_1 = x_2$. Thus, $g \circ f$ is one-to-one. $\qquad\square$

A function $f \colon A \to B$ is **onto** or **surjective** if $f(A) = B$. This means that for any $b \in B$, there exists $a \in A$ such that $f(a) = b$.

Proposition 1.2.2. *The composition of two surjective functions is surjective.*

Proof. Suppose that $f \colon A \to B$ and $g \colon B \to C$ are onto. Then

$$f(A) = B \quad \text{and} \quad g(B) = C.$$

It follows that

$$g \circ f(A) = g(f(A)) = g(B) = C.$$

This proves that $g \circ f$ is onto. $\qquad\square$

A function $f \colon A \to B$ is called a **one-to-one correspondence** or a **bijective** function if f is both one-to-one and onto. Now Propositions 1.2.1 and 1.2.2 together give us the following result.

Corollary 1.2.3. *The composition of two bijective functions is bijective.*

There is a special property regarding bijectivity unique to finite sets.

Proposition 1.2.4. *Let S be a finite set and let $f\colon S \to S$ be a function.*

(a) *If f is onto then f is one-to-one and hence bijective.*

(b) *If f is one-to-one then f is onto and hence bijective.*

This proposition is basically a restatement of

Pigeonhole Principle. *Let $m > n$ be two positive integers. If one is to put m pigeons into n pigeonhole, then there is at least one hole containing two or more pigeons. In other words, let A be a set of m elements and B be a set of n elements. Then there are no injective functions from A to B.*

We leave the proof of Proposition 1.2.4 as an exercise (see Exercise 5).

At last we give a different classification of bijective maps.

Let A and B be sets. We define the **identity function** or the **identity map** from A to A, denoted 1_A, to be the function sending $a \in A$ to a itself. Let $f\colon A \to B$ be a function. Then

$$(f \circ 1_A)(x) = f(1_A(x)) = f(x),$$
$$(1_B \circ f)(x) = 1_B(f(x)) = f(x)$$

for all $x \in A$. It follows that

(1.2.2) $$f \circ 1_A = f = 1_B \circ f.$$

If $f\colon A \to B$ is a function, then we say $g\colon B \to A$ is an **inverse function**, or simply an **inverse**, of f if $g \circ f = 1_A$ and $f \circ g = 1_B$. It is an easy exercise to show that the inverse function is unique if it exists (see Exercise 2). Hence, we can denote the inverse function of f by f^{-1}.

Theorem 1.2.5. *The function $f\colon A \to B$ is bijective if and only if f has an inverse function.*

Proof. To show the "if" part, let g be the inverse function of f. Suppose $f(a) = f(a')$. Then $a = g(f(a)) = g(f(a')) = a'$. This shows that f is one-to-one. On the other hand, let $b \in B$. Then $b = f(g(b))$ is in the image of f. Hence f is also onto.

To show the "only if" part, assume that f is bijective. Define the function $g\colon B \to A$ as follows. For each $b \in B$, there is exactly one $a \in A$ such that $f(a) = b$. Define $g(b) = a$.

Let $a \in A$. Then $(g \circ f)(a) = g(f(a)) = g(b) = a$ if $f(a) = b$. Hence we have $g \circ f = 1_A$. On the other hand, let $b \in B$. Find $a \in A$ such that $f(a) = b$. Then $(f \circ g)(b) = f(g(b)) = f(a) = b$. This implies that $f \circ g = 1_B$. We have shown that g is the inverse of f. $\qquad\square$

Sometimes when one wants to show a function is bijective, it is easier to construct its inverse function than actually show that the function is one-to-one and onto.

To simplify notation, $f \circ g$ is often denoted as fg when no confusion arises.

Exercises 1.2

1. Let $f\colon A \to B$ and $g\colon B \to C$ be two functions.

   (a) If $g \circ f$ is one-to-one, show that f is one-to-one.

   (b) Give an example of f and g such that $g \circ f$ is one-to-one but g is not.

   (c) If $g \circ f$ is onto, show that g is onto.

   (d) Give an example of f and g such that $g \circ f$ is onto but f is not.

2. Show that the inverse of a function is unique if it exists.

3. Suppose that f and g are bijective functions from S into itself satisfying $g \circ f(x) = x$ for all $x \in S$. Show that $f \circ g(x) = x$ for all $x \in S$.

4. Suppose given a function $f\colon S \to T$ and let $A \subseteq S$ and $B \subseteq T$. We define the **image** of A under f to be

$$f(A) = \{\, b \in T : b = f(a) \text{ for some } a \in A \,\},$$

and the **preimage** of B under f to be

$$f^{-1}(B) = \{\, a \in S : f(a) \in B \,\}.$$

In particular, $f^{-1}(\{b\})$ is also written as $f^{-1}(b)$ when no confusion arises.[1] Prove the following assertions:

(a) $A \subseteq A' \Rightarrow f(A) \subseteq f(A')$ and $B \subseteq B' \Rightarrow f^{-1}(B) \subseteq f^{-1}(B')$.

(b) $A \subseteq f^{-1}(f(A))$ and $f(f^{-1}(B)) \subseteq B$.

(c) $f^{-1}(B \cup B') = f^{-1}(B) \cup f^{-1}(B')$ and $f^{-1}(B \cap B') = f^{-1}(B) \cap f^{-1}(B')$.

(d) $f(A \cup A') = f(A) \cup f(A')$ and $f(A \cap A') \subsetneq f(A) \cap f(A')$.

5. Use Pigeonhole Principle to prove Proposition 1.2.4.

1.3 Equivalence Relations and Partitions

A **relation** R on a set A is a subset of $A \times A$. If R is a relation, it is customary to write aRb when $(a, b) \in R$.

The notion of "relations" is rather general. For example, let $A = \{1, 3, 5, 6\} \subseteq \mathbb{Z}$. The common notion of the relation $<$ on A is the set

$$\{(1,3),\ (1,5),\ (1,6),\ (3,5),\ (3,6),\ (5,6)\}.$$

Of course, the more customary expression for "$(1,3) \in <$" is "$1 < 3$". Similarly, the relation $\leq$ on A is the set

$$\{(1,1),\ (1,3),\ (1,5),\ (1,6),\ (3,3),\ (3,5),\ (3,6),\ (5,5),\ (5,6),\ (6,6)\}.$$

For another example let $S = \{0, 1\}$. The relation $\supseteq$ on 2^S is the set

$$\Big\{ (\varnothing, \varnothing),\ (\{0\}, \varnothing),\ (\{1\}, \varnothing),\ (\{0,1\}, \varnothing),\ (\{0\}, \{0\}),$$

$$(\{0,1\}, \{0\}),\ (\{1\}, \{1\}),\ (\{0,1\}, \{1\}),\ (\{0,1\}, \{0,1\}) \Big\}.$$

One can make up all kinds of relations on a set. For example, $=$, $\leq$, $\geq$, $>$ and $<$ are relations on real numbers $\mathbb{R}$. Even "no relation" is a relation if one takes the relation to be the empty set.

---

[1] The notation $f^{-1}(b)$ may also stand for the value of the inverse function of f at b. However, in this case f must be bijective.

Definition 1.3.1. An **equivalence relation** E on a set A is a relation on A such that the following conditions are satisfied for all $a, b, c \in A$:

(i) (reflexivity) aEa;

(ii) (symmetry) $aEb \implies bEa$;

(iii) (transitivity) aEb and $bEc \implies aEc$.

The equality $=$ on numbers is an equivalence relation while $\leq, \geq, >$ and $<$ are not equivalence relations. For example, the relation $\leq$ satisfies reflexivity and transitivity but not symmetry. The statement $2 \leq 3$ is true but the statement $3 \leq 2$ is false. In fact, an equivalence relation is a generalization of equality. The symbols $\sim, \approx, \equiv, \simeq$ or $\cong$ are often used to denote an equivalence relation.

Definition 1.3.2. Let E be an equivalence relation on S. Let $a \in S$. Define
$$[a] = \{\, b \in S : bEa \,\}.$$

The set $[a]$ is called the **equivalence class** of a (with respect to E) and a is called a **representative** of this class. We often use S/E to denote the set of the equivalence classes with respect to E.

Example 1.3.3. Let $a, b \in \mathbb{Z}$. Define $a \equiv b \Leftrightarrow 2 \,|\, a + b$. Then $\equiv$ is an equivalence relation on $\mathbb{Z}$. Let a, b and c be arbitrary integers. Since $2 \,|\, a + a = 2a$, we have $a \equiv a$. If $a \equiv b$, then $2 \,|\, a + b = b + a$ and we have $b \equiv a$. If $a \equiv b$ and $b \equiv c$, then $2 \,|\, a + b$ and $2 \,|\, b + c$. This implies that $2 \,|\, (a + b) + (b + c) = a + c + 2b$. It follows that $2 \,|\, a + c$ and $a \equiv c$. We conclude that $\equiv$ is an equivalence relation.

Under $\equiv$, the set $\mathbb{Z}$ is divided into exactly two equivalence classes: the set of even integers and the set of odd integers.

An equivalence relation naturally partitions a set.

Definition 1.3.4. A **partition** of a set S is a collection of *disjoint nonempty* subsets of S whose union is S. More precisely, a collection of subsets $\{A_i\}_{i \in \Lambda}$ of S is a partition if the following three conditions are satisfied:

(i) $A_i \neq \varnothing$ for all $i \in \Lambda$;

(ii) $A_i \cap A_j = \varnothing$ for $i \neq j$ in Λ;

(iii) $S = \bigcup_{i \in \Lambda} A_i$.

For example, let $S = \{1, 2, 3, 4, 5, 6\}$. Then $\Big\{\{1, 3, 5\}, \ \{2, 4\}, \ \{6\}\Big\}$ is a partition of S while $\Big\{\{1, 3, 5\}, \ \{2, 4\}, \ \{6, 3\}\Big\}$ is not.

If a family of subsets forms a partition of the set S, every element of S belongs to exactly one member of this family. If one takes two subsets A and B from a partition of S, then either $A = B$ or $A \cap B = \varnothing$.

Proposition 1.3.5. *Let E be an equivalence relation on S. Then the equivalence classes with respect to E form a partition of S.*

Proof. Take any $a \in S$. Then $a \in [a]$. It follows that $[a] \neq \varnothing$ for all $a \in A$. It is also clear that $S = \bigcup_{a \in A} [a]$. It remains to show that if we take any $a, b \in S$, then either $[a] = [b]$ or $[a] \cap [b] = \varnothing$. Suppose $[a] \cap [b] \neq \varnothing$. Find $c \in [a] \cap [b]$. Then cEa and cEb. If $d \in [a]$, then dEa. By symmetry of E, we have aEc. By transitivity of E, we have dEb and thus $d \in [b]$. We have shown that $[a] \subseteq [b]$. Similarly, we also have $[b] \subseteq [a]$. Hence $[a] = [b]$. $\square$

In Exercise 3 we will see that a partition will give rise to an equivalence relation. "Equivalence relation" and "Partition" are the two faces of the same notion (see Exercise 4).

Exercises 1.3

1. Let $f \colon S \to T$ be a function. Then f defines a relation $\sim_f$ on S by letting
$$a \sim_f b \iff f(a) = f(b), \qquad \text{for } a, b \in S.$$
   Show that $\sim_f$ is an equivalence relation. Describe the equivalence classes.

2. Define a relation on $\mathbb{Z}$ by letting aRb if and only if $ab \geq 0$. Is R an equivalence relation on $\mathbb{Z}$? If yes, describe its equivalence classes.

3. Let π be a partition of S. Define a relation $\sim_\pi$ on S by letting $a \sim_\pi b$ if and only if a and b belong to the same set in π. Show that $\sim_\pi$ is an equivalence relation on S.

4. Let E be an equivalence relation on S and let π_E be the partition given in Proposition 1.3.5. Show that $\sim_\pi = E$ where $\pi = \pi_E$. Let π be an arbitrary partition of S. Show that $\pi_E = \pi$ where $E = \sim_\pi$.

1.4   A Note on Natural Numbers

Most students probably believe that they are extremely familiar with the set of natural numbers $\mathbb{N}$. However, it is quite tedious to go through the formal construction of the natural numbers and establish the subsequent facts and properties regarding them. And many students starting the study of abstract algebra will be puzzled why on earth would anyone need to go through all this trouble. Hence, we will skip this part of work, and hopefully one day the reader would acquire enough mathematical maturity to feel the need to fill the gap herself or himself.

The interesting thing is that even mathematicians have different opinions regarding the notation $\mathbb{N}$. Some take it as the set of all positive integers $\{1, 2, 3, \ldots\}$, while some take it as the set $\{0, 1, 2, 3, \ldots\}$. To set theorists, it is more natural to use the latter. However, it does not really matter in most situations, as long as one knows where one stands. In this book, we will assume

$$\mathbb{N} = \{1, 2, 3, 4, 5, \ldots\}$$

is the set of positive integers. Hence, the notations $\mathbb{N}$ and $\mathbb{Z}_+$ stand for the same set.

In this section, we would like to review some of the most important properties of the natural numbers without giving the proofs. Interested readers are encouraged to read books on set theory for more details. These properties will be used in many proofs in this book. If not for the discussions in this section, many students would probably never realize that they are using them.

Theorem 1.4.1 (The Well-ordering Principle for Natural Numbers). *Every nonempty subset of $\mathbb{N}$ contains a least number.*

Theorem 1.4.2 (The First Principle of Mathematical Induction). *Let S be a subset of $\mathbb{N}$ which contains 1. Suppose S is such that $n \in S$ implies $n + 1 \in S$. Then $S = \mathbb{N}$.*

We present here another commonly used form of mathematical induction.

Theorem 1.4.3 (The Second Principle of Mathematical Induction). *Let S be a subset of $\mathbb{N}$ which contains 1. Suppose S is such that $x \in S$ for all $1 \leq x \leq n$ implies $n + 1 \in S$. Then $S = \mathbb{N}$.*

We will assume that readers are familiar with addition and multiplication as well as the order in $\mathbb{N}$. Finally, we add a few words on *cardinality*.

Definition 1.4.4. We say that two sets A, B are of the same **cardinality**, denoted $|A| = |B|$, if there is a bijective map from A onto B. We use $|A|$ to denote the cardinality of A. When A is a finite set, $|A|$ is simply the number of the elements in A.

It is not always easy to construct a bijective map between two sets of the same cardinality.

Definition 1.4.5. If there is an injective map from A to B, then we say B **dominates** A, and we will write $A \preceq B$.

Theorem 1.4.6 (Schröder-Bernstein Theorem). *If $A \preceq B$ and $B \preceq A$ then $|A| = |B|$.*

Hence, to show two sets A and B are of the same cardinality, one can do so by constructing an injective map from A to B and an injective map from B to A.

Definition 1.4.7. We say that an infinite set S is **countable** if $|S| = |\mathbb{N}|$, that is, there is a bijective map from $\mathbb{N}$ to S. Otherwise, we say S is **uncountable**.

It is well-known that the set of rational numbers is countable, while the set of real numbers is uncountable.

Exercises 1.4

1. Let S be a set which satisfies the well-ordering principle and let A be a subset of S. Show that A also satisfies the well-ordering principle.

2. (Long division in $\mathbb{N} \cup \{0\}$) Show that for any $a, b \in \mathbb{N} \cup \{0\}$ where $b \neq 0$, there exist $q \in \mathbb{N} \cup \{0\}$ and $0 \leq r < b$ such that $a = bq + r$.

3. Let $S = \{5m + 8n \in \mathbb{N} : m, n \in \mathbb{N}\}$. Find the largest natural number which is not contained in S.

4. Let $\{A_i\}_{i \in \mathbb{N}}$ be a family of countable subsets of a set S. Show that $\bigcup_{i \in \mathbb{N}} A_i$ is countable.

Review Exercises for Chapter 1

1. Let f, g and h be three functions from a set into itself.

   (a) Suppose that f is one-to-one and $f \circ g = f \circ h$. Show that $g = h$.

   (b) Suppose that f is onto and $g \circ f = h \circ f$. Show that $g = h$.

2. Suppose that f and g are both functions from S into itself satisfying $g \circ f(x) = x$ for all $x \in S$. Show that $f \circ g(x) = x$ for all $x \in S$ if S is finite. Is this assertion still true in general?

3. Let f be a bijective function from $S = \{a, b, c\}$ onto itself. Prove that
$$f^6(x) = x$$
for all $x \in S$. Here $f^n = f \circ f \circ \cdots \circ f$ (n times).

4. Let $z = a + bi$ be a complex number where $a, b \in \mathbb{R}$. Remember that the **absolute value** of z is $|z| = \sqrt{a^2 + b^2}$. Define a relation on $\mathbb{C}$ by letting $z \sim w$ if $|z| = |w|$. Show that $\sim$ is an equivalence relation.

5. Let S be the set of all infinite sequences of $\mathbb{N}$, that is,
$$S = \{(a_i)_{i=1}^{\infty} : a_i \in \mathbb{N}\}.$$

   Is S countable?

CHAPTER 2

Algebraic Structure of Numbers

In 1882, Walther von Dyck (1856–1934) and Heinrich Weber (1842–1913) independently gave clear definitions of an abstract group for the first time. However, long before the development of group theory, there are numbers such as integers, rational numbers, real numbers and complex numbers which form natural groups, rings or fields under the operations of the usual addition and multiplication. These algebraic structures of numbers are employed as axioms for abstract groups, rings and fields. In this chapter we review the algebraic structures of integers and rational numbers.

2.1   The Set of Integers

Remember that we use $\mathbb{Z}$ to denote the set of integers. It consists of 0, the positive integers $1, 2, 3, \ldots$, and the negative integers $-1, -2, -3, \ldots$.[1] The set of integers $\mathbb{Z}$ can be thought of as an extension of the set of natural numbers. The purpose is to add all the additive inverses as needed. The reader should be familiar with the operations of addition and multiplication inside $\mathbb{Z}$. These two operations have many nice properties. These operations and properties make up the algebraic structure of $\mathbb{Z}$. Below is a list of properties that we are particularly interested in.

(1) **Associativity** of addition and multiplication. For k, ℓ and $m \in \mathbb{Z}$, we have

$$(k + \ell) + m = k + (\ell + m) \qquad \text{and} \qquad (k\ell)m = k(\ell m).$$

(2) The existence of **identity elements**. For all $k \in \mathbb{Z}$,

$$k + 0 = 0 + k = k \qquad \text{and} \qquad k \cdot 1 = 1 \cdot k = k.$$

The element 0 is called the *additive* identity and 1 is called the *multiplicative* identity.

(3) The existence of additive **inverses**. For each $k \in \mathbb{Z}$,

$$k + (-k) = (-k) + k = 0.$$

The element $-k$ is called the *additive* inverse of k. Note that one must have an identity element first before one tries to find the inverse of an element. For example, if k is a multiplicative inverse of 2, then $k \cdot 2 = 2k = 1$. However, there is no such k. Thus, multiplicative inverses do not exist in general for elements in $\mathbb{Z}$. The only integers with multiplicative inverses are ± 1.

---

[1] Formally, the set of integers is constructed as below. Define an equivalence relation on $\mathbb{N} \times \mathbb{N}$ by letting $(m_1, m_2) \sim (n_1, n_2)$ if and only if $m_1 + n_2 = m_2 + n_1$. Then let $\mathbb{Z} = (\mathbb{N} \times \mathbb{N})/\sim$. Let $m > n$. The equivalence class $[(n, n)]$ is denoted as 0. The equivalence class $[(m, n)]$ is called a positive integer and eventually identified as the natural number $m - n$. The equivalence class $[(n, n)]$ is called a negative integer and is often denoted $-(m - n)$. We leave it to the reader to try to define addition and multiplication in $\mathbb{Z}$.

(4) **Cancellation law**. This is a compensation for having no multiplicative inverses. For all k, ℓ and $m \in \mathbb{Z}$ such that $k \neq 0$,

$$k\ell = km \implies \ell = m.$$

(5) **Commutativity** of addition and multiplication. For k and $\ell \in \mathbb{Z}$,

$$k + \ell = \ell + k \quad \text{and} \quad k\ell = \ell k.$$

(6) **Distributive law**. For any integers k, ℓ and m, we have

$$k(\ell + m) = k\ell + km.$$

In this case, we say that $\cdot$ is distributive with respect to $+$.

The properties above are essential in the study of abstract algebra. When we are given a set with one or two operations, the first thing we do is to check if these properties are satisfied.

For any integer n, let

$$n\mathbb{Z} = \{ nk : k \in \mathbb{Z} \}$$

be the set of integer multiples of n. The set $n\mathbb{Z}$ is a subset of $\mathbb{Z}$ and inherits addition and multiplication from $\mathbb{Z}$. If we take any two elements from $n\mathbb{Z}$, say, nk and $n\ell$ where $k, \ell \in \mathbb{Z}$, we can see that

$$nk + n\ell = n(k + \ell) \quad \text{and} \quad (nk)(n\ell) = n(kn\ell)$$

by distributivity and associativity. Hence the sum and the product of any two elements in $n\mathbb{Z}$ are still in $n\mathbb{Z}$. In this case, we say $n\mathbb{Z}$ is **closed** under addition and multiplication. We can now view addition and multiplication as operations in $n\mathbb{Z}$ and study the algebraic structure in $n\mathbb{Z}$.

Addition and multiplication are both associative in $n\mathbb{Z}$ and multiplication is distributive with respect to addition. In $n\mathbb{Z}$, 0 is the additive identity. However, there is no multiplicative identity in $n\mathbb{Z}$ unless $n = \pm 1$. For any element $nk \in n\mathbb{Z}$, $n(-k) = -(nk)$ is the additive inverse of nk. Again, there are no multiplicative inverse for nk unless $n = \pm 1$ and $k = \pm 1$.

Since $n\mathbb{Z}$ is a subset of $\mathbb{Z}$, inherits the same operations from $\mathbb{Z}$ and have many algebraic properties in common, we may view $n\mathbb{Z}$ as a "sub*structure*" of $\mathbb{Z}$. In the course of studying this textbook, the reader will gradually understand what this means.

A set S equipped with an operation $*$ is often denoted $(S, *)$ to emphasize the algebraic structure. For example, we may use $(\mathbb{Z}, +)$ or $(\mathbb{Z}, \cdot)$ depending on which operation we want to study. Or we may write $(\mathbb{Z}, +, \cdot)$ when we want to study the set $\mathbb{Z}$ with both operations. We may even write $(\mathbb{Z}, +, 0)$ or $(\mathbb{Z}, +, \cdot, 0, 1)$ when we want to emphasize the identities.

Exercises 2.1

1. Let $\mathbb{Z}^*$ be the set of nonzero integers. Answer the following questions.

   (a) Is $\mathbb{Z}^*$ closed under addition? If yes, decide whether the addition is associative and commutative? Is there an additive identity in $\mathbb{Z}^*$?

   (b) Is $\mathbb{Z}^*$ closed under multiplication? If yes, decide whether the multiplication is associative and commutative? Is there a multiplicative identity in $\mathbb{Z}^*$ and is there a multiplicative inverse for every element of $\mathbb{Z}^*$?

   In general, suppose given $(S, +)$ with 0 as the additive identity. We often use the notation S^* to denote the set $S \setminus \{0\}$.

2. Let $U = \{1, -1\} \subseteq \mathbb{Z}$. Is $(U, \cdot)$ closed under multiplication? If yes, decide whether the multiplication is associative and commutative? Is there a multiplicative identity in U and is there a multiplicative inverse for every element of U?

3. Let
$$\mathbb{Z}[i] = \{a + bi : a, b \in \mathbb{Z}\}$$
   be the set of **Gaussian integers**.

   (a) Define an addition in $\mathbb{Z}[i]$ by letting
$$(a + bi) + (c + di) = (a + c) + (b + d)i.$$

Show that $+$ is associative and commutative. Show that there is an additive identity in $\mathbb{Z}[i]$ and there is an additive inverse for every element of $\mathbb{Z}[i]$.

(b) Define an multiplication in $\mathbb{Z}[i]$ by letting

$$(a + bi)(c + di) = (ac - bd) + (ad + bc)i.$$

Show that the multiplication is associative and commutative. Show that there is an multiplicative identity in $\mathbb{Z}[i]$.

(c) Find a subset S of 4 elements in $\mathbb{Z}[i]$ so that S is closed under multiplication and inverses (that is, every element of S has an multiplicative inverse in S).

4. Let a and b be integers. Show that the set

$$S = \{ am + bn : m, n \in \mathbb{Z} \}$$

is closed under $+$. Show that $+$ is associative and commutative in S. Show that there is an additive identity in S and there is an additive inverse for every element of S.

5. Let S be a set.

(a) Is the binary operation $\cup$ associative and commutative on 2^S. Is there an identity in 2^S and is there an inverse for every element of 2^S?

(b) Is the binary operation $\cap$ associative and commutative on 2^S. Is there an identity in 2^S and is there an inverse for every element of 2^S?

(c) Is $\cup$ distributive with respect to $\cap$? Is $\cap$ distributive with respect to $\cup$?

2.2   Congruences of Integers

Let n be a positive integer. Two integers a and b are **congruent modulo** n, denoted

$$a \equiv b \pmod{n},$$

if they have the same remainder when divided by n. To be more formal, we define

$$a \equiv b \pmod{n} \quad \Longleftrightarrow \quad n \mid a - b \quad \Longleftrightarrow \quad a - b \in n\mathbb{Z}.$$

For example, we have the following congruences

$$5 \equiv 7 \pmod{2}, \qquad 7 \equiv 19 \pmod{3},$$
$$-4 \equiv 16 \pmod{5} \quad \text{and} \quad -7 \equiv 19 \pmod{13}.$$

Below are some simple properties concerning congruences.

Proposition 2.2.1. *Let n be a positive integer.*

(a) *If $a \equiv b \pmod{n}$ and $c \equiv d \pmod{n}$, then*

$$a + c \equiv b + d \pmod{n} \quad \text{and} \quad ac \equiv bd \pmod{n}.$$

(b) *If $ab \equiv ac \pmod{n}$ and a is relatively prime to n, then*

$$b \equiv c \pmod{n}.$$

Proof. (a) This part is easy, and we leave it as an exercise (see Exercise 1).

(b) Since a is relatively prime to n, there exist integers p and q such that $ap + nq = 1$. It follows that $ap \equiv 1 \pmod{n}$. Hence from $ab \equiv ac$ $\pmod{n}$, we obtain that $b \equiv pab \equiv pac \equiv c \pmod{n}$. $\qquad\square$

We leave it as an exercise to verify that $\equiv$ (congruence modulo n) is an equivalence relation (see Exercise 1). Thus, when n is a positive integer we may define $\mathbb{Z}_n$ to be $\mathbb{Z}/\equiv$ as a set. Remember that each integer belongs to exactly one equivalence class. We have that

$$[k] = [\ell] \quad \Longleftrightarrow \quad k \equiv \ell \pmod{n}.$$

Hence the elements of $\mathbb{Z}_n$ can be represented by the residues of integers modulo n,

$$\begin{aligned}
[0] &= \{\, m \in \mathbb{Z} : m \equiv 0 \pmod{n} \,\} = n\mathbb{Z}, \\
[1] &= \{\, m \in \mathbb{Z} : m \equiv 1 \pmod{n} \,\} = 1 + n\mathbb{Z}, \\
[2] &= \{\, m \in \mathbb{Z} : m \equiv 2 \pmod{n} \,\} = 2 + n\mathbb{Z},
\end{aligned}$$

$$\vdots \qquad\qquad\qquad\qquad \vdots$$

$$[n-1] = \{\, m \in \mathbb{Z} : m \equiv n - 1 \pmod{n} \,\} = (n-1) + n\mathbb{Z}.$$

Here $k + n\mathbb{Z} = \{ k + ni : i \in \mathbb{Z} \}$. The equivalence class $[k]$ in $\mathbb{Z}_n$ is also called the **residue class** of k, and is often denoted as $\overline{k}$. However, when there is no confusion, we can also simply write k for $\overline{k}$. Hence we may write

$$\mathbb{Z}_n = \{ \overline{0}, \overline{1}, \overline{2}, \ldots, \overline{n-1} \} \quad \text{or} \quad \mathbb{Z}_n = \{ 0, 1, 2, \ldots, n-1 \}.$$

The major difference of $\mathbb{Z}_n$ from the other sets of numbers that one has encountered so far is that an element of $\mathbb{Z}_n$ is in fact an equivalence class which can be represented by any other element in it. For example, in $\mathbb{Z}_5$, we have $1 = 6 = -34$. This makes it a little bit trickier to define a function or an operation on $\mathbb{Z}_n$. We have to make sure that the definition works for all representatives.

For example, when we want to define addition on $\mathbb{Z}_n$, we start by taking $\overline{a}, \overline{b} \in \mathbb{Z}_n$. Suppose $\overline{a} = \overline{c}$ and $\overline{b} = \overline{d}$ in $\mathbb{Z}_n$. This is equivalent to saying that $a \equiv c \pmod{n}$ and $b \equiv d \pmod{n}$. Proposition 2.2.1(a) tells us that $a + b \equiv c + d \pmod{n}$, and hence $\overline{a+b} = \overline{c+d}$. Thus, we may define addition on $\mathbb{Z}_n$ by letting

(2.2.1) $\overline{a} + \overline{b} = \overline{a+b} \in \mathbb{Z}_n, \qquad \overline{a}, \overline{b} \in \mathbb{Z}_n.$

This definition would create no contradiction since even when we change the representatives for $\overline{a}$ and $\overline{b}$ to $\overline{c}$ and $\overline{d}$, the resulting class of the sum would remain the same. Similarly, we may use Proposition 2.2.1(a) to define multiplication on $\mathbb{Z}_n$,

(2.2.2) $\overline{a} \cdot \overline{b} = \overline{ab} \in \mathbb{Z}_n, \qquad \overline{a}, \overline{b} \in \mathbb{Z}_n.$

For example when $n = 4$, we have the following tables for addition and multiplication.

+	0	1	2	3
0	0	1	2	3
1	1	2	3	0
2	2	3	0	1
3	3	0	1	2

·	0	1	2	3
0	0	0	0	0
1	0	1	2	3
2	0	2	0	2
3	0	3	2	1

Table 2.1: Additive and multiplicative tables for $\mathbb{Z}_4$

It is easy to check that $+$ and $\cdot$ are associative and commutative in $\mathbb{Z}_n$. The additive identity is $\bar{0}$ and the multiplicative identity is $\bar{1}$. For any $\bar{a} \in \mathbb{Z}_n$, the additive inverse is $\overline{-a} = \overline{n-a}$. However, the element $\bar{0}$ has no multiplicative inverse in $\mathbb{Z}_n$. If we observe Table 2.1, we can see $\bar{2}$ has no multiplicative inverse in $\mathbb{Z}_4$ either. So, when does an element in $\mathbb{Z}_n$ have a multiplicative inverse? The following proposition answers this question.

Proposition 2.2.2. *The subset of elements with a multiplicative inverse in $\mathbb{Z}_n$ is the set*

$$U(\mathbb{Z}_n) = \left\{ \bar{a} \in \mathbb{Z}_n : 1 \le a \le n, \ a \text{ is relatively prime to } n \right\}.$$

Every element in $U(\mathbb{Z}_n)$ can be represented by a unique positive integer less than n and relatively prime to n. Furthermore, $U(\mathbb{Z}_n)$ is closed under $\cdot$.

Proof. The first part of the statement is true since

$$\text{An element } \bar{a} \text{ has an inverse in } \mathbb{Z}_n$$
$$\iff \quad \text{there exists } \bar{b} \in \mathbb{Z}_n \text{ such that } \overline{ab} = 1$$
$$\iff \quad ab \equiv 1 \pmod{n}$$
$$\iff \quad ab = 1 + kn \quad \text{for some } k \in \mathbb{Z}$$
$$\iff \quad a \text{ is relatively prime to } n.$$

Take $\bar{a}$ and $\bar{b} \in U(\mathbb{Z}_n)$. We may choose the representatives so that $0 < a, b < n$. Then a and b are both relatively prime to n. Hence ab is also relatively prime to n and $\overline{ab} \in U(\mathbb{Z}_n)$. $\qquad\qquad\qquad\qquad\qquad$ $\square$

We can see now $U(\mathbb{Z}_n)$ inherits multiplication (but not addition) and multiplicative identity from $\mathbb{Z}_n$. In addition, every element in $U(\mathbb{Z}_n)$ has a multiplicative inverse.

$U(\mathbb{Z}_4)$	1	3
1	1	3
3	3	1

$U(\mathbb{Z}_5)$	1	2	3	4
1	1	2	3	4
2	2	4	1	3
3	3	1	4	2
4	4	3	2	1

Table 2.2: Multiplicative tables for $U(\mathbb{Z}_4)$ and $U(\mathbb{Z}_5)$

From the tables above, we see that the inverse of 3 is 3 in $U(\mathbb{Z}_4)$ and the inverse of 3 is 2 in $U(\mathbb{Z}_5)$.

Let p be a positive prime integer. Then $U(\mathbb{Z}_p) = \mathbb{Z}_p^* = \{1, 2, \ldots, p-1\}$ contains $p-1$ elements. In general, we define the **Euler φ-function**, denoted $\varphi(n)$, to be the number of elements in $U(\mathbb{Z}_n)$. In other words, $\varphi(n)$ is the number of positive integers no greater than n which is relatively prime to n. We make the following remarks on φ.

(1) Define $\varphi(1) = 1$.

(2) For any positive prime integer p and any positive integer n,

$$\varphi(p^n) = p^{n-1}(p-1).$$

In particular, $\varphi(p) = p - 1$. Among the p^n integers between 1 and p^n, there are p^{n-1} integers which are multiples of p. Hence, for $1 \le k \le p^n$, there are $p^n - p^{n-1}$ integers k which are relatively prime to p.

We leave it as an exercise to check the following general formula.

(3) If $n = p_1^{\alpha_1} p_2^{\alpha_2} \cdots p_s^{\alpha_s}$ where the p_i's are distinct positive prime integers, then

$$(2.2.3) \qquad \varphi(n) = \varphi(p_1^{\alpha_1})\varphi(p_2^{\alpha_2}) \cdots \varphi(p_s^{\alpha_s})$$

$$= n\left(1 - \frac{1}{p_1}\right)\left(1 - \frac{1}{p_2}\right) \cdots \left(1 - \frac{1}{p_s}\right).$$

We now apply the properties of $U(\mathbb{Z}_p)$ to prove some classical results.

Theorem 2.2.3 (Fermat's Little Theorem). *Let a be an integer not divisible by the positive prime integer p. Then*

$$a^{p-1} \equiv 1 \pmod{p}.$$

Proof. Define $f : U(\mathbb{Z}_p) \to U(\mathbb{Z}_p)$ by letting $f(\bar{j}) = \overline{aj}$ (verify this function is well-defined). If $\overline{aj} = \overline{ak}$ then $aj \equiv ak \pmod{p}$. By Proposition 2.2.1(b), we have $j \equiv k \pmod{p}$, that is, $\bar{j} = \bar{k}$. Hence this function is one-to-one. We obtain that f is bijective by Proposition 1.2.4. It follows that $\left\{\bar{a}, \overline{2a}, \ldots, \overline{(p-1)a}\right\} = \left\{\bar{1}, \bar{2}, \ldots, \overline{p-1}\right\}$, and therefore

$$(p-1)! = 1 \cdot 2 \cdots (p-1) \equiv a \cdot 2a \cdots (p-1)a = a^{p-1}(p-1)! \pmod{p}.$$

Since $(p-1)!$ is relatively prime to p, we may cancel $(p-1)!$ from both sides using Proposition 2.2.1(b) again to obtain

$$a^{p-1} \equiv 1 \pmod{p}. \qquad \square$$

The same method can be modified to prove the following generalization.

Theorem 2.2.4 (Euler's Generalization of Fermat's Little Theorem). *Let a be an integer relatively prime to a positive integer n. Then*

$$a^{\varphi(n)} \equiv 1 \pmod{n}.$$

Fermat's little theorem and its generalization can be proved in an entirely different fashion. It is an easy application of a basic theorem in group theory (see Corollary 6.1.5).

Theorem 2.2.6 is another classical result. Here, we prove this theorem using what we know about integers, but it can also be treated as a result regarding finite fields (*Cf.* Exercise 6, §16.2).

Lemma 2.2.5. *Let p be a prime integer. Then*

$$x^2 \equiv 1 \pmod{p} \quad \Longrightarrow \quad x \equiv \pm 1 \pmod{p}.$$

Proof. To solve this equation, we have that

$$
\begin{aligned}
& x^2 - 1 \equiv 0 \pmod{p} \\
& \Longrightarrow p \mid (x-1)(x+1) \qquad \text{in } \mathbb{Z} \\
& \Longrightarrow p \mid x-1 \text{ or } p \mid x+1 \qquad \text{in } \mathbb{Z} \\
& \Longrightarrow x-1 \in p\mathbb{Z} \quad \text{or} \quad x+1 \in p\mathbb{Z} \\
& \Longrightarrow x \equiv \pm 1 \pmod{p}.
\end{aligned}
$$

Note that this congruence equation has two distinct solutions if p is an odd prime. $\qquad \square$

Theorem 2.2.6 (Wilson's Theorem). *Let p be a positive prime integer. Then*

$$(p-1)! \equiv -1 \pmod{p}.$$

Proof. The result is trivial when $p = 2$. We will assume $p \geq 3$.

For each $a \in \mathbb{Z}_p^*$, there exists $b \in \mathbb{Z}_p^*$ (for example, we can take $b = a^{p-2}$ using Fermat's Little Theorem) such that $ab = 1$. Note that if $ab = ac = 1$ in $\mathbb{Z}_p^*$, then

$$b = b(ab) = b(ac) = (ba)c = (ab)c = c.$$

Thus, there is exactly one inverse a^{-1} for each element a in $\mathbb{Z}_p^*$. The elements $a \in \mathbb{Z}_p^*$ with $a = a^{-1}$ can be found by solving the congruence equation $x^2 \equiv 1 \pmod{p}$ in $\mathbb{Z}$. By Lemma 2.2.5, except for $\overline{1}$ and $\overline{p-1}$, the rest of the elements in $\mathbb{Z}_p^*$ can be grouped into $(p-3)/2$ pairs of elements which are inverse to each other. It follows that

$$(p-1)! \equiv 1 \times \left(\text{products of } \frac{p-3}{2} \text{ pairs} \right) \times (p-1) \equiv -1 \pmod{p}. \quad \square$$

Exercises 2.2

1. Let n be a positive integer. Prove the following statements.

   (a) The relation $\equiv$ (congruence modulo n) is an equivalence relation.

   (b) If $a \equiv b \pmod{n}$ and $c \equiv d \pmod{n}$ then $a + c \equiv b + d \pmod{n}$ and $ac \equiv bd \pmod{n}$.

2. In $\mathbb{Z}_{12}^* = \{1, 5, 7, 11\}$, give the table of multiplication and then find the inverses of 5, 7 and 11.

3. Solve the following congruences.

   (a) $3x \equiv 1 \pmod{5}$

   (b) $7x \equiv 2 \pmod{11}$

   (c) $563x \equiv 37 \pmod{13}$

4. Prove the general formula for the Euler φ-function in (2.2.3). (Hint: Use induction on s, the number of distinct prime factors of n.)

5. Prove Theorem 2.2.4.

6. Suppose that a, b and c are integers. If a is a divisor of bc and a is relatively prime to b, then a is a divisor of c.

7. Solve the following quadratic equations in $\mathbb{Z}_5$.

   (a) $x^2 - x + 1 = 0$

   (b) $x^2 + 2x - 3 = 0$

   (c) $x^2 - 2x - 1 = 0$

   (d) $x^2 - 3x + 4 = 0$

8. Solve $x^2 - 5x = 0$ in $\mathbb{Z}_6$. Be careful, there are more than two solutions.

2.3   Rational Numbers

Define a relation $\sim$ on $\mathbb{Z} \times \mathbb{Z}^*$ by letting $(a, b) \sim (c, d)$ if $ad = bc$. The reflexivity and symmetry of $\sim$ is easy to prove. We now check the transitivity of $\sim$. Let $(a, b) \sim (c, d)$ and $(c, d) \sim (e, f)$. Then $ad = bc$ and $cf = de$ and we have $acdf = bcde$. Remember that $d \neq 0$. If $c \neq 0$, then $cd \neq 0$. We have $af = be$ by cancellation law in $\mathbb{Z}$, and thus $(a, b) \sim (e, f)$. If $c = 0$, then $ad = bc = 0$ and $de = cf = 0$. This implies that $a = e = 0$ since $d \neq 0$. Thus $af = 0 = be$ and we have $(a, b) \sim (e, f)$. In conclusion, $\sim$ is an equivalence relation. We may now define the set of **rational numbers**, denoted $\mathbb{Q}$, to be $(\mathbb{Z} \times \mathbb{Z}^*)/\sim$. Traditionally, we use a/b or $\dfrac{a}{b}$ to denote the equivalence class of (a, b). Thus a rational number a/b has infinitely many different expressions (representatives). For example, we have

$$\frac{1}{2} = \frac{2}{4} = \frac{3}{6} = \frac{-4}{-8}, \quad \frac{2}{3} = \frac{8}{12} = \frac{-6}{-9} = \frac{-12}{-18} = \frac{-72}{-108}.$$

We define the addition and multiplication in $\mathbb{Q}$ as

$$\frac{a}{b} + \frac{c}{d} = \frac{ad + bc}{bd}, \qquad \frac{a}{b} \cdot \frac{c}{d} = \frac{ac}{bd}.$$

We have to make sure these definitions are independent of the representatives of rational numbers. First, we need to check that

$$\frac{a}{b} = \frac{a_1}{b_1} \quad \text{and} \quad \frac{c}{d} = \frac{c_1}{d_1} \quad \Longrightarrow \quad \frac{a}{b} + \frac{c}{d} = \frac{a_1}{b_1} + \frac{c_1}{d_1}.$$

That is, we need to show that

$$(ad + bc)b_1d_1 = (a_1d_1 + b_1c_1)bd$$

under the assumptions (i) $ab_1 = a_1b$ and (ii) $cd_1 = c_1d$.

Multiplying dd_1 to identity (i) and bb_1 to identity (ii), we obtain

$$ab_1dd_1 = a_1bdd_1 \qquad \text{and} \qquad cd_1bb_1 = c_1dbb_1.$$

Adding the identities together, we have

$$(ad + bc)b_1d_1 = ab_1dd_1 + cd_1bb_1 = a_1bdd_1 + c_1dbb_1 = (a_1d_1 + b_1c_1)bd.$$

Similarly, to check the multiplication in $\mathbb{Q}$ is well-defined, we need to check that

$$\frac{a}{b} = \frac{a_1}{b_1} \quad \text{and} \quad \frac{c}{d} = \frac{c_1}{d_1} \quad \Longrightarrow \quad \frac{a}{b} \cdot \frac{c}{d} = \frac{a_1}{b_1} \cdot \frac{c_1}{d_1}.$$

We leave this part to the reader.

The set of rational numbers can be considered as an extension of the set of integers. An integer k is identified with the rational number $k/1$. Basically speaking, the set of rational numbers is constructed so that every nonzero integer (as well as any nonzero rational number) can have a multiplicative inverse. To be more thorough, we make a list of algebraic properties of $\mathbb{Q}$ below.

(1) **Associativity.** For rational numbers r_1, r_2 and r_3, we have

$$(r_1 + r_2) + r_3 = r_1 + (r_2 + r_3) \qquad \text{and} \qquad (r_1r_2)r_3 = r_1(r_2r_3).$$

We demonstrate the associativity of $+$ here. Let

$$r_j = \frac{a_j}{b_j}, \qquad j = 1, 2, 3.$$

On the one hand, we have

$$(r_1 + r_2) + r_3 = \left(\frac{a_1}{b_1} + \frac{a_2}{b_2}\right) + \frac{a_3}{b_3}$$
$$= \frac{a_1b_2 + a_2b_1}{b_1b_2} + \frac{a_3}{b_3} = \frac{(a_1b_2 + a_2b_1)b_3 + a_3b_1b_2}{b_1b_2b_3}$$
$$= \frac{a_1b_2b_3 + a_2b_1b_3 + a_3b_1b_2}{b_1b_2b_3}.$$

On the other hand, we have

$$r_1 + (r_2 + r_3) = \frac{a_1}{b_1} + \left(\frac{a_2}{b_2} + \frac{a_3}{b_3}\right)$$

$$= \frac{a_1}{b_1} + \frac{a_2 b_3 + a_3 b_2}{b_2 b_3} = \frac{a_1 b_2 b_3 + (a_2 b_3 + a_3 b_2) b_1}{b_1 b_2 b_3}$$

$$= \frac{a_1 b_2 b_3 + a_2 b_1 b_3 + a_3 b_1 b_2}{b_1 b_2 b_3}.$$

Hence the associativity of addition follows. We leave it to the reader to verify that the associativity of multiplication.

(2) **Commutativity**. For rational numbers r_1 and r_2, we have

$$r_1 + r_2 = r_2 + r_1 \qquad \text{and} \qquad r_1 r_2 = r_2 r_1.$$

We leave the verification of this part to the reader.

(3) The existence of **identity elements**. The element $0/1$ is the additive identity in $\mathbb{Q}$. For $a/b \in \mathbb{Q}$, we have

$$\frac{a}{b} + \frac{0}{1} = \frac{a \cdot 1 + b \cdot 0}{b \cdot 1} = \frac{a \cdot 1}{b \cdot 1} = \frac{a}{b}.$$

The rational number $0/1$ is usually simply denoted 0. Often times, an additive identity is denoted as 0. Similarly, $1/1$ is the multiplicative identity in $\mathbb{Q}$ and is often denoted as 1.

(4) The existence of **inverses**. In $\mathbb{Q}$, the additive inverse of a/b is $(-a)/b$ since

$$\frac{a}{b} + \frac{-a}{b} = \frac{a \cdot b + b(-a)}{b \cdot b} = \frac{0}{b \cdot b} = \frac{0}{1}.$$

The element 0 still fails to possess a multiplicative inverse. Next we show that any nonzero rational number has a multiplicative inverse. A rational number $b/a = 0/1$ if and only if $b = 0$. Hence the set of nonzero rational numbers is

$$\mathbb{Q}^* = \mathbb{Q} \setminus \{0\} = \left\{\frac{b}{a} \in \mathbb{Q} \,\middle|\, a, b \neq 0\right\}.$$

Thus, $\mathbb{Q}^*$ is closed under multiplication. More importantly, for any nonzero rational number b/a,

$$\frac{a}{b} \cdot \frac{b}{a} = \frac{b}{a} \cdot \frac{a}{b} = \frac{ab}{ab} = \frac{1}{1}.$$

The nonzero rational number a/b is the multiplicative inverse of b/a.

(5) *Distributive Law*. For rational numbers r_1, r_2 and r_3, we have

$$r_1(r_2 + r_3) = r_1r_2 + r_1r_3.$$

At this point, this should be easy to verify.

In this chapter, we have seen several different algebraic structures which have many good properties in common. For example, $(\mathbb{Z}, +)$ is *associative* with an *identity* and its elements all have *inverses*. We thus say $(\mathbb{Z}, +)$ is a **group**. In particular, since $(\mathbb{Z}, +)$ is commutative, we say it is an **abelian group**. Similarly, $(n\mathbb{Z}, +)$, $(\mathbb{Z}_n, +)$ and $(\mathbb{Q}, +)$ are also abelian groups. Furthermore, $(\mathbb{Q}^*, \cdot)$ is also an example of an abelian group.

Note that in all the structures we have studied so far, the **zero element** (the additive identity) is always without a multiplicative inverse. Even when we only consider the nonzero elements, the algebraic structure $(\mathbb{Z}^*, \cdot)$ still fails to become an abelian group since most elements in $\mathbb{Z}^*$ are without multiplicative inverses. For an integer $n \geq 2$, $(\mathbb{Z}_n^*, \cdot)$ is not necessarily a group by Proposition 2.2.2. Instead, we should look at $U(\mathbb{Z}_n)$, which is always an abelian group and is called the **group of units** in $\mathbb{Z}_n$. Note that $U(\mathbb{Z}_n) = \mathbb{Z}_n^*$ if and only if n is a positive prime integer. Hence, $(\mathbb{Z}_n^*, \cdot)$ is an abelian group if and only if n is a positive prime integer.

We have seen that $(\mathbb{Q}, +)$ and $(\mathbb{Q}^*, \cdot)$ are both abelian groups. With distribution law linking the two operations together, we say $(\mathbb{Q}, +, \cdot)$ is a **field**. In comparison, we say $(\mathbb{Z}, +, \cdot)$ is a **commutative ring** instead of a field, because $\mathbb{Z}$ is missing one property: the existence of multiplicative inverses for all nonzero elements. We only require the associativity and the existence of an identity for the multiplication in a ring. Similarly, we say $(\mathbb{Z}_n, +, \cdot)$, $n \in \mathbb{Z}_+$, is a commutative ring and $(\mathbb{Z}_p, +, \cdot)$, p a positive prime, is a field.

There are lots and lots of examples of fields and rings consisting of numbers. Let

$$\begin{cases} \mathbb{Q}[\sqrt{2}] = \left\{ a + b\sqrt{2} \in \mathbb{R} : a, b \in \mathbb{Q} \right\} \subseteq \mathbb{R}; \\ \mathbb{Z}[\sqrt{2}] = \left\{ a + b\sqrt{2} \in \mathbb{R} : a, b \in \mathbb{Z} \right\} \subseteq \mathbb{R}. \end{cases}$$

It is easy to check that $(\mathbb{Q}[\sqrt{2}], +, \cdot)$ is a field and $(\mathbb{Z}[\sqrt{2}], +, \cdot)$ is a commutative ring. See Exercise 1.

Groups, rings and fields are the main objects of interest in this book. From next chapter on, we begin the study of groups.

Exercises 2.3

1. Verify that $(\mathbb{Z}[\sqrt{2}], +, \cdot)$ is a commutative ring and $(\mathbb{Q}[\sqrt{2}], +, \cdot)$ is a field.

2. Let $\omega = (-1 + \sqrt{3}i)/2$ and

$$\mathbb{Z}[\omega] = \{\, a + b\omega \in \mathbb{C} : a, b \in \mathbb{Z} \,\}.$$

   Prove that $\mathbb{Z}[\omega]$ is an abelian group under the addition of complex numbers. Find a subset T of $\mathbb{Z}[\omega]$ so that T is a group under the operation of multiplication. Is $\mathbb{Z}[\omega]$ a field?

3. Let $\mathbb{Q}[x]$ be the set of polynomials in the variable x with rational coefficients,

$$\mathbb{Q}[x] = \left\{ \sum_{j=0}^{n} a_j x^j \;\middle|\; a_j \in \mathbb{Q}, \; j = 0, 1, 2, 3, \ldots \right\}.$$

   Prove that $\mathbb{Q}[x]$ is a commutative ring but not a field.

4. Let $\mathbb{Q}(x)$ be the set of **rational functions** over $\mathbb{Q}$,

$$\mathbb{Q}(x) = \left\{ \frac{p(x)}{q(x)} \;\middle|\; p(x), q(x) \in \mathbb{Q}[x], \; q(x) \neq 0 \right\}.$$

   Is $\mathbb{Q}(x)$ a field?

5. Let p be a positive prime integer and

$$\mathbb{Z}_p[i] = \{\, a + bi \mid a, b \in \mathbb{Z}_p \,\}$$

   with addition and multiplication defined by

$$\begin{cases} (a + bi) + (c + di) = (a + c) + (b + d)i, \\ (a + bi)(c + di) = (ac - bd) + (ad + bc)i. \end{cases}$$

   Is $\mathbb{Z}_2[i]$ a field? Is $\mathbb{Z}_3[i]$ a field?

Review Exercises for Chapter 2

1. Solve the following equations in $\mathbb{Z}_5$.

    (a) $4x = 3$

    (b) $3x - 1 = x + 3$

    (c) $x^2 = 1$

    (d) $x^2 = -1$

2. Solve the system of equations

$$\begin{cases} x \equiv 5 \pmod 7 \\ x \equiv 11 \pmod{17} \end{cases}$$

   in $\mathbb{Z}$.

3. Let n be a positive integer with $n \geq 3$ and

$$\zeta = \cos \frac{2\pi}{n} + i \sin \frac{2\pi}{n}.$$

   Suppose that $S = \{\, 1, \zeta, \zeta^2, \ldots, \zeta^{n-1} \,\}$. Prove that S is an abelian group under the multiplication of complex numbers.

4. In this problem, we describe the process of constructing the set $\mathbb{R}$ of real numbers.

   An infinite sequence $\{\, a_n \,\}_{n=1}^{\infty}$ of rational numbers is a **Cauchy sequence** if for any given $\epsilon > 0$, there exists a positive integer N such that $|a_m - a_n| < \epsilon$ whenever $m, n \geq N$. Let S be the set of all Cauchy sequences of rational numbers. Define a relation on S as

$$\{\, a_n \,\} \sim \{\, b_n \,\} \quad \text{if and only if} \quad \lim_{n \to \infty} a_n = \lim_{n \to \infty} b_n.$$

    (a) Show that $\sim$ is an equivalence relation.

    (b) The set of real numbers $\mathbb{R}$ is defined to be $S/\sim$. The addition and multiplication are then defined as

$$\begin{cases} [\{\, a_n \,\}] + [\{\, b_n \,\}] = [\{\, a_n + b_n \,\}], \\ [\{\, a_n \,\}] \cdot [\{\, b_n \,\}] = [\{\, a_n b_n \,\}]. \end{cases}$$

    Show that $+$ and $\cdot$ are both well-defined.

(c) Show that $(\mathbb{R}, +, \cdot)$ is a field.

(d) Define a map f from $\mathbb{Q}$ to $\mathbb{R}$ by letting $f(r)$ be the equivalence class of $\{a_n\}$ where $a_n = r$ for all n. Show that f is one-to-one and thus we can consider $\mathbb{Q}$ as a subset of $\mathbb{R}$. Moreover, show that the addition and multiplication in $\mathbb{Q}$ are compatible with the addition and multiplication in $f(\mathbb{Q})$. In other words, show that

$$f(r_1 + r_2) = f(r_1) + f(r_2) \quad \text{and} \quad f(r_1 r_2) = f(r_1) f(r_2)$$

for all $r_1, r_2 \in \mathbb{Q}$.

CHAPTER 3

Basic Notions of Groups

A group is a nonempty set with a binary operation such that (i) the operation is associative, (ii) there is an identity and (iii) each element has an inverse. In this chapter we shall see more examples of groups besides those we had already seen in previous chapters. Furthermore, basic properties of groups will be investigated.

3.1   Definitions and Examples

Definition 3.1.1. A **binary operation** $*$ on a set S is a function from $S \times S$ to S. The image of (a, b) is denoted as $a * b$.

We use $(S, *)$ to denote a set with a binary operation $*$ on S.

For example, $+$ and $\cdot$ are both binary operations on $\mathbb{Z}$ (or $\mathbb{Q}$, etc.). Let S be a set. Then $\cup$ and $\cap$ are binary operations on 2^S. Of course, if one defines a binary operation too sloppily, the operation might not have good properties and become difficult to handle. The binary operations we are interested in should at least have the following property.

Definition 3.1.2. We say a binary operation $*$ on a set S is **associative** if

$$(a * b) * c = a * (b * c)$$

for all $a, b, c \in S$.

A binary operation tells us how to combine two elements to produce a third element. When three or more elements are involved, there are more than one way to combine them. For example, there are five types of products of 4 elements

$$((a*b)*c)*d, \quad (a*b)*(c*d), \quad (a*(b*c))*d, \quad a*((b*c)*d), \quad a*(b*(c*d)),$$

and there is a chance that these products are not of the same value. An associative binary operation guarantees that no matter in what order one combines the elements the result would remain the same. Without this property, computing the product of elements would become extremely complicated and there will be many many parentheses involved.

Presently, we are ready to give the definition of a group.

Definition 3.1.3. We say a *nonempty* set $(G, *)$ is a **group** if the following conditions are satisfied.

(G1) The operation $*$ is **associative**.

(G2) There exists an **identity** e such that $e * a = a * e = a$ for all $a \in G$.

(G3) There is an **inverse** for every $a \in G$, that is, for every $a \in G$ there exists $b \in G$ such that $a * b = b * a = e$.

Besides the examples we have seen in Chapter 2, here we give some more classical examples of groups.

Example 3.1.4. Suppose that m and n are positive integers. Let $M_{m \times n}(\mathbb{R})$ be the set of $m \times n$ matrices with entries in $\mathbb{R}$. For $A = \left[a_{ij} \right]$ and $B = \left[b_{ij} \right]$ in $M_{m \times n}(\mathbb{R})$, define
$$A + B = \left[a_{ij} + b_{ij} \right].$$
Then $(M_{m \times n}(\mathbb{R}), +)$ is a group. The operation $+$ is easily checked to be associative. The additive identity is the zero matrix, the $m \times n$ matrix whose entries are all 0. The additive inverse of $A = \left[a_{jk} \right]$ is

$$-A = \left[-a_{ij} \right].$$

Example 3.1.5. Let

$$GL_n(\mathbb{R}) = \{ A \in M_{n \times n}(\mathbb{R}) : \det A \neq 0 \}$$

be the set of non-singular $n \times n$ matrices with entries in $\mathbb{R}$. With respect to the usual matrix multiplication, $GL_n(\mathbb{R})$ is a group, called the **general linear group** over $\mathbb{R}$. The **special linear group** $SL_n(\mathbb{R})$ is defined as

$$SL_n(\mathbb{R}) = \{ A \in M_{n \times n}(\mathbb{R}) : \det A = 1 \}.$$

Note that $GL_n(\mathbb{R})$ and $SL_n(\mathbb{R})$ are closed under the matrix multiplication since
$$\det(AB) = (\det A)(\det B).$$

Also a square matrix A is invertible if and only if $\det A \neq 0$. Indeed the inverse of $A = \left[a_{ij} \right]$ is given by

$$\frac{1}{\det A} \left[A_{ji} \right]_{i,j}$$

where A_{ji} is the **cofactor** of A at the (j, i)-entry, i.e., A_{ji} is the product of $(-1)^{i+j}$ and the determinant of the matrix obtained by deleting the j-th row and i-th column in A.

Example 3.1.6. Let p_n be the function from $\mathbb{Z}$ into $\mathbb{Z}$ defined by

$$p_n(x) = x + n$$

and $G = \{\, p_n \mid n \in \mathbb{Z} \,\}$. Then G is a group with respect to the composition of functions

$$p_m \circ p_n = p_{m+n}.$$

Composition of functions is associative. The function p_0 is the identity and the inverse of p_n is p_{-n} since

$$p_n \circ p_{-n} = p_0 = p_{-n} \circ p_n.$$

Example 3.1.7. Let $V = \{\, e, a, b, c \,\}$ be a set of 4 elements. Define the operation $*$ on V by the following table.

$*$	e	a	b	c
e	e	a	b	c
a	a	e	c	b
b	b	c	e	a
c	c	b	a	e

Then V is a group under the operation $*$. It is called the **Klein 4-group**.

Example 3.1.8. Let $\mathbb{Q}[x]$ be the set of polynomials in the variable x with rational coefficients. For

$$p(x) = a_0 + a_1 x + \cdots + a_n x^n \quad \text{and} \quad q(x) = b_0 + b_1 x + \cdots + b_m x^m,$$

define

$$p(x) + q(x) = q(x) + p(x) = (a_0 + b_0) + (a_1 + b_1)x + \cdots$$
$$+ (a_m + b_m)x^m + a_{m+1}x^{m+1} + \cdots + a_n x^n$$

if $n \geq m$. Then $(\mathbb{Q}[x], +)$ is a group. The operation $+$ is associative. The identity is the zero polynomial and the inverse of $p(x)$ is

$$-p(x) = (-a_0) + (-a_1)x + \cdots + (-a_n)x^n.$$

Definition 3.1.9. A group G is a **finite group** if it contains only finitely many elements. The number of elements in G is called the **order** of G and is denoted as $|G|$.

For example, $(\mathbb{Z}_n, +)$ is a finite group of order n and $(U(\mathbb{Z}_n), \cdot)$ is a finite group of order $\varphi(n)$.

We may simply use G to denote the group $(G, *)$ if $*$ is understood. For example, although $+$ and $\cdot$ are both binary operations on $\mathbb{Z}_n$, $(\mathbb{Z}_n, \cdot)$ is not a group. Hence, the notation $\mathbb{Z}_n$ usually refers to the additive group. Similarly, $U(\mathbb{Z}_n)$ usually refers to the multiplicative group.

Here are some of the smallest groups.

Example 3.1.10. Let $G = \{e\}$ be a set with a single element. Then G is a group with the operation

$$e * e = e.$$

Clearly, this is the only possible operation in G. A group of one element is called the **trivial** group.

Example 3.1.11. Let $G = \{e, a\}$ be a set with two elements. Define

$$e * e = e, \quad e * a = a * e = a, \quad a * a = e.$$

Then $(G, *)$ is a group of order 2. This is also the only possible way to define an operation in G so that G becomes a group with e as the identity. If $a * a = a$, then

$$a = e * a = (a^{-1} * a) * a = a^{-1} * (a * a) = a^{-1} * a = e,$$

which is a contradiction. Thus, there is basically only one way to construct a group of order 2.

A group G is called an **abelian** group if $a * b = b * a$ for all a, $b \in G$. In this case, we say the operation $*$ is **commutative**.

The trivial group and the group of order 2 are both abelian. The groups $\mathbb{Z}$, $\mathbb{Z}_n$ and $U(\mathbb{Z}_n)$ are all abelian groups while $GL_n(\mathbb{R})$ is not an abelian group for $n \geq 2$ (see Exercise 1).

Example 3.1.12. Let $S = \{a^j : j \in \mathbb{Z}\}$ be the set of powers of a nonzero rational number a. Then $(S, \cdot)$ is an abelian group in which the identity is $a^0 = 1$ and the inverse of a^j is a^{-j}.

Example 3.1.13. Let x be an indeterminate and let $T = \{x^j : j \in \mathbb{Z}\}$ be the set of powers of x. The multiplication in T is defined as

$$x^j \cdot x^k = x^{j+k}, \qquad \text{for } j, k \in \mathbb{Z}.$$

Then $(T, \cdot)$ is an abelian group. The identity of T is x^0 and the inverse of x^j is x^{-j}.

At the end of this section, we give some examples of non-groups.

Example 3.1.14. Neither $(\mathbb{N}, +)$ nor $(\mathbb{N}, \cdot)$ is a group.

Example 3.1.15. Subtraction is a binary operation in $\mathbb{Z}$, but $(\mathbb{Z}, -)$ is not a group since
$$(a - b) - c \neq a - (b - c)$$
in general.

Example 3.1.16. Let $M_n(\mathbb{R})$ be the set of $n \times n$ matrices with entries in $\mathbb{R}$. Then $M_n(\mathbb{R})$ is not a group with the matrix multiplication. If we let $M_n(\mathbb{R})^*$ be the subset of nonzero matrices in $M_n(\mathbb{R})$, it is still not a group under the matrix multiplication. A square matrix has a multiplicative inverse if and only if its determinant is nonzero. There are plenty of nonzero matrices whose determinant is zero such as

$$\begin{pmatrix} 0 & \cdots & 0 & 1 \\ 0 & \cdots & 0 & 0 \\ \vdots & \ddots & \vdots & \vdots \\ 0 & \cdots & 0 & 0 \end{pmatrix}$$

Example 3.1.17. Let $G = \{a, b, c\}$ be a set of three elements. Define
$$\alpha * \beta = \alpha$$
for any α and $\beta \in G$. Then $(G, *)$ is not a group. The operation $*$ is associative but unfortunately there is no identity in this structure.

Exercises 3.1

1. Show that the group $GL_n(\mathbb{R})$ is non-abelian for $n \geq 2$.

2. Let S be a set. Show that $(2^S, +)$ is a group where $+$ is the Boolean sum defined in Exercise 5, §1.1.

3. Let $\mathbb{Q}_+$ be the set of positive rational numbers. Define $*$ by letting

$$a * b = \frac{ab}{2}.$$

Show that $(\mathbb{Q}_+, *)$ is a group with 2 as the identity.

4. Let $G = \mathbb{R} \setminus \{-1\}$. Define the operation $*$ on G as

$$a * b = a + b + ab.$$

(a) Show that $(G, *)$ is a group.

(b) Solve the equation $(-2) * x * 3 = -7$.

5. Let $V = \{e, a, b, c\}$ be the Klein 4-group. Solve the following equations in V.

(a) $ax = b$

(b) $bx = c$

(c) $x^2 = e$

(d) $x^2 = b$

6. Let $S = \{\cos\theta + i\sin\theta \in \mathbb{C} : \theta \in \mathbb{R}\} \subseteq \mathbb{C}$. Show that $(S, \cdot)$ is a group.

7. If G is a finite group and $a \in G$, show that there exists a positive integer n such that $a^n = e$.

3.2   Basic Properties

Let $(S, *)$ be a set with a binary operation $*$. Usually we write ab for $a * b$ and S for $(S, *)$ when no confusion arises.

Proposition 3.2.1. *The identity of $(S, *)$ is unique if it exists. If $(S, *)$ is associative with an identity, then the inverse of a in S is unique when it exists.*

Proof. Suppose e and e' are both identities of S. Then $e = e * e' = e'$. Now suppose e is the identity in S and b and b' are both inverses of a in S. Then $b = b * e = b * (a * b') = (b * a) * b' = e * b' = b'$. $\qquad\square$

It is customary to use a^{-1} to denote the inverse of an element a when it is unique. However, for the binary operation $+$, it is customary to use $-a$ to denote the inverse of a.

Proposition 3.2.2. *In a group G, the following statements are true for all $a, b \in G$:*

(a) $(a^{-1})^{-1} = a;$

(b) $(ab)^{-1} = b^{-1}a^{-1}.$

Proof. (a) Since $aa^{-1} = e = a^{-1}a$, a is the inverse of a^{-1}, that is, $a = (a^{-1})^{-1}$.

(b) First note that

$$(ab)(b^{-1}a^{-1}) = a(bb^{-1})a^{-1} = aea^{-1} = aa^{-1} = e.$$

Similarly, $(b^{-1}a^{-1})(ab) = e$. Hence the inverse of ab is $b^{-1}a^{-1}$. $\square$

Here are some remarks.

(1) The reader must note that $(ab)^{-1} = b^{-1}a^{-1}$ instead of $a^{-1}b^{-1}$. In general, we have

$$(a_1a_2\cdots a_n)^{-1} = a_n^{-1}\cdots a_2^{-1}a_1^{-1}.$$

(2) Define $a^0 = e$. For a positive integer k,

$$a^k = \underbrace{a \cdot a \cdots a}_{k \text{ times}}$$

From (1), it is easy to see by induction on k that

$$(a^k)^{-1} = \underbrace{a^{-1} \cdot a^{-1} \cdots a^{-1}}_{k \text{ times}}$$

The element $(a^k)^{-1}$ is usually denoted as a^{-k}. With this notation, for any element a in any group G we have

$$a^i a^j = a^{i+j}, \qquad \text{for all } i, j \in \mathbb{Z}.$$

Furthermore, by induction one can show that

$$(a^i)^j = a^{ij}, \qquad \text{for all } i, j \in \mathbb{Z}.$$

(3) Suppose G is a group with the binary operation $+$. Then G is called
an **additive** group. For example, $\mathbb{Z}$ is an additive group. As a convention, additive groups are usually abelian. In an additive group,
the identity element is usually denoted as 0 and the inverse of an
element a is denoted as $-a$. It is customary to write

$$
\begin{cases}
0a = 0, \\
ka = \underbrace{a + \cdots + a}_{k \text{ times}}, \\
(-k)a = -(ka) = \underbrace{(-a) + \cdots + (-a)}_{k \text{ times}}
\end{cases}
$$

for $a \in G$ and $k \in \mathbb{N}$. In this book, the results regarding groups are
mostly given in multiplicative form. The reader is asked to interpret
all respective results in additive form (*Cf.* Exercise 1).

Proposition 3.2.3 (Cancellation Law). *In a group G, the following properties hold for all $a, b, c \in G$.*

(a) *If $ab = ac$, then $b = c$.*

(b) *If $ba = ca$, then $b = c$.*

Proof. (a) Suppose $ab = ac$. Then $b = eb = (a^{-1}a)b = a^{-1}(ab) = a^{-1}(ac) = (a^{-1}a)c = ec = c$.

(b) Suppose $ba = ca$. Then $b = be = b(aa^{-1}) = (ba)a^{-1} = (ca)a^{-1} = c(aa^{-1}) = ce = c$. $\square$

In $(S, *)$, we say e is a *left identity* in S if $ea = a$ for all $a \in S$, and we
say b is a **left inverse** of a with respect to e if $ba = e$. Right identities and
right inverses are defined accordingly.

If you check the proof of Proposition 3.2.3 carefully, you will see that
when proving left cancellation law in (a), one only needs the existence
of a left identity and the left inverses. Similarly, when proving the right
cancellation law, one only needs the existence of a right identity and the
right inverses.

Proposition 3.2.4. *Let G be a set with an associative binary operation.
Suppose e is a left identity and every element in G has a left inverse with*

respect to e. Then e is the *identity and the left inverse of an element is* *indeed* the *inverse. Thus, in this situation G is a group.*

Proof. In this proof, all the left inverses are with respect to e. For $a \in G$, let b be a left inverse of a. Then $bab = eb = b$. Find a left inverse c for b. Then $ab = eab = cbab = ceb = cb = e$. We have shown that b is also a right inverse, hence an inverse of a with respect to e. Moreover, $ae = a(ba) = (ab)a = ea = a$. Since this is true for all $a \in G$, e is an identity in G.

Of course, the identity and inverses are unique by Proposition 3.2.1. □

Similarly, a set G with an associative operation is a group if it possesses a right identity and a right inverse for every element in G.

Exercises 3.2

1. Let G be an additive group. For all $a \in G$ and $m, n \in \mathbb{Z}$, show that $ma + na = (m + n)a$ and $m(na) = (mn)a$.

2. Suppose that a is an element in a group G. Show that the function $L_a \colon G \to G$ defined by $L_a(g) = ag$ is a bijective function from G into G. Also show that the function $R_a \colon G \to G$ defined by $R_a(g) = ga$ is a bijective function from G into G.

3. An element a in a group G is called an **idempotent** if $a^2 = a$. Prove that the identity e is the only idempotent in a group.

4. Let G be a group of order $2n$. Show that G contains an element x such that $x \neq e$ and satisfies the equation $x^2 = e$.

5. Let G be a group in which $(ab)^{-1} = a^{-1}b^{-1}$ for all $a, b \in G$. Show that G is abelian.

6. Let G be a group with the identity e. Suppose that $a^2 = e$ for all $a \in G$. Show that G is an abelian group.

7. Let G be a group in which $(ab)^3 = a^3b^3$ and $(ab)^5 = a^5b^5$ for all $a, b \in G$. Show that G is abelian.

8. Let S be a set with $|S| \geq 2$. Define a binary operation $*$ by letting $a * b = a$. Let b_0 be an arbitrary element in S.

   (a) Show that $*$ is associative.

   (b) Show that b_0 is a right identity of $(S, *)$.

   (c) Show that every element in S has a left inverse with respect to the right identity b_0.

   (d) Show that $(S, *)$ is *not* a group.

9. Let G be a group and $g \in G$. Define a function f from G into G by letting $f(x) = gxg^{-1}$. Prove that f is bijective.

3.3   Subgroups

A nonempty subset H of a group G is a **subgroup** of G if H is a group under the same operation of G. For example, $(\mathbb{Z}, +)$ is a subgroup of $(\mathbb{Q}, +)$ and $(\mathbb{Q}^*, \cdot)$ is a subgroup of $(\mathbb{R}^*, \cdot)$. Let

$$\zeta = \cos \frac{2\pi}{n} + i \sin \frac{2\pi}{n}.$$

In a previous exercise (see Exercise 3, P. 33) we have seen that the set

$$S = \left\{ 1, \zeta, \zeta^2, \ldots, \zeta^{n-1} \right\}$$

is a subgroup of $\mathbb{C}^*$ under the multiplication of complex numbers.

However, to check whether a set is a subgroup, we don't really have to check all the conditions for groups. We have the following less tedious criteria for subgroups.

Theorem 3.3.1 (The 3-Step Test for Subgroups). *Let G be a group. A set $H \subseteq G$ is a subgroup of G if and only if it satisfies the following conditions:*

   (i) *H is nonempty;*

   (ii) *H is closed under the operation of G: $a, b \in H \Longrightarrow ab \in H$;*

   (iii) *H is closed under inverses: $a \in H \Longrightarrow a^{-1} \in H$.*

We often check condition (i) by showing the identity of G is in H, but any other element will suffice.

Proof. The "only if" part follows directly from the definition of groups. We concentrate on showing that H is a group if the three conditions are satisfied.

A group is *nonempty* by definition and this is achieved by condition (i). Condition (ii) guarantees that H can inherit the binary operation from G. The associativity of the operation in H then follows from the associativity of the same operation in G. Since H is nonempty, find $a \in H$. By (iii), $a^{-1} \in H$. By (ii), the identity e of G is $aa^{-1} \in H$. It is easy to see that e is the identity of H. Now (iii) guarantees that H is a group. $\square$

From this proof, we can see that a subgroup also inherits the identity and the inverses from its mother group.

This 3-step test for subgroups can be shortened to the following test.

Theorem 3.3.2 (The 2-Step Test for Subgroups). *A subset H of a group G is a subgroup, if the following conditions are satisfied:*

(i) H *is nonempty;*

(ii) $ab^{-1} \in H$ *for a, $b \in H$.*

Proof. We will show that the two tests are equivalent to each other. Obviously, the 3-step test implies the 2-step test. Assuming the two conditions in the 2-step test hold, we only need to verify H is closed under the inherited binary operation and inverses. Since H is nonempty, find $a \in H$. Then $e = aa^{-1} \in H$ by (ii). Now suppose given arbitrary elements $a, b \in H$. Then $a^{-1} = ea^{-1} \in H$ by (ii). This shows that H is closed under inverses. Since $b^{-1} \in H$, we have that $ab = a(b^{-1})^{-1} \in H$. $\square$

These two tests are equally easy to use. You can simply choose the one you prefer for the following examples.

Example 3.3.3. For any integer m, the set $m\mathbb{Z} = \{ma \mid a \in \mathbb{Z}\}$ is a subgroup of $\mathbb{Z}$ under the operation of addition. The set $m\mathbb{Z}$ is definitely nonempty. The sum of two multiples of m is a multiple of m and so is the inverse of a multiple of m.

Let n be a positive integer. Similarly, the set

$$m\mathbb{Z}_n = \left\{\, \overline{m}\,\overline{k} = \overline{mk} \in \mathbb{Z}_n : k \in \mathbb{Z} \,\right\}$$

is also a subgroup of $\mathbb{Z}_n$. The element $\overline{0} \in m\mathbb{Z}_n$. For $\overline{mk}, \overline{m\ell} \in m\mathbb{Z}_n$, the sum is $\overline{m(k+\ell)}$ and the inverse of $\overline{mk}$ is $\overline{m(-k)}$.

Example 3.3.4. Let $V = \{\,e, a, b, c\,\}$ be the Klein 4-group. Then V contains the following subgroups

$$A = \{\,e, a\,\}, \quad B = \{\,e, b\,\} \quad \text{and} \quad C = \{\,e, c\,\}.$$

The set of identity $\{\,e\,\}$ is a subgroup of G and is called the **trivial** subgroup of G. The group G itself is also a subgroup, and is called the **improper** subgroup of G. The other subgroups are called **proper nontrivial** subgroups of G. A group without any proper nontrivial subgroup is an example of a **simple group** (for the true definition of simple groups, see Definition 8.1.7).

Example 3.3.5. Let H be a nontrivial subgroup of $\mathbb{Z}_3$. Then H contains elements other than 0. If $1 \in H$, then $2 = 1 + 1 \in H$ and thus $H = \mathbb{Z}_3$. On the other hand, if $2 \in H$, then $1 = 2 + 2 \in H$ and $H = \mathbb{Z}_3$. In either case, we have $H = \mathbb{Z}_3$. Thus, $\mathbb{Z}_3$ is a simple group.

In the case of finite subgroups, the subgroup tests can be further simplified to the following test.

Theorem 3.3.6 (Finite Subgroups Test). *A finite subset H of a group G is a subgroup if and only if the following conditions hold:*

(i) *H is nonempty;*

(ii) *H is closed under the operation:* $a,\, b \in H \implies ab \in H$.

We leave this theorem as an exercise. See Exercise 5.

Example 3.3.7. Consider the subset $\{1, -1, i, -i\}$ in $\mathbb{C}$. This subset is clearly closed under multiplication. By the finite subgroups test, it is a subgroup of $(\mathbb{C}^*, \cdot)$.

Exercises 3.3

1. Show that $\{\, A \in GL_n(\mathbb{R}) : \det A = \pm 1 \,\}$ is a subgroup of $GL_n(\mathbb{R})$ under multiplication.

2. Show that $G = \left\{\, \begin{bmatrix} \cos\theta & -\sin\theta \\ \sin\theta & \cos\theta \end{bmatrix} \,\middle|\, \theta \in \mathbb{R} \,\right\}$ is a subgroup of $SL_2(\mathbb{R})$ under multiplication. Is G equal to $SL_2(\mathbb{R})$?

3. Let A and B be subgroups of G. Is $A \cap B$ is also a subgroup of G?

4. Let H and K be subgroups of an abelian group G. Define

$$HK = \{\, hk : h \in H, k \in K \,\}.$$

   Show that HK is a subgroup of G.

5. Prove Theorem 3.3.6.

6. Let G be a group. Define the **center** of G, denoted $Z(G)$, to be the set $\{\, a \in G : ab = ba \text{ for all } b \in G \,\}$.

   (a) Show that G is abelian if and only if $Z(G) = G$.

   (b) Show that $Z(G)$ is a subgroup of G.

   (c) Show that $Z(G)$ is an abelian group.

3.4   Generating Sets

To describe a subgroup in a group, there is really no need to specify every element in it. It will be sufficient just to describe certain key elements. For example, suppose H is a subgroup of $\mathbb{Z}$ and we know $2 \in H$. Without speaking, H must also contain $0, 2, 4, 6, \ldots$ and $-2, -4, -6, \ldots$. Now let's explore this idea further.

Proposition 3.4.1. *Let $\{H_i\}_{i \in \Lambda}$ be a family of subgroups of G. Then $\bigcap_{i \in \Lambda} H_i$ is also a subgroup of G.*

Proof. Write $H = \bigcap_{i \in \Lambda} H_i$. We use Theorem 3.3.2 to prove this proposition.

Since the identity e of G is contained in every H_i, it is also in H. Hence H is nonempty. Now assume $a, b \in H$. Then $a, b \in H_i$ for all $i \in \Lambda$. Since all the H_i's are subgroups of G, we have that $ab^{-1} \in H_i$ for all $i \in \Lambda$. This implies that $ab^{-1} \in H$. We conclude that H is a subgroup of G. $\qquad\square$

Now let S be a subset of a group G. By the proposition above we can see that

$$\bigcap_{\substack{S \subseteq H \\ H \text{ subgroup of } G}} H$$

is a subgroup of G containing S. Obviously, this subgroup is also contained in every subgroup of G containing S. Thus, this subgroup is *the* smallest subgroup in G containing S.

Let S be a subset of a group G. The smallest subgroup of G containing S is called the **subgroup generated by** S, and is denoted as $\langle S \rangle$. If $H = \langle S \rangle$, we say that S **generates** H or S is **a set of generators** or a **generating set** for H.

If $S = \{a_1, \ldots, a_n\}$ is finite, it is customary to write $\langle a_1, \ldots, a_n \rangle$ for $\langle S \rangle$. If $G = \langle a_1, \ldots, a_n \rangle$, we say that $a_1, \ldots, a_n$ generate G and G is a **finitely generated** group.

There are two things worth noting.

(1) Let G be a group with e as the identity element. Then $\langle \varnothing \rangle = \{e\}$ because it is the smallest subgroup in G.

(2) To show that $\langle S \rangle \subseteq H$, it suffices to show that $S \subseteq H$.

Example 3.4.2. The additive group $\mathbb{Z} = \langle 1 \rangle = \langle -1 \rangle$. The additive group E of even integers $= \langle 2 \rangle = \langle -2 \rangle$. The additive group $\mathbb{Z}_n = \langle \bar{1} \rangle$. All the groups in this example are finitely generated; they can all be generated by one element.

The following proposition gives an explicit description of the elements in the subgroup generated by a given set. We will use the convention that the product of an empty collection of elements is defined to be the identity element.

Proposition 3.4.3. *Let S be a subset of a group G. Then*

$$(3.4.1) \qquad \langle S \rangle = \left\{ a_{i_1}^{k_1} a_{i_2}^{k_2} \cdots a_{i_n}^{k_n} : n = 0, 1, 2, \ldots, \ a_{i_j} \in S, \ k_j \in \mathbb{Z} \right\}.$$

In particular, $\langle a \rangle = \left\{ a^k \in G : k \in \mathbb{Z} \right\}$.

This proposition says that an element is in $\langle S \rangle$ if and only if it is a product of elements or inverses of elements in S. For example, let a, b be elements in a group G. Then ab, aba, $abab$, $ababa$, $ababab$, $\ldots$ are all elements in $\langle a, b \rangle$, and these elements may or may not be distinct from each other. Other elements such as $ba^{-3}b^{-5}a^9$ or $a^{46}bab^3aba^{-100001}$ are also elements in $\langle a, b \rangle$.

Proof. We will use H to denote the set on the right-hand side of (3.4.1). Obviously, every element in H is in the subgroup $\langle S \rangle$. It remains to show that $\langle S \rangle \subseteq H$. Since clearly $S \subseteq H$, to complete the proof, we need to show that H is a subgroup in G.

By convention, the identity element e is the product of 0 elements and is thus contained in H. We have that H is nonempty. It is also clear that the product of two elements in H remains a product of elements or inverses of elements in S. Hence, the product of two elements in H is still in H. By the first remark to Proposition 3.2.2 this is also true for the inverse of any element in H. Hence, H is a subgroup of G by Theorem 3.3.1. $\qquad\square$

Corollary 3.4.4. *Let S be a subset of an* abelian *group G. Then*

$$\langle a_1, \ldots, a_n \rangle = \left\{ a_1^{k_1} \cdots a_n^{k_n} \in G : k_1, \ldots, k_n \in \mathbb{Z} \right\}.$$

Exercises 3.4

1. Let H be a subgroup of G. Show that $\langle H \rangle = H$.

2. Let G be an additive (abelian) group. For $a_1, \ldots, a_n \in G$, show that

$$\langle a_1, \ldots, a_n \rangle = \{ k_1 a_1 + \cdots + k_n a_n \in G : k_1, \ldots, k_n \in \mathbb{Z} \}.$$

3. For $k \in \mathbb{Z}$, show that $\langle k \rangle = k\mathbb{Z} = \langle -k \rangle$ in $\mathbb{Z}$.

4. Show that $\langle -8, 18 \rangle \subseteq \mathbb{Z}$ can be generated by one single element. Find a generator for this subgroup.

5. List the elements of $\langle \overline{15}, \overline{12} \rangle \subseteq \mathbb{Z}_{21}$. Show that this subgroup can be generated by one element.

6. Let H, H' be two subgroups of a group G.

   (a) Give an example of H, H' and G such that $H \cup H'$ is not a subgroup of G (*Cf.* Proposition 3.4.1).

   (b) Show that the smallest subgroup containing both H and H' in G is $\langle H \cup H' \rangle$.

   (c) If $H = \langle S \rangle$ and $H' = \langle S' \rangle$, show that the smallest subgroup containing both H and H' in G is $\langle S \cup S' \rangle$.

7. Suppose the group $G = \langle S \rangle$ such that $ab = ba$ for all $a, b \in S$. Show that G is an abelian group.

Review Exercises for Chapter 3

1. If G is a finite abelian group with n elements $a_1, a_2, \ldots, a_n$, then the element $x = a_1 a_2 \cdots a_n$ must satisfy $x^2 = e$.

2. Define a relation on a group G by letting $a \sim b$ if $b = x^{-1}ax$ for some $x \in G$. Determine whether $\sim$ is an equivalence relation.

3. Let G be an abelian group and let d be an integer. Show that $G_d = \{ a \in G : a^d = e \}$ is a subgroup of G.

4. Let A and B be subgroups of a group G. Define the relation $\sim$ by $a \sim b$ if $a = xby$ for some $x \in A$ and $y \in B$. Prove that $\sim$ is an equivalence relation and describe its equivalence class containing a.

5. Let A and B be subgroups of a group G. Show that $A \cup B$ is a subgroup of G if and only if $A \subseteq B$ or $B \subseteq A$. If C is also a subgroup, does a similar necessary and sufficient condition hold for $A \cup B \cup C$ to be a subgroup of G.

6. Let G be a finite group. Show that there are only finitely many subgroups in G.

7. Let G be a group and $a, b \in G$. Suppose that $ab = ba^{-1}$ and $ba = ab^{-1}$. Show that $a^4 = b^4 = e$.

8. We say that $(S, *)$ is a **semigroup** if $*$ is an associative binary operation.

   Suppose that $(S, *)$ is a semigroup such that for all $a \in S$ there exists a unique $b \in S$ such that $a * b * a = a$.

   (a) Show that if $a * b * a = a$ then $a * b$ is an idempotent in S.

   (b) Show that there is exactly one idempotent in S.

   (c) Show that the idempotent in S is the identity in S.

   (d) Show that the b such that $a * b * a = a$ is the inverse of a.

   (e) Show that $(S, *)$ is a group.

CHAPTER 4

Cyclic Groups

One of the purposes of this book is to familiarize ourselves with the basic concepts regarding groups. We will start by demonstrating the ability to "dissect" small groups, that is, to understand reasonably small and uncomplicated groups from inside out. In this chapter, we will study in full details the simplest kind of groups: cyclic groups.

Cyclic groups are groups that can be generated by one single element. Basically there are only two types of cyclic groups: $\mathbb{Z}$ and $\mathbb{Z}_n$.

4.1   Cyclic Groups

In §3.4 we discussed the concept of "generating sets". The idea is that the fewer generators a group needs, the easier it is to understand its elements. In this respect there are no groups simpler than *cyclic groups*.

A **cyclic** group is a group that can be generated by one single element. This element is called a **generator** of the cyclic group.

Suppose g is a generator of the group G. Then

$$G = \langle g \rangle = \{\, g^n : n \in \mathbb{Z} \,\} = \{e, g^{\pm 1}, g^{\pm 2}, g^{\pm 3}, \dots \}.$$

For additive cyclic groups,

$$G = \langle g \rangle = \{\, ng : n \in \mathbb{Z} \,\} = \{0, \pm g, \pm 2g, \pm 3g, \dots \}.$$

Example 4.1.1.   (1) The group $\mathbb{Z} = \langle 1 \rangle$ is cyclic, and so is the group $\mathbb{Z}_n = \langle \overline{1} \rangle$.

(2) Remember that $U(\mathbb{Z}_6) = \{\overline{1}, \overline{5}\}$. Since $\langle \overline{5} \rangle = \{\overline{1}, \overline{5}\}$, we have that $U(\mathbb{Z}_6)$ is cyclic.

(3) In $U(\mathbb{Z}_7) = \mathbb{Z}_7^*$,

$$\langle \overline{3} \rangle = \{\overline{1},\ \overline{3},\ \overline{3}^2 = \overline{2},\ \overline{3}^3 = \overline{6},\ \overline{3}^4 = \overline{4},\ \overline{3}^5 = \overline{5}\} = U(\mathbb{Z}_7).$$

Hence $U(\mathbb{Z}_7)$ is cyclic.

(4) The group $U(\mathbb{Z}_8) = \{\overline{1}, \overline{3}, \overline{5}, \overline{7}\}$ is not cyclic, since

$$\langle \overline{1} \rangle = \{\overline{1}\}, \quad \langle \overline{3} \rangle = \{\overline{1}, \overline{3}\}, \quad \langle \overline{5} \rangle = \{\overline{1}, \overline{5}\}, \quad \langle \overline{7} \rangle = \{\overline{1}, \overline{7}\}.$$

To check if a group is cyclic, we need to check if any element can be a generator. Next we introduce a useful concept which will facilitate this job.

Let a be an element in a group. The **order** of a, denoted $o(a)$ or $|a|$, is defined to be $\inf \{\, n \in \mathbb{N} : a^n = e \,\}$. (Remember that the infimum of an empty set is defined to be ∞.)

By definition, $o(e) = 1$. In general, if we want to compute the order of an element a, we simply list the elements $a, a^2, a^3, \dots$. The first positive integer n we encounter with $a^n = e$ is the order of a. If $a^n \neq e$ for all $n > 0$, we say the order of a is ∞.

For example, inside the additive group $\mathbb{Z}$, $o(0) = 1$ while $o(1) = o(-1) = o(2) = \infty$. In fact, any nonzero element in $\mathbb{Z}$ is of order ∞. As another easy example, inside the Klein 4-group, every element but the identity is of order 2.

Example 4.1.2. Consider the additive group $\mathbb{Z}_6$. To find the order of $\overline{4}$ we make the computation that

$$\overline{4} = \overline{4}, \quad 2 \cdot \overline{4} = \overline{2}, \quad 3 \cdot \overline{4} = \overline{0}.$$

We conclude that $o(\overline{4}) = 3$. In this manner, we can find the orders of all the elements in $\mathbb{Z}_6$:

$$o(\overline{0}) = 1; \quad o(\overline{1}) = o(\overline{5}) = 6; \quad o(\overline{2}) = o(\overline{4}) = 3; \quad o(\overline{3}) = 2.$$

Proposition 4.1.3. *Let a be an element of finite order in a group.*

(a) *If $a^n = e$ then $o(a) \,\big|\, n$.*

(b) *If $a^i = a^j$ then $o(a) \,\big|\, i - j$.*

Proof. To prove (a), assume $a^n = e$. Find $q, r \in \mathbb{Z}$ such that

$$n = o(a)q + r, \qquad 0 \leq r < o(a).$$

Then $a^r = a^{n - o(a)q} = a^n \big(a^{o(a)}\big)^{-q} = e$. By definition, $o(a)$ is the smallest positive integer k such that $a^k = e$. Hence $r = 0$ and $o(a) \,\big|\, n$.

Now (b) follows from (a). $\qquad\qquad\qquad\qquad\qquad\qquad\qquad\qquad$ $\square$

Proposition 4.1.4. *Let a be an element in a group.*

(a) *If $o(a) = \infty$, then $a^i = a^j$ if and only if $i = j$.*

(b) *If $o(a) = n$ is finite, then for any $i \in \mathbb{Z}$, $a^i = a^k$ for some unique k with $0 \leq k \leq n - 1$.*

Proof. (a) Assume $o(a) = \infty$. Let $a^i = a^j$ where $i \geq j$. Then $a^{i-j} = a^i(a^j)^{-1} = a^i(a^i)^{-1} = e$. By definition, $i - j = 0$, that is, $i = j$. The "if" part is trivial.

(b) Assume $o(a) = n$ is finite. Suppose given any integer i. Find $q, k \in \mathbb{Z}$ such that

$$i = nq + k, \qquad 0 \leq k \leq n - 1.$$

Then
$$a^i = a^{nq+k} = (a^n)^q a^k = e^q a^k = a^k.$$

If $0 \le j \le i \le n-1$ and $a^i = a^j$, then by Proposition 4.1.3(b) we have $n \mid i - j$ while $0 \le i - j < n$. This implies that $i - j = 0$, that is, $i = j$. $\square$

From the proposition above, we see that if $o(a) = \infty$, then

$$\langle a \rangle = \{ \ldots, a^{-3}, a^{-2}, a^{-1}, e, a, a^2, a^3, \ldots \}$$

consists of infinitely many elements, and if $o(a) = n$, then

$$\langle a \rangle = \{ e, a, a^2, \ldots, a^{n-1} \}$$

consists of n elements. Now we can draw the following conclusions.

Corollary 4.1.5. *For any element a in a group, $|\langle a \rangle| = o(a)$.*

Corollary 4.1.6. *Let G be a finite group.*

(a) *An element g in G is a generator if and only if $o(g) = |G|$.*

(b) *The group G is cyclic if and only if there exists $g \in G$ such that $o(g) = |G|$.*

Suppose that $G = \langle g \rangle$ is a cyclic group. Proposition 4.1.4 tells us that if $o(g) = \infty$, there is a natural one-to-one correspondence

$$\begin{array}{ccc} G & \rightsquigarrow & \mathbb{Z} \\ g^i & \longleftrightarrow & i \end{array}$$

between elements of G and $\mathbb{Z}$. Moreover, $g^i g^j = g^{i+j}$ corresponds to $i + j$: the binary operations also corresponds to each other. If $o(g) = n$ is finite, then by Proposition 4.1.4 again we have a natural one-to-one correspondence

$$\begin{array}{ccc} G & \rightsquigarrow & \mathbb{Z}_n \\ g^i & \longleftrightarrow & \bar{i}. \end{array}$$

Again, $g^i g^j = g^{i+j}$ corresponds to $\overline{i+j} = \bar{i} + \bar{j}$. We can see that the algebraic structure of a cyclic group is "identical" to that of $\mathbb{Z}$ or $\mathbb{Z}_n$ depending on the order of the cyclic group. (We will be back with a more detailed discussion on this in Chapter 7.) Hence, there are only two types of cyclic groups. In the next section, we will study $\mathbb{Z}$ and $\mathbb{Z}_n$ in full details.

Exercises 4.1

1. Show that all cyclic groups are abelian. Show that the Klein 4-group V is abelian but not cyclic.

2. Find $o(\bar{3})$ and $o(\bar{7})$ in $U(\mathbb{Z}_{20})$. Find out if $U(\mathbb{Z}_{20})$ is a cyclic group.

3. Let a be an element of order 15 in a group.

   (a) Find $o(a^9)$, $o(a^{10})$ and $o(a^8)$.

   (b) What are the possible orders for the elements in $\langle a \rangle$?

4. Find all generators of the additive group $\mathbb{Z}_{12}$.

5. Let a be an element in a group. Show that $\langle a^{-1} \rangle = \langle a \rangle$. Hence $o(a^{-1}) = o(a)$.

4.2   Subgroups of Cyclic Groups

To understand a group, it is necessary to understand all its subgroups. A **subgroup lattice** of a group G is a diagram which depicts all the subgroups of G with an edge connecting any subgroups A, B with $A \subseteq B$. It is customary to put the smaller subgroup in a lower position. In this section, we study the subgroups of cyclic groups.

Proposition 4.2.1. *Any subgroup of a cyclic group is still cyclic.*

Proof. Let $G = \langle g \rangle$ be cyclic and H be a subgroup of G. If H is trivial there is nothing to prove. We now assume $|H| \geq 2$. Consider the set

$$S = \{\, i \in \mathbb{N} : g^i \in H \,\}.$$

Note that there must be at least one nonzero integer i such that $g^i \in H$. If $i > 0$ then $i \in S$. Otherwise, $g^{-i} = (g^i)^{-1} \in H$, which implies that $-i \in S$.

In either case, $S \cap \mathbb{N}$ is nonempty. By the well-ordering principle on $\mathbb{N}$, S contains a least element k. We claim that $H = \langle g^k \rangle$ is cyclic.

Clearly, $\langle g^k \rangle \subseteq H$ since $g^k \in H$. Take any element in H. It is of the form g^i for some $i \in \mathbb{Z}$. Find $q, r \in \mathbb{Z}$ such that $i = kq + r$ where $0 \le r < k$. Then $g^r = g^{i-kq} = g^i(g^k)^{-q} \in H$. The choice of k forces r to be 0. Thus, $g^i = (g^k)^q \in \langle g^k \rangle$. We have shown that $H \subseteq \langle g^k \rangle$. $\qquad\square$

From the proof above, we can also see that k is a g.c.d. (greatest common divisor) of all the exponents of the elements in H. In fact, since we are considering factorization in $\mathbb{Z}$, the g.c.d. of a collection of integers is not unique. For example, 1 and -1 can be both considered as the g.c.d. of 2 and -3. However, the two g.c.d.'s only differ by a multiple of -1. We will use the notation $(a, b) \sim d$ to mean that d and $-d$ are the g.c.d.'s of a and b. Similarly, we will use $[a, b] \sim m$ to mean that m and $-m$ are the l.c.m.'s (least common multiples) of a and b.

We have the following basic results.

Proposition 4.2.2. *Let a be an element in a group.*

(a) *If $o(a) = \infty$, then $o(a^k) = \infty$ for all $k \neq 0$. In fact, $\langle a^k \rangle = \langle a \rangle$ if and only if $k = \pm 1$.*

(b) *If $o(a) = n$, then $\langle a^k \rangle = \langle a^d \rangle$ where $d > 0$ and $d \sim (n, k)$. In this case, $o(a^k) = o(a^d) = n/d$.*

Proof. (a) If $(a^k)^i = e$ then $ki = 0$ by Proposition 4.1.4(a). If $k \neq 0$ then $i = 0$. This proves that $o(a^k) = \infty$ for $k \neq 0$.

In general, we have $\langle a^k \rangle \subseteq \langle a \rangle$ since $a^k \in \langle a \rangle$. If $\langle a^k \rangle = \langle a \rangle$, then $a \in \langle a^k \rangle$. In other words, $a = (a^k)^i$ for some $i \in \mathbb{Z}$. By Proposition 4.1.4(a) again, we have $1 = ki$. This implies that $k = \pm 1$.

(b) Since $d \mid k$ we have that $\langle a^k \rangle \subseteq \langle a^d \rangle$. To show the other inclusion, find $i, j \in \mathbb{Z}$ so that $d = ni + kj$. Then $a^d = a^{ni+kj} = (a^n)^i(a^k)^j = (a^k)^j \in \langle a^k \rangle$. Hence $\langle a^d \rangle \subseteq \langle a^k \rangle$.

By Corollary 4.1.5, $o(a^k) = o(a^d)$. We leave it to the reader to verify that $o(a^d) = n/d$. $\qquad\square$

Corollary 4.2.3. *Let G be a cyclic group of order n. Then the order of every element of G is a divisor of n.*

We are now ready to describe all the subgroups inside a given cyclic group. Let $G = \langle g \rangle$ be a cyclic group.

(1) If $|G| = \infty$, then each subgroup of G is of the form $\langle g^k \rangle$ for a unique $k \geq 0$. Moreover, when $i, j \neq 0$, $\langle g^i \rangle \subseteq \langle g^j \rangle$ if and only if j is a divisor of i (see Exercise 1).

(2) Let $G = \langle g \rangle$ be a cyclic group of order n. Then Proposition 4.2.2 tells us that each subgroup of G is of the form $\langle g^k \rangle$ for a unique $k \mid n$ and $k > 0$. Let i and j both be divisors of n. We leave it to the reader to verify that $\langle g^i \rangle \subseteq \langle g^j \rangle$ if and only if $j \mid i$ (see Exercise 2).

We are now ready to construct the subgroup lattice for $\mathbb{Z}_{12}$.

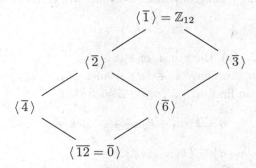

Part (a) of Proposition 4.2.2 tells us that in an infinite cyclic group there are exactly two generators. On the other hand, part (b) says that if $G = \langle g \rangle$ is a cyclic group of order n, then

$$g^k \text{ is a generator of } G \quad \Longleftrightarrow \quad (n, k) \sim 1.$$

In particular, $\overline{k} \in \mathbb{Z}_n$ (as an additive group) is a generator if and only if $(n, k) \sim 1$. We also conclude that in a cyclic group of order n, the number of generators for G is given by the Euler φ-function $\varphi(n)$ (see Formula (2.2.3)).

Example 4.2.4. The generators for $\mathbb{Z}_{30}$ are $1, 7, 11, 13, 17, 19, 23, 29$. Note that $30 = 2 \cdot 3 \cdot 5$ and $\varphi(30) = 30(1/2)(2/3)(4/5) = 8$.

Example 4.2.5. Consider the multiplicative group

$$U(\mathbb{Z}_{50}) = \{\, 1, 3, 7, 9, 11, 13, 17, 19, 21, 23, 27, 29, 31, 33, 37, 39, 41, 43, 47, 49 \,\}.$$

Note that $50 = 2 \cdot 5^2$ and $\varphi(50) = 50(1/2)(4/5) = 20$. The group $U(\mathbb{Z}_{50})$ is of order 20. This group is cyclic if we can find an element of order 20 in it. With patience, we can list the powers of 3 as

3, 9, 27, 31, 43, 29, 37, 11, 33, 49, 47, 41, 23, 19, 7, 21, 13, 39, 17, 1

We conclude that $o(3) = 20$. Hence 3 is a generator for $U(\mathbb{Z}_{50})$. Since $\varphi(20) = 20(1/2)(4/5) = 8$, there are altogether 8 generators for $U(\mathbb{Z}_{50})$. The generators of $U(Z_{50})$ are 3^k with $(k, 20) \sim 1$:

3, $3^3 = 27$, $3^7 = 37$, $3^9 = 33$, $3^{11} = 47$, $3^{13} = 23$, $3^{17} = 13$, $3^{19} = 17$.

Example 4.2.6. Let $k_1, \ldots, k_r \in \mathbb{Z}^*$. Then Proposition 4.2.1 tells us that $\langle k_1, \ldots, k_r \rangle$, being a subgroup of $\mathbb{Z}$, is cyclic. We claim that

$$\langle k_1, \ldots, k_r \rangle = \langle d \rangle,$$

where $d \sim (k_1, \ldots, k_r)$, the g.c.d. of $k_1, \ldots, k_r$.

Since for each i, we have $k_i \in \langle d \rangle$. Hence $\langle k_1, \ldots, k_r \rangle \subseteq \langle d \rangle$. On the other hand, we can find $m_1, \ldots, m_r \in \mathbb{Z}$ such that

$$d = k_1 m_1 + k_2 m_2 + \cdots + k_r m_r.$$

Hence we also have $\langle d \rangle \subseteq \langle k_1, \ldots, k_r \rangle$.

Let's also try to find a generator for the group $\langle k_1 \rangle \cap \cdots \cap \langle k_r \rangle$. This time we claim that

$$\langle k_1 \rangle \cap \cdots \cap \langle k_r \rangle = \langle m \rangle,$$

where $m \sim [k_1, \ldots, k_r]$, the l.c.m. of $k_1, \ldots, k_r$.

For each i we have $m \in \langle k_i \rangle$ since $k_i \mid m$. Thus $\langle m \rangle \subseteq \langle k_1 \rangle \cap \cdots \cap \langle k_r \rangle$. Conversely, let $n \in \langle k_1 \rangle \cap \cdots \cap \langle k_r \rangle$. Then $n \in \langle k_i \rangle$, that is, $k_i \mid n$ for each i. Hence n is a common multiple of $k_1, \ldots, k_r$. Therefore $m \mid n$ and $n \in \langle m \rangle$. We have shown that $\langle k_1 \rangle \cap \cdots \cap \langle k_r \rangle \subseteq \langle m \rangle$.

We can now use the knowledge on cyclic groups to derive an interesting identity on the Euler φ-function.

Proposition 4.2.7. *If n is a positive integer, then*

(4.2.1)
$$\sum_{d \mid n} \varphi(d) = n.$$

Proof. Let $G = \langle g \rangle$ be a cyclic group of order n. Suppose given a positive divisor d of n. Then $G_d = \langle g^{n/d} \rangle$ is the only subgroup of order d inside G. It follows that there are exactly $\varphi(d)$ generators for G_d. In other words, there are exactly $\varphi(d)$ elements of order d in G. The group G consists of elements of order d with $d \,|\, n$ by Corollary 4.2.3. Hence

$$n = |G| = \sum_{d|n} \varphi(d). \qquad \qquad \square$$

In number theory, the identity (4.2.1) is nothing but a special case of the well-known *Möbius inversion formula*

$$F(n) = \sum_{d|n} f(d) \iff f(n) = \sum_{d|n} \mu(d) F\left(\frac{n}{d}\right),$$

where $\mu(n)$ is the **Möbius function** defined by

$$\mu(n) = \begin{cases} 1, & \text{if } n = 1; \\ (-1)^k, & \text{if } n = p_1 p_2 \cdots p_k \text{ is a product of distinct positive primes}; \\ 0, & \text{otherwise}. \end{cases}$$

Note that Möbius function has the property that $\mu(m)\mu(n) = \mu(mn)$ if m and n are relatively prime. Thus

$$\varphi(p^\alpha) = p^\alpha - p^{\alpha-1} = \mu(1)p^\alpha + \mu(p)\frac{p^\alpha}{p} = \sum_{d|p^\alpha} \mu(d)\frac{p^\alpha}{d}.$$

In general, let $n = p_1^{\alpha_1} \cdots p_s^{\alpha_s}$. By induction on s we have that

$$\varphi(n) = \varphi(p_1^{\alpha_1})\varphi\left(\frac{n}{p_1^{\alpha_1}}\right) = \left(\mu(1)p_1^\alpha + \mu(p_1)\frac{p_1^\alpha}{p_1}\right)\left(\sum_{d \,\big|\, \frac{n}{p_1^{\alpha_1}}} \mu(d)\frac{n}{dp_1^{\alpha_1}}\right)$$

$$= \sum_{d \,\big|\, \frac{n}{p_1^{\alpha_1}}} \mu(d)\frac{n}{d} + \sum_{d \,\big|\, \frac{n}{p_1^{\alpha_1}}} \mu(dp_1)\frac{n}{dp_1} = \sum_{d|n} \mu(d)\frac{n}{d}.$$

Now the inversion formula gives the identity (4.2.1) if we choose $f(n)$ to be $\varphi(n)$ and $F(n)$ to be n.

Corollary 4.2.8. *Suppose that G is a finite abelian group of order n and the equation $x^d = e$ has at most d solutions in G for any positive divisor d of n. Then G is a cyclic group.*

Proof. Let $G_d = \{ x \in G \mid x^d = e \}$. Then G_d is a subgroup of G (see Exercise 3, P. 51) containing all elements of order d in G. By assumption, $|G_d| \le d$.

(1) If $|G_d| < d$, then G contains no elements of order d.

(2) If $|G_d| = d$ and G_d contains an element of order d, then G_d is cyclic and G contains exactly $\varphi(d)$ elements of order d. Otherwise G contains no elements of order d at all.

Let m_d be the number of elements of order d in G. We have shown that $m_d \le \varphi(d)$ for all $d \mid n$. Since $n = \sum_{d\mid n} m_d = \sum_{d\mid n} \varphi(d)$, we have that $m_d = \varphi(d)$ for all $d \mid n$. In particular, $m_n = \varphi(n) > 0$. Hence, G contains an element of order n and it is cyclic. $\square$

Exercises 4.2

1. Let a be an element of infinite order in a group. Show that $\langle a^i \rangle \subseteq \langle a^j \rangle$ if and only if j is a divisor of i. Moreover, $\langle a^i \rangle = \langle a^j \rangle$ if and only if $i = \pm j$.

2. Let $\langle g \rangle$ be a cyclic group of order n and let i and j both divide n. Show that $\langle g^i \rangle \subseteq \langle g^j \rangle$ if and only if $j \mid i$.

3. Construct the subgroup lattice of $\mathbb{Z}_{36}$.

4. Prove that a group of order 3 must be cyclic. Is a group of order 4 always cyclic as well?

5. Find all the generators for the additive group $\mathbb{Z}_{56}$.

6. Find the generators for $U(\mathbb{Z}_{29})$.

7. Let a be an element of order n in a group. Show that $\langle a^i \rangle = \langle a^j \rangle$ if and only if $(n, i) \sim (n, j)$.

8. Let $\overline{k}_1, \ldots, \overline{k}_r \in \mathbb{Z}_n$. Show that $\langle \overline{k}_1, \ldots, \overline{k}_r \rangle = \langle \overline{d} \rangle$ where $d \sim (n, k_1, \ldots, k_r)$.

9. Let a be an element of order n in a group. Show that $\langle a^i \rangle \cap \langle a^j \rangle = \langle a^m \rangle$ where $m \sim [(n, i), (n, j)]$.

10. Find a generator for $\langle \overline{24}, \overline{9}, \overline{21} \rangle \cap \langle \overline{8}, \overline{12}, \overline{20}, \overline{36} \rangle \cap \langle \overline{20}, \overline{10} \rangle$ in $\mathbb{Z}_{120}$.

Review Exercises for Chapter 4

1. Let n be a positive integer and let

$$G = \left\{ \cos \frac{k\pi}{12} + i \sin \frac{k\pi}{12} \,\middle|\, k = 0, 1, \ldots, 23 \right\}.$$

Construct the subgroup lattice of G.

2. Let $G = \langle a \rangle$ be a cyclic group of infinite order. Find all the generators for the subgroup $\langle a^8 \rangle$.

3. Let $G = \langle a \rangle$ be a cyclic group of order 20. Find all the generators for the subgroup $\langle a^8 \rangle$.

4. Let a and b be elements in an abelian group G. Suppose that $o(a)$ and $o(b)$ are relatively prime. Show that $o(ab) = o(a)o(b)$.

5. Let a be an element in a group G. Suppose the order of a is mn where the positive integers m and n are relatively prime. Show that there are b, c in G such that $o(b) = m$, $o(c) = n$ and $a = bc$.

6. Show that $U(\mathbb{Z}_{2^n})$ is not cyclic for $n \geq 3$.

7. Show that any group of order 4 is abelian.

8. Show that $H = \left\{ \begin{bmatrix} 1 & n \\ 0 & 1 \end{bmatrix} \in M_2(\mathbb{R}) \,\middle|\, n \in \mathbb{Z} \right\}$ is a cyclic subgroup of $GL_2(\mathbb{R})$.

9. Let $G = \{ ax^2 + bx + c : a, b, c \in \mathbb{Z}_3 \}$. Define

$$(a_1 x^2 + b_1 x + c_1) + (a_2 x^2 + b_2 x + c_2) = (a_1 + a_2)x^2 + (b_1 + b_2)x + (c_1 + c_2)$$

for $a_i, b_i, c_i \in \mathbb{Z}_3$. Show that G is an additive group. Is the additive group G cyclic?

10. Is the additive group $\mathbb{Q}$ cyclic? Is the multiplicative group $\mathbb{Q}^*$ cyclic?

11. Let G be a group of order mn such that $m, n > 1$. Show that G contains at least one proper nontrivial subgroup.

12. Give an example of an infinite group which contains a nontrivial finite cyclic subgroup.

CHAPTER 5

Permutation Groups

Permutation groups can be reckoned as the prototype of all groups. A permutation, in group language, is a bijective function. A permutation group is a group consisting of permutations on a given set S. The group of all permutations on S is called the symmetric group of S. The study of groups arose out of study of symmetry of geometric objects, which can be interpreted naturally as permutations. We will see that all groups are essentially permutation groups in Chapter 7.

5.1   Symmetric Groups

Let S be a set and let $\operatorname{Sym} S$ denote the set of all bijective functions from S onto itself. We will also call an element in $\operatorname{Sym} S$ a **permutation** on S. In §1.2 we have carefully investigated many properties of this set. In particular, Corollary 1.2.3 tells us that "$\circ$", composition of functions, is a binary operation on $\operatorname{Sym} S$. From (1.2.1), we see that the composition of functions is associative. From (1.2.2), we see that 1_S is the identity in $(\operatorname{Sym} S, \circ)$ (we usually simply write 1 for 1_S). Finally, by Theorem 1.2.5, every $f \in \operatorname{Sym} S$ has an inverse with respect to $\circ$. Thus, $(\operatorname{Sym} S, \circ)$ is a group and we call it the **symmetric group** on S.

We use S_n to denote $\operatorname{Sym} S$ where $S = \{1, 2, \ldots, n\}$ and call it the **symmetric n-group**. In general, when S is a finite set of n elements, we can index the elements in S by the set $\{1, 2, \ldots, n\}$. In this way we can think of $\operatorname{Sym} S$ as S_n. Note that $\operatorname{Sym} S$ has exactly $n!$ elements. For $\sigma \in S_n$, we write σ as

$$\begin{bmatrix} 1 & 2 & \cdots & n \\ \sigma(1) & \sigma(2) & \cdots & \sigma(n) \end{bmatrix},$$

which is seen naturally as a *permutation* of n objects. For example when $n = 3$, S_3 has 6 elements as listed below:

$$\begin{bmatrix} 1 & 2 & 3 \\ 1 & 2 & 3 \end{bmatrix}, \quad \begin{bmatrix} 1 & 2 & 3 \\ 1 & 3 & 2 \end{bmatrix}, \quad \begin{bmatrix} 1 & 2 & 3 \\ 3 & 2 & 1 \end{bmatrix},$$

$$\begin{bmatrix} 1 & 2 & 3 \\ 2 & 1 & 3 \end{bmatrix}, \quad \begin{bmatrix} 1 & 2 & 3 \\ 2 & 3 & 1 \end{bmatrix}, \quad \begin{bmatrix} 1 & 2 & 3 \\ 3 & 1 & 2 \end{bmatrix}.$$

It is alright to switch the columns in a permutation, as long as in the first row of a permutation is a list of all elements in $\{1, 2, \ldots, n\}$ and the image of each element is listed right below itself in the second row. For example,

$$\begin{bmatrix} 1 & 2 & 3 \\ 1 & 3 & 2 \end{bmatrix} = \begin{bmatrix} 2 & 1 & 3 \\ 3 & 1 & 2 \end{bmatrix}.$$

The composition of functions can then be performed as

$$\begin{bmatrix} 1 & 2 & 3 \\ 1 & 3 & 2 \end{bmatrix} \begin{bmatrix} 1 & 2 & 3 \\ 2 & 1 & 3 \end{bmatrix} = \begin{bmatrix} 2 & 1 & 3 \\ 3 & 1 & 2 \end{bmatrix} \begin{bmatrix} 1 & 2 & 3 \\ 2 & 1 & 3 \end{bmatrix} = \begin{bmatrix} 1 & 2 & 3 \\ 3 & 1 & 2 \end{bmatrix}.$$

Such kind of notation is quite cumbersome and so we develop the **cycle notation** for S_n. The permutation

$$\begin{bmatrix} 1 & 2 & 3 \\ 1 & 3 & 2 \end{bmatrix}$$

fixes 1 and exchanges 2 and 3; we write it as a 2-cycle (2 3) or (3 2). The permutation

$$\begin{bmatrix} 1 & 2 & 3 \\ 2 & 3 & 1 \end{bmatrix}$$

maps 1 to 2, then 2 to 3 and then 3 back to 1; we write it as a 3-cycle (1 2 3), (2 3 1) or (3 1 2).

In general, when we write an r-cycle $(a_1\ a_2\ \cdots\ a_r)$, it is required that all the a_i's be distinct. This cycle represents the permutation sending a_1 to a_2, a_2 to a_3, ..., a_{r-1} to a_r, and eventually a_r back to a_1, while leaving all other elements in $\{1, 2, \ldots, n\}$ fixed. When several cycles are put together in a product, it is viewed as an composition of the corresponding functions. A 1-cycle (a) is simply the identity function and so it can be present or omitted in a product.

In cycle notation, the elements in S_3 can be represented as

$$\{1, (1\ 2), (1\ 3), (2\ 3), (1\ 2\ 3), (1\ 3\ 2)\},$$

where 1 is the identity function.

In general, we can write a permutation as a product of **disjoint cycles**. We say two cycles are **disjoint** if no integers appear in both cycles. In a product of disjoint cycles, every integer can appear in at most one cycle. For example,

$$\begin{bmatrix} 1 & 2 & 3 & 4 & 5 & 6 & 7 & 8 \\ 2 & 1 & 4 & 5 & 6 & 3 & 8 & 7 \end{bmatrix} = (1\ 2)(3\ 4\ 5\ 6)(7\ 8)$$

is expressed as a product of disjoint cycles.

Next, we summarize some basic properties regarding disjoint cycles.

(1) We have different expressions for the same cycle. For example, we have

$$(1\ 2\ 3\ 4) = (2\ 3\ 4\ 1) = (3\ 4\ 1\ 2) = (4\ 1\ 2\ 3)$$

and $(1\ 2) = (2\ 1)$.

(2) Disjoint cycles commute with each other.

(3) The order of a k-cycle in S_n is k.

(4) Every permutation in S_n is a product of disjoint cycles and the decomposition is unique up to different orderings of the disjoint cycles.

(5) Let $\sigma \in S_n$. Suppose that σ can be decomposed into a product of disjoint cycles of length $m_1, m_2, \ldots, m_k$. Then the order of σ is the (positive) least common multiple of $m_1, m_2, \ldots, m_k$.

Here we shall describe briefly the reasoning behind the decompositions of permutations into products of disjoint cycles without really getting into the tedious details. Let $\sigma \in S_n$ be a permutation of order m. Then

$$\langle \sigma \rangle = \left\{ e, \sigma, \sigma^2, \ldots, \sigma^{m-1} \right\}$$

is a subgroup of S_n of order m. Define a relation $\sim$ on $S = \{1, 2, \ldots, n\}$ by

$$j \sim k \quad \Longleftrightarrow \quad k = \sigma^\ell(j) \quad \text{for some } \ell.$$

Then $\sim$ is an equivalence relation and induces a partition of S. For each $1 \le j \le n$, the equivalence class containing j is some r-element set

$$\left\{ j, \sigma(j), \sigma^2(j), \ldots, \sigma^{r-1}(j) \right\}$$

where r is a divisor of m. This set would form one of the disjoint cycles

$$(j \; \sigma(j) \; \sigma^2(j) \; \cdots \; \sigma^{r-1}(j))$$

appearing in the decomposition.

Example 5.1.1. Now we are able to list all elements in S_4 in cycle notation:

- the identity function 1,

- the 2-cycles: (1 2), (1 3), (1 4), (2 3), (2 4), (3 4),

- the product of two 2-cycles: (1 2)(3 4), (1 3)(2 4), (1 4)(2 3),

- the 3-cycles: (1 2 3), (1 3 2), (2 3 4), (2 4 3), (1 3 4), (1 4 3), (1 2 4), (1 4 2),

- the 4-cycles: $(1\ 2\ 3\ 4)$, $(1\ 3\ 4\ 2)$, $(1\ 2\ 4\ 3)$, $(1\ 4\ 3\ 2)$, $(1\ 3\ 2\ 4)$, $(1\ 4\ 2\ 3)$.

The total number of elements in S_4 is $1 + 6 + 3 + 8 + 6 = 24 = 4!$.

Example 5.1.2. Let $\sigma = (1\ 2\ 4\ 3)$ and $\tau = (1\ 2\ 3)$. Write $\sigma\tau$ and $\tau\sigma$ as products of disjoint cycles.

Solution. We have

$$\sigma = \begin{bmatrix} 1 & 2 & 3 & 4 \\ 2 & 4 & 1 & 3 \end{bmatrix} \quad \text{and} \quad \tau = \begin{bmatrix} 1 & 2 & 3 & 4 \\ 2 & 3 & 1 & 4 \end{bmatrix}.$$

Hence

$$\sigma\tau = \begin{bmatrix} 2 & 3 & 1 & 4 \\ 4 & 1 & 2 & 3 \end{bmatrix} \begin{bmatrix} 1 & 2 & 3 & 4 \\ 2 & 3 & 1 & 4 \end{bmatrix} = \begin{bmatrix} 1 & 2 & 3 & 4 \\ 4 & 1 & 2 & 3 \end{bmatrix} = (1\ 4\ 3\ 2).$$

However, with some practice we can speed up the calculation. Remember in the product

$$\sigma\tau = (1\ 2\ 4\ 3)(1\ 2\ 3)$$

we go from right to left. Note that

$$1 \mapsto 2 \mapsto 4, \quad 4 \mapsto 3, \quad 3 \mapsto 1 \mapsto 2, \quad 2 \mapsto 3 \mapsto 1.$$

Thus, $\sigma\tau = (1\ 4\ 3\ 2)$. Similarly,

$$\tau\sigma = (1\ 2\ 3)(1\ 2\ 4\ 3) = (1\ 3\ 2\ 4).$$

We now see that $\sigma\tau \neq \tau\sigma$ and hence S_4 is a non-abelian group.   ◇

Symmetric groups, especially S_n, are very important in group theory. However, these groups tend to be too large to study. Nevertheless, they do contain many interesting subgroups. A subgroup of a symmetric group is called a **permutation group**. In the next section we will study a useful class of permutation groups.

Exercises 5.1

1. In S_3, we let $a = (1\ 2\ 3)$ and $b = (1\ 2)$.

(a) Write $b^{-1}ab$ and $a^{-1}ba$ as products of disjoint cycles.

(b) Show that $a^3 = b^2 = 1$.

2. Decompose the following permutations into products of disjoint cycles.

(a) $\begin{bmatrix} 1 & 2 & 3 & 4 & 5 & 6 \\ 6 & 5 & 4 & 3 & 2 & 1 \end{bmatrix}$

(b) $\begin{bmatrix} 1 & 2 & 3 & 4 & 5 & 6 & 7 \\ 4 & 1 & 2 & 3 & 6 & 7 & 5 \end{bmatrix}$

(c) $\begin{bmatrix} 1 & 4 & 6 & 2 & 7 & 5 & 3 \\ 7 & 5 & 2 & 4 & 1 & 6 & 3 \end{bmatrix}^{-1}$

(d) $\begin{bmatrix} 1 & 2 & 3 & 4 & 5 \\ 2 & 3 & 4 & 5 & 1 \end{bmatrix} \begin{bmatrix} 1 & 2 & 3 & 4 & 5 \\ 3 & 4 & 5 & 1 & 2 \end{bmatrix}$

(e) $\begin{bmatrix} 1 & 2 & 3 & 4 & 5 & 6 \\ 6 & 3 & 2 & 5 & 4 & 1 \end{bmatrix} \begin{bmatrix} 1 & 2 & 3 & 4 & 5 & 6 \\ 3 & 2 & 1 & 6 & 5 & 4 \end{bmatrix}$

(f) $\begin{bmatrix} 1 & 2 & 3 & 4 & 5 \\ 3 & 5 & 4 & 2 & 1 \end{bmatrix}^{-1} \begin{bmatrix} 1 & 2 & 3 & 4 & 5 \\ 1 & 3 & 4 & 2 & 5 \end{bmatrix} \begin{bmatrix} 1 & 2 & 3 & 4 & 5 \\ 3 & 5 & 4 & 2 & 1 \end{bmatrix}$

3. Rewrite the following products of cycles into products of disjoint cycles and find the order of each product.

(a) $(2\ 3\ 4\ 1)(1\ 2\ 3)$

(b) $(1\ 2\ 3)(2\ 3\ 4)(3\ 4\ 1)$

(c) $(1\ 2\ 3\ 4)(2\ 3\ 5)(4\ 5)$

(d) $(1\ 2\ 3\ 4)(2\ 4\ 3\ 6)$

(e) $(1\ 2\ 3)(2\ 3\ 4)(4\ 5\ 6)$

(f) $(1\ 2)(1\ 3)(1\ 4)(1\ 5)$

(g) $(1\ 2)(2\ 3)(3\ 4)(4\ 1)$

(h) $\begin{bmatrix} 1 & 2 & 3 & 4 & 5 & 6 & 7 \\ 3 & 4 & 5 & 6 & 7 & 1 & 2 \end{bmatrix}$

(i) $\begin{bmatrix} 1 & 2 & 3 & 4 & 5 & 6 \\ 6 & 5 & 4 & 3 & 2 & 1 \end{bmatrix}$

$$(j) \quad \begin{bmatrix} 1 & 2 & 3 & 4 & 5 & 6 & 7 & 8 & 9 & 10 \\ 4 & 3 & 9 & 5 & 1 & 6 & 10 & 2 & 8 & 7 \end{bmatrix}$$

4. Show that S_n is non-abelian for $n \geq 3$.

5. Inside S_3, find three subgroups of order 2 and a subgroup of order 3.

6. Show that if σ is any permutation, then

$$\sigma(i_1 \ i_2 \ \cdots \ i_r)\sigma^{-1} = (\sigma(i_1) \ \sigma(i_2) \ \cdots \ \sigma(i_r)).$$

5.2   Dihedral Groups

Certain important examples of groups arise out of symmetry of geometric objects. In this section we study a classical example.

Let $n \geq 3$. There are two kinds of motions which display the symmetry of regular n-gons: **rotations** and **reflections**.

Rotations: Rotate a regular n-gon by the angle of a multiple of $2\pi/n$.

Reflections: On a regular n-gon, we can find n-lines along which the reflections of the regular n-gon remain the same from the outset even though the vertices may have been moved to different positions.

We demonstrate this for the cases $n = 3$ and 4. The rotations of an equilateral triangle are shown in Figure 5.1. If we label the three vertices

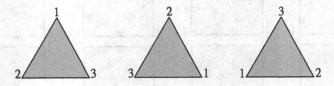

Figure 5.1: Rotations of an equilateral triangle

by 1, 2 and 3, we can express the three rotations by the three permutations

$$1 = \begin{bmatrix} 1 & 2 & 3 \\ 1 & 2 & 3 \end{bmatrix}, \quad \begin{bmatrix} 1 & 2 & 3 \\ 2 & 3 & 1 \end{bmatrix}, \quad \begin{bmatrix} 1 & 2 & 3 \\ 3 & 1 & 2 \end{bmatrix}.$$

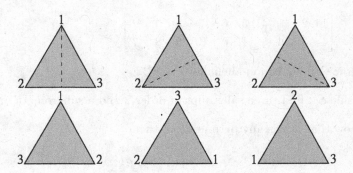

Figure 5.2: Reflections of an equilateral triangle

Moreover, we can "flip" over the triangle along the axes given in Figure 5.2 without changing its appearance. These reflections can be expressed by the following three permutations

$$\begin{bmatrix} 1 & 2 & 3 \\ 1 & 3 & 2 \end{bmatrix}, \quad \begin{bmatrix} 1 & 2 & 3 \\ 3 & 2 & 1 \end{bmatrix}, \quad \begin{bmatrix} 1 & 2 & 3 \\ 2 & 1 & 3 \end{bmatrix}.$$

As we shall see, the rotations and the reflections of the equilateral triangles form a subgroup of S_3, which is called the **dihedral 3-group** and is denoted D_3. However, note that $D_3 = S_3$ since D_3 contains 6 elements.

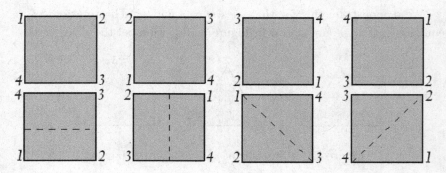

Figure 5.3: Rotations and reflections of a square

On a square we also have the 4 rotations and the 4 reflections as shown in Figure 5.3. Again, these 8 permutations form the **dihedral 4-group**

D_4. Note that in this case, $D_4 \subsetneq S_4$.

In general, we have the n rotations on the regular n-gon:

$$\rho^0 = 1$$

$$\rho = \begin{bmatrix} 1 & 2 & \cdots & n-2 & n-1 & n \\ 2 & 3 & \cdots & n-1 & n & 1 \end{bmatrix}$$

$$\rho^2 = \begin{bmatrix} 1 & 2 & \cdots & n-2 & n-1 & n \\ 3 & 4 & \cdots & n & 1 & 2 \end{bmatrix}$$

$$\vdots$$

$$\rho^{n-1} = \begin{bmatrix} 1 & 2 & \cdots & n-2 & n-1 & n \\ n & 1 & \cdots & n-3 & n-2 & n-1 \end{bmatrix}.$$

It is clear that $o(\rho) = n$ such that

$$\rho^i \rho^j = \rho^{i+j} \qquad \text{and} \qquad (\rho^i)^{-1} = \rho^{n-i}$$

for $0 \leq i \leq n-1$. Hence $\{1, \rho, \rho^2, \ldots, \rho^{n-1}\}$ is an abelian group of order n under multiplication (composition). The rotations form the cyclic group $\langle \rho \rangle$.

We also have the n reflections on the regular n-gon:

$$\gamma_1 = \begin{bmatrix} 1 & 2 & 3 & \cdots & n-2 & n-1 & n \\ 1 & n & n-1 & \cdots & 4 & 3 & 2 \end{bmatrix}$$

$$\gamma_2 = \begin{bmatrix} 1 & 2 & 3 & \cdots & n-2 & n-1 & n \\ 2 & 1 & n & \cdots & 5 & 4 & 3 \end{bmatrix}$$

$$\vdots$$

$$\gamma_n = \begin{bmatrix} 1 & 2 & 3 & \cdots & n-2 & n-1 & n \\ n & n-1 & n-2 & \cdots & 3 & 2 & 1 \end{bmatrix}.$$

It is also obvious that $\gamma_i^2 = 1$, and so $\gamma_i^{-1} = \gamma_i$.

If you observe the rotations and the reflections in Figures 5.1–5.3, we can see that a rotation does not change the direction of the indices around the regular n-gon while a reflection does change the direction of the indices from clockwise to counterclockwise and vice versa. Thus, one can easily check that

$$(\text{rotation})(\text{rotation}) = \text{rotation}$$
$$(\text{rotation})(\text{reflection}) = \text{reflection}$$
$$(\text{reflection})(\text{rotation}) = \text{reflection}$$
$$(\text{reflection})(\text{reflection}) = \text{rotation}$$
$$(\text{rotation})^{-1} = \text{rotation}$$
$$(\text{reflection})^{-1} = \text{reflection}.$$

The n rotations and n reflections form a group of order $2n$ called the **dihedral n-group D_n**. The group D_n is a relatively small subgroup inside S_n. Hence dihedral groups are very useful as examples.

Let α be an arbitrary rotation of order n and let β be an arbitrary reflection in D_n. Remember that $1, \alpha, \ldots, \alpha^{n-1}$ are n distinct rotations in D_n. Note that for $0 \leq i, j \leq n - 1$,

$$\alpha^i \beta = \alpha^j \beta \implies \alpha^i = \alpha^j \implies i = j.$$

Thus $\beta, \alpha\beta, \ldots, \alpha^{n-1}\beta$ are n distinct reflections in D_n. Therefore we conclude that

(5.2.1) $$D_n = \left\{ 1, \alpha, \ldots, \alpha^{n-1}, \beta, \alpha\beta, \ldots, \alpha^{n-1}\beta \right\}.$$

We see that $D_n = \langle \alpha, \beta \rangle$. Since $\alpha\beta$ is a reflection, we have that $\alpha\beta\alpha\beta = 1$. Thus, we have the following important relation in D_n:

$$\beta\alpha\beta^{-1} = \beta\alpha\beta = \alpha^{-1}.$$

This further implies that

$$\beta\alpha^i\beta = (\beta\alpha\beta)^i = \alpha^{-i},$$

which can be rewritten as

$$\beta\alpha^i = \alpha^{-i}\beta.$$

Together with the relations

$$\alpha^n = \beta^2 = 1$$

in D_n, we can rewrite any product of α and β in the form $\alpha^i \beta^j$ where $0 \le i \le n-1$ and $j = 0, 1$. For example, in D_5,

$$\alpha \beta^3 \alpha^{-3} = \alpha \beta \alpha^2 = \alpha(\alpha^{-2}\beta) = \alpha^{-1}\beta = \alpha^4 \beta.$$

It is useful to write D_n as

(5.2.2) $\quad \left\{ 1, \alpha, \ldots, \alpha^{n-1}, \beta, \alpha\beta, \ldots, \alpha^{n-1}\beta : \alpha^n = \beta^2 = 1, \ \beta\alpha = \alpha^{-1}\beta \right\}$

to constantly remind us of the essential relations in D_n. Remember that it suffices to choose an arbitrary rotation of order n for α and an arbitrary reflection for β.

Exercises 5.2

1. Let α be a rotation of order n and β a reflection in D_{10}. Rewrite the following products in the form $\alpha^i \beta^j$ where $0 \le i \le 9$ and $j = 0, 1$.

    (a) $(\alpha\beta)(\alpha^2\beta)$

    (b) $(\beta\alpha)(\beta\alpha^2)$

    (c) $\beta\alpha\beta\alpha\beta\alpha\beta$

    (d) $\beta\alpha^{-2}\beta$

2. Write down every element in D_4 and D_5 in cycle notation.

3. Construct the subgroup lattices of D_4 and D_5.

4. Show that D_n is non-abelian for $n \ge 3$.

5. Find a non-abelian group of order 6. Find a non-abelian group of order 8. (These two groups are in fact the two smallest non-abelian subgroups there are.)

6. Let α be a rotation of order n and β be an arbitrary reflection in D_n. Show that $D_n = \left\{ 1, \alpha, \ldots, \alpha^{n-1}, \beta, \beta\alpha, \ldots, \beta\alpha^{n-1} \right\}$. Show that any element in D_n can be expressed uniquely as $\beta^i \alpha^j$ where $i = 0, 1$ and $0 \le j \le n-1$.

5.3   Alternating Groups

A 2-cycle in S_n is also called a **transposition**. Each transposition is of order 2. We can decompose a cycle into a product of transpositions, which are very likely not disjoint. For example, we have

$$(1\ 2\ 3) = (1\ 3)(1\ 2) \quad \text{and} \quad (1\ 2\ 3\ 4) = (1\ 4)(1\ 3)(1\ 2).$$

In general, we have

(5.3.1) $$(i_1\ i_2\ \cdots\ i_k) = (i_1\ i_k)(i_1\ i_{k-1})\cdots(i_1\ i_2).$$

Since any element in S_n can be expressed as a product of cycles, it can be rewritten as a product of transpositions using (5.3.1). Hence

$$S_n = \langle\,(i\ j)\,\big|\,1 \le i < j \le n\,\rangle$$

is generated by all the transpositions in S_n.

It is worth noting that there are many distinct ways to write a permutation as a product of transpositions. For example, $(2\ 3) = (1\ 2)(1\ 3)(1\ 2)$ are obviously two different expressions. What remains invariant is the *parity* of the number of transpositions in any such product.

Proposition 5.3.1. *Let $\sigma \in S_n$ and write*

$$\sigma = \tau_1\tau_2\cdots\tau_s = \tau_1'\tau_2'\cdots\tau_t'$$

as two products of transpositions. Then s and t are either both even or both odd.

Proof. Let $f(x_1,\ldots,x_n)$ be a polynomial in the variables $x_1,\ldots,x_n$ with integer coefficients. We will now let σ perform an action on the polynomial f to obtain a new polynomial $\sigma(f)$ by permuting the variables in f using σ while leaving the coefficients unchanged. For example, if $f = x_1 - 2x_2 + 3x_3$ and $\sigma = (1\ 2\ 3)$, then $\sigma(f) = x_2 - 2x_3 + 3x_1$. In general, for any two permutations σ and π in S_n, we leave it as a routine exercise to check that

(5.3.2) $$(\pi\sigma)(f) = \pi(\sigma(f)).$$

Let

$$A = \begin{bmatrix} 1 & x_1 & x_1^2 & \cdots & x_1^{n-1} \\ 1 & x_2 & x_2^2 & \cdots & x_2^{n-1} \\ \vdots & & & & \vdots \\ 1 & x_n & x_n^2 & \cdots & x_n^{n-1} \end{bmatrix}.$$

Consider the formal product

$$\chi = \det A = \prod_{i > j} (x_i - x_j).$$

This is the well-known *Vandermonde determinant*.[1] It is clear that

$$\sigma(\chi) = \det \begin{bmatrix} 1 & x_{\sigma(1)} & x_{\sigma(1)}^2 & \cdots & x_{\sigma(1)}^{n-1} \\ 1 & x_{\sigma(2)} & x_{\sigma(2)}^2 & \cdots & x_{\sigma(2)}^{n-1} \\ \vdots & & & & \vdots \\ 1 & x_{\sigma(n)} & x_{\sigma(n)}^2 & \cdots & x_{\sigma(n)}^{n-1} \end{bmatrix} = \prod_{i > j} (x_{\sigma(i)} - x_{\sigma(j)}).$$

Obviously, since $\sigma(\chi)$ is the determinant of a matrix obtained by permuting the rows in A, we have that $\sigma(\chi) = \chi$ or $-\chi$. In particular, if $\tau = (i\ j)$ is a transposition, then $\tau(\chi)$ is the determinant of the matrix obtained by exchanging the i-th and the j-th rows in A. Therefore $\tau(\chi) = -\chi$. Now we can use (5.3.2) to compute that

$$\begin{aligned} \sigma(\chi) &= (\tau_1 \tau_2 \cdots \tau_s)(\chi) = (-1)^s \chi \\ &= (\tau_1' \tau_2' \cdots \tau_t')(\chi) = (-1)^t (\chi). \end{aligned}$$

Hence $(-1)^s = (-1)^t$. We conclude that s and t are both even or both odd. $\qquad\square$

If $\sigma \in S_n$ is a product of an even number of transpositions, we say that σ is an **even** permutation. Otherwise, we say σ is an **odd** permutation. Clearly, we have the following properties on permutations.

(1) The identity permutation is even.

---

[1] We will give a more detailed discussion on determinant in §19.2. Students will be asked to work out the Vandermonde determinant in Exercise 4, §19.2.

(2) The product of two even permutations is even.

(3) The product of two odd permutations is even.

(4) The product of an even permutation and an odd one (or vice versa) is odd.

If $\sigma = \tau_1 \tau_2 \cdots \tau_n$ is a product of transpositions, then $\sigma^{-1} = \tau_n \tau_{n-1} \cdots \tau_1$. Hence the following statements are also true.

(5) The inverse of an even permutation is even.

(6) The inverse of an odd permutation is odd.

Let

$$A_n = \{\, \sigma \in S_n : \sigma \text{ is an even permutation} \,\}.$$

By properties (1), (2) and (5) above, the set A_n is a subgroup of S_n. We call A_n the n-th **alternating group**.

Now let S be the set of all the odd permutations in S_n where $n \geq 2$. If $\sigma \in A_n$, then $(1\ 2)\sigma$ is odd. If $\sigma \in S$, then $(1\ 2)\sigma$ is even. Thus we have the two mappings

$$
\begin{array}{ccc}
A_n \xrightarrow{\varphi} S \\
\sigma \longmapsto (1\ 2)\sigma
\end{array}
\quad \text{and} \quad
\begin{array}{ccc}
S \xrightarrow{\psi} A_n \\
\sigma \longmapsto (1\ 2)\sigma.
\end{array}
$$

It is easy to check that $\psi\varphi = 1_{A_n}$ and $\varphi\psi = 1_S$. Hence φ is a bijective map and we conclude that there are as many even permutations in S_n as there are odd permutations. We now have the following proposition.

Proposition 5.3.2. *The alternating group A_n is a group of order $n!/2$ for $n \geq 2$.*

By using (5.3.1), we can see that an r-cycle can be written as a product of $r - 1$ transpositions. Hence,

$$
\text{an } r\text{-cycle is } \begin{cases} \text{even,} & \text{if } r \text{ is odd;} \\ \text{odd,} & \text{if } r \text{ is even.} \end{cases}
$$

We can then use properties (2)–(4) to determine whether an arbitrary permutation is even or odd.

Exercises 5.3

1. Determine whether each of the following permutations is even or odd.

   (a) $(1\ 2\ 3\ 4\ 5)(2\ 3\ 4)$

   (b) $(1\ 4\ 5)(4\ 2\ 3)(3\ 5\ 6)$

   (c) $(1\ 2)(1\ 3)(1\ 4)(1\ 5)$

   (d) $(2\ 3\ 4\ 5)(2\ 4\ 6\ 7)$

2. Write down explicitly the elements in A_3.

3. Write down explicitly the elements in A_4.

4. Show that $S_n = \langle\, (1\ 2),\ (1\ 3), \ldots, (1\ n)\,\rangle$.

5. Show that $S_n = \langle\, (1\ 2),\ (2\ 3), \ldots, (n-1\ n)\,\rangle$.

6. Show that $S_n = \langle\, (1\ 2),\ (1\ 2\ \cdots\ n-1\ n)\,\rangle$.

7. Show that for $n \geq 3$, A_n is generated by the 3-cycles.

8. Let H be a subgroup of S_n. Show that $H \cap A_n$ is either H or contains exactly half of the elements in H.

Review Exercises for Chapter 5

1. Let T be the set of linear functions from $\mathbb{R}$ into itself,

$$\{\, L_{a,b} \mid L_{a,b}(x) = ax + b,\ a,b \in \mathbb{R},\ a \neq 0\,\}.$$

   (a) Show that $L_{a,b} \circ L_{c,d} = L_{ac,ad+b}$.

   (b) Find $L_{a,b}^{-1}$.

   (c) Show that T is a subgroup of $\mathrm{Sym}(\mathbb{R})$.

2. Let T_n be the function from $\mathbb{Z}$ to $\mathbb{Z}$ defined by

$$T_n(a) = n + a.$$

Suppose that
$$B = \{ T_n \mid n \in \mathbb{Z} \}.$$

Show that B is a subgroup of $\mathrm{Sym}(\mathbb{Z})$.

3. Let σ and τ be in S_n. Show that $o(\sigma\tau\sigma^{-1}) = o(\tau)$.

4. Find $Z(S_n)$ for $n \geq 3$.

5. Find $Z(D_n)$ for $n \geq 3$.

6. Find $Z(A_n)$ for $n \geq 4$.

7. Let $\alpha \in S_n$. Let

$$\alpha A_n = \{ \alpha\pi \in S_n : \pi \in A_n \}$$

and
$$A_n\alpha = \{ \pi\alpha \in S_n : \pi \in A_n \}.$$

Show that $\alpha A_n = A_n\alpha$.

CHAPTER 6

Counting Theorems

Any theorem involving "counting" can be very useful. In this chapter we present two of them.

For each subgroup H of a group G, by using a carefully chosen equivalence relation, we are able to decompose G into a disjoint union of *right cosets* (or *left cosets*, respectively) of H in G. In particular, when G is a finite group, all these cosets, H being one of them, have the same number of elements in them. Hence the order of H is a divisor of the order of G. This is the well-known Lagrange's theorem.

By another carefully chosen equivalence relation, we can establish the *class equation* of a finite group. The class equation is important in analyzing the structure of a group.

6.1 Lagrange's Theorem

Lagrange's theorem asserts that the order of a subgroup of a finite group is a divisor of the order of the group. For example, if G is a group of order 6, G cannot contain a subgroup of order 4. To prove Lagrange's theorem, we have to introduce an equivalence relation on a group.

Let H be a subgroup of a group G. For a, b in G, we define

$$a \sim b \quad \text{if} \quad ab^{-1} \in H.$$

We claim that this relation $\sim$ is an equivalence relation on the group G. To see this, first note that $a \sim a$ since $aa^{-1} = e \in H$. Suppose that $a \sim b$. Then $ab^{-1} \in H$. Since H is a subgroup of G, we have $ba^{-1} = (ab^{-1})^{-1} \in H$. This implies $b \sim a$. At last, suppose that $a \sim b$ and $b \sim c$. Then $ab^{-1} \in H$ and $bc^{-1} \in H$. Thus $(ab^{-1})(bc^{-1}) = a(b^{-1}b)c^{-1} = ac^{-1} \in H$. This shows that $a \sim c$.

Note that the equivalence class $[a]$ of a is

$$Ha := \{\, ha \in G : h \in H \,\}.$$

Suppose that $u \in [a]$. Then $u \sim a$ and $ua^{-1} \in H$. It follows that $ua^{-1} = h$ for some $h \in H$ and $u = ha \in Ha$. This proves that $[a] \subseteq Ha$. Conversely, for $v \in Ha$, $v = ha$ for some $h \in H$. We have that $va^{-1} = h \in H$ and $v \sim a$. Hence $v \in [a]$. This shows that $Ha \subseteq [a]$. We have that $[a] = Ha$.

The set Ha is called a **right coset** of H represented by a. We also say that a is a **representative** of the right coset Ha. Remember that a coset can be represented by any of its members.

By Proposition 1.3.5, we have that the right cosets of H form a partition of G. In particular, if G contains only finitely many right cosets of H (especially when G is finite), then we can find representatives $a_1, \ldots, a_s \in G$ such that

$$G = Ha_1 \,\dot{\cup}\, Ha_2 \,\dot{\cup} \cdots \dot{\cup}\, Ha_s.$$

Remember that $\dot{\cup}$ denotes the disjoint union such that $Ha_j \cap Ha_k = \varnothing$ if $j \neq k$. We call the number of distinct right cosets of H in G the **index** of H in G, and we denote it by $[G : H]$.

Presently we are ready to prove a well-known theorem due to Lagrange on finite groups.

Theorem 6.1.1 (Lagrange's Theorem). *Let H be a subgroup of a finite group G. Then $|H|$ is a divisor of $|G|$. Furthermore, $[G : H] = |G|/|H|$.*

Proof. From our previous discussion, it remains to show that all the right cosets Ha have the same number of elements as H does. We do this by constructing a bijective map between H and Ha.

Define $\varphi\colon H \to Ha$ by letting $\varphi(h) = ha$. The map φ is onto by construction. Suppose that $\varphi(h_1) = \varphi(h_2)$. This means that $h_1 a = h_2 a$. By cancellation law, we get $h_1 = h_2$. Hence φ is also one-to-one. This implies that $|Ha| = |H|$ for all $a \in G$. Now we can conclude that

$$|G| = |Ha_1| + |Ha_2| + \cdots + |Ha_s| = s|H|$$

for a choice of representatives $a_1, \ldots, a_s$ in G. It follows that $|H|$ is a divisor of $|G|$ and $[G : H] = s = |G|/|H|$. $\square$

There is a variation on the proof of Lagrange's theorem if we had started with the equivalence relation

(6.1.1) $$a \sim b \quad \text{if} \quad a^{-1}b \in H.$$

The equivalence class of a is then

$$aH = \{\, ah \mid h \in H \,\}.$$

We leave the verification as an exercise (see Exercise 1).

The set aH is called a **left coset** of H represented by a in G. If G is a finite group, the number of left cosets of H in G is also given by $[G : H] = |G|/|H|$.

Next we give a few examples to describe the coset decompositions of groups.

Example 6.1.2. Remember that in (5.2.1) we have that

$$D_n = \{\, e, \alpha, \alpha^2, \ldots, \alpha^{n-1}, \beta, \alpha\beta, \ldots, \alpha^{n-1}\beta \,\}.$$

where α is a rotation of order n and β is a reflection. Note that

$$H = \langle \alpha \rangle = \{\, e, \alpha, \alpha^2, \ldots, \alpha^{n-1} \mid \alpha^n = e \,\}$$

is a subgroup of G and we have that

$$D_n = H \overset{.}{\cup} H\beta.$$

Similarly we have that

$$D_n = H \overset{.}{\cup} \beta H$$

since H contains the n distinct rotations and βH contains the n distinct reflections. In this case $H\beta = D_n \setminus H = \beta H$.

Example 6.1.3. Remember that $S_3 = D_3$. Let $\alpha = (1\ 2\ 3)$, a rotation of order 3, and $\beta = (1\ 2)$, a reflection. Thus

$$S_3 = \left\{ e, \alpha, \alpha^2, \beta, \alpha\beta, \alpha^2\beta \right\} = \left\{ e, \alpha, \alpha^2, \beta, \beta\alpha, \beta\alpha^2 \right\}.$$

Let $K = \langle \alpha \rangle$. Then

$$S_3 = K \overset{.}{\cup} K\beta = K \overset{.}{\cup} \beta K.$$

Let $H = \langle \beta \rangle = \{ e, \beta \}$. Then

$$S_3 = H \overset{.}{\cup} H\alpha \overset{.}{\cup} H\alpha^2 = H \overset{.}{\cup} \alpha H \overset{.}{\cup} \alpha^2 H.$$

However, note that $H\alpha \neq \alpha H$ since $\beta\alpha = \alpha^{-1}\beta = \alpha^2\beta$ is in $H\alpha$ but not in αH. In general, the left cosets and the right cosets are not necessarily the same.

As immediate results of Lagrange's Theorem, we have the following corollaries.

Corollary 6.1.4. *Let G be a finite group and $a \in G$. Then $o(a)$ is a divisor of $|G|$ and in particular, $a^{|G|} = e$.*

Proof. Suppose that $m = o(a)$. Then $\langle a \rangle$ is a subgroup of order m in G. Thus m is a divisor of the order of G. The rest is trivial. $\qquad\square$

Corollary 6.1.5 (Euler's Generalization of Fermat's Little Theorem). *Let a and n be relatively prime positive integers. Then $a^{\varphi(n)} \equiv 1 \pmod{n}$.*

Proof. It follows from the fact that the multiplicative group $U(\mathbb{Z}_n)$ is of order $\varphi(n)$. $\qquad\square$

Corollary 6.1.6. *Let G be a group of prime order p. Then G is cyclic and G has no proper nontrivial subgroups. In fact, G is generated by any nontrivial element in G.*

Proof. Let H be a subgroup of G. Then $|H|$ is a divisor of p by Lagrange's theorem. This implies that $|H| = 1$ or $|H| = p$. Thus H is either trivial or G. Let $a \in G$ and $a \neq e$. Then $\langle a \rangle$ is a nontrivial subgroup of G and thus $G = \langle a \rangle$ is cyclic. □

Example 6.1.7. In this example we see that A_4 arises as a symmetry group of a regular tetrahedron, each face of which is an equilateral triangle. We use $\mathscr{S}$ to denote this symmetry group, which is a proper subgroup of S_4. By rotating the side faces of the tetrahedron in Figure 6.1 while leaving the vertex 1 unmoved, we see that (2 3 4) and (2 4 3) are in $\mathscr{S}$. Similarly, (1 2 3), (1 3 2), (1 2 4), (1 4 2), (1 3 4), (1 4 3) are also in the symmetry group $\mathscr{S}$. Since $\mathscr{S}$ contains all the 3-cycles of S_4, $\mathscr{S}$ contains A_4 by Exercise 7, §5.3.

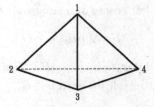

Figure 6.1: The motions of a regular tetrahedron

By Lagrange's theorem, $|\mathscr{S}| = 12$ or 24. However, any permutation in $\mathscr{S}$ must leave one of the vertices of the tetrahedron fixed. We conclude that $\mathscr{S} \neq S_4$ and hence $\mathscr{S} = A_4$.

Exercises 6.1

1. Verify that the relation $\sim$ in (6.1.1) is an equivalence relation on the set G. Verify that the equivalence class of $a \in G$ is aH.

2. Let $V = \{e, a, b, c\}$ be the Klein 4-group and $H = \{e, a\}$. Find all the left cosets of H in V.

3. Find all the cosets of the subgroup $6\mathbb{Z}$ in $2\mathbb{Z}$.

4. Let H be a subgroup of a group G and a, $b \in G$. Prove or disprove the following statements.

   (a) If $Ha = Hb$, then $b \in Ha$ and $a \in Hb$.

   (b) If $Ha = Hb$, then $aH = bH$.

   (c) If $aH = bH$, then $Ha^{-1} = Hb^{-1}$.

   (d) If $aH = bH$, then $a^2 H = b^2 H$.

5. Let H be a subgroup of a group G. Show that the following statements are equivalent:

   (i) $ghg^{-1} \in H$ for all $h \in H$ and all $g \in G$;

   (ii) $gH = Hg$ for all $g \in G$.

6. Let S be the set of left cosets of a subgroup H in a group G, that is,

$$S = \{ gH \mid g \in G \}.$$

Define $\varphi_a \colon S \to S$ by $\varphi_a(gH) = agH$. Prove that φ_a is a bijective function from S onto itself for each $a \in G$.

7. Let H be a subgroup of a group G. Let Γ_L be the collection of left cosets of H in G and let Γ_R be the collection of right cosets of H in G. Let $\varphi \colon \Gamma_L \to \Gamma_R$ be the function sending aH to Ha^{-1} for $a \in G$. Show that φ is well-defined and bijective. Hence Γ_L and Γ_R are of the same cardinality. Thus, the index of H in G can be defined to be the cardinality of Γ_L or of Γ_R.

6.2   Conjugacy Classes of a Group

Let G be a group, we define

$$a \sim b \quad \text{if} \quad b = xax^{-1} \quad \text{for some } x \in G.$$

The relation $\sim$ is an equivalence relation on G as shown below.

(1) As $a = eae^{-1}$ or aaa^{-1}, we have $a \sim a$.

(2) If $a \sim b$, then $b = xax^{-1}$ for some $x \in G$. It follows that $a = x^{-1}bx = x^{-1}b(x^{-1})^{-1}$ and hence $b \sim a$.

(3) If $a \sim b$ and $b \sim c$, then $b = xax^{-1}$ and $c = yby^{-1}$ for some x and y in G. It follows that $c = yby^{-1} = y(xax^{-1})y^{-1} = (yx)a(yx)^{-1}$. Consequently, we have $a \sim c$.

When $a \sim b$ we say that a is **conjugate to** b in G. We also say that b is a **conjugate** of a in G. Conjugation plays an important role in group theory. The equivalence class of a under conjugation is called the **conjugacy class** containing a in G and it is given by

$$[a] = \left\{ gag^{-1} \in G : g \in G \right\}.$$

Proposition 6.2.1. *Let G be a group and $a \in G$. The set*

$$C_G(a) = \{ g \in G : ga = ag \} = \{ g \in G : gag^{-1} = a \}$$

is a subgroup of G. In case G is finite, we have that $\left| [a] \right| = \dfrac{|G|}{|C_G(a)|}.$

The subgroup $C_G(a)$ (or simply $C(a)$ when G is understood) is called the **centralizer** of a in G.

Proof. We leave the verification that $C(a)$ is a subgroup of G as an exercise (see Exercise 1). To prove this proposition, we shall establish a bijective map φ between the left cosets of $C(a)$ in G and the elements in the conjugacy class $[a]$.

Define $\varphi(gC(a)) = gag^{-1}$. We first check that φ is well-defined. Suppose $gC(a) = g_1C(a)$. Then $g^{-1}g_1 \in C(a)$. This implies that

$$(g^{-1}g_1)a(g^{-1}g_1)^{-1} = a.$$

Hence,

$$g_1ag_1^{-1} = g(g^{-1}g_1)a(g^{-1}g_1)^{-1}g^{-1} = gag^{-1}.$$

Clearly, φ is onto. Now suppose that

$$\varphi(gC(a)) = gag^{-1} = \varphi(g_1C(a)) = g_1ag_1^{-1}.$$

Then $g_1^{-1}gag^{-1}g_1 = a$ and thus $g_1^{-1}g \in C(a)$. It follows that $g_1C(a) = gC(a)$ and φ is one-to-one. We now have that

$$\big|[a]\big| = [G : C(a)] = |G|/|C(a)|. \qquad \square$$

If G is a finite group, then we can find representatives $a_1, \ldots, a_t$ in G such that

$$G = [a_1] \,\dot\cup\, [a_2] \,\dot\cup \cdots \dot\cup\, [a_t].$$

Remember that the center $Z(G)$ of a group G is a subgroup of G (see Exercise 6, §3.3). It is easy to see that $a \in Z(G)$ if and only if $C(a) = G$ if and only if $[a] = \{\, a \,\}$ using Proposition 6.2.1. Thus $Z(G)$ is the union of all the conjugacy classes which are singleton sets. Without loss of generality, we assume that $Z(G) = \{\, a_{s+1}, \ldots, a_t \,\}$. Then

$$G = Z(G) \,\dot\cup\, [a_1] \,\dot\cup\, [a_2] \,\dot\cup \cdots \dot\cup\, [a_s], \qquad \text{where } \big|[a_i]\big| \geq 2.$$

Consequently, we obtain the **class equation** for G:

$$(6.2.1) \qquad\qquad |G| = |Z(G)| + \sum_{i=1}^{s} \frac{|G|}{|C(a_i)|},$$

where $C(a_i)$ is a proper subgroup of G for each i.

Example 6.2.2. Remember that $S_3 = D_3 = \{\, e, \alpha, \alpha^2, \beta, \alpha\beta, \alpha^2\beta \,\}$ with $\alpha = (1\ 2\ 3)$ and $\beta = (1\ 2)$. Since we have that

$$C(\alpha) = \{\, e, \alpha, \alpha^2 \,\}, \quad C(\beta) = \{\, e, \beta \,\} \quad \text{and} \quad Z(S_3) = \{\, e \,\},$$

the class equation for S_3 is

$$6 = |S_3| = |Z(S_3)| + \frac{|S_3|}{|C(\alpha)|} + \frac{|S_3|}{|C(\beta)|} = 1 + 2 + 3.$$

To determine the class equation for S_n in general, we need to know exactly when two permutations are conjugate to each other. Remember that every permutation can be written as a product of disjoint cycles

$$\pi = \gamma_1 \gamma_2 \cdots \gamma_t,$$

where the length of γ_i is r_i. Since disjoint cycles commute with each other, we can arrange it so that

$$r_1 \geq r_2 \geq \cdots \geq r_s > r_{s+1} = \cdots = r_t = 1$$

and $\sum_{i=1}^{t} r_i = n$. We call π a permutation of type $(r_1, \ldots, r_t)$. The type may be simplified as $(r_1, \ldots, r_s)$ where the dropped 1's are understood implicitly. For example, in S_6 we say the permutation $(1\ 2\ 3)(4\ 5)$ is of type $(3, 2, 1)$ or simply $(3, 2)$.

We have the following results regarding conjugates and types.

Lemma 6.2.3. *Let $\pi = (i_1\ i_2\ \cdots\ i_r)(j_1\ j_2\ \cdots\ j_s)\cdots(k_1\ k_2\ \cdots\ k_t)$. Then for all $\sigma \in S_n$, $\sigma\pi\sigma^{-1}$ is the product*

$$\big(\sigma(i_1)\ \sigma(i_2)\ \cdots\ \sigma(i_r)\big)\big(\sigma(j_1)\ \sigma(j_2)\ \cdots\ \sigma(j_s)\big)\cdots\big(\sigma(k_1)\ \sigma(k_2)\ \cdots\ \sigma(k_t)\big).$$

Moreover, a permutation is of the same type as any of its conjugates.

Proof. First note that if $\pi = (i_1\ i_2\ \cdots\ i_r)$ is an r-cycle, then

$$\sigma\pi\sigma^{-1} = (\sigma(i_1)\ \sigma(i_2)\ \cdots\ \sigma(i_r))$$

is also an r-cycle (see Exercise 6, §5.1). In general, if $\pi = \gamma_1\gamma_2\cdots\gamma_s$ where the γ_i's are cycles, then

$$\sigma\pi\sigma^{-1} = (\sigma\gamma_1\sigma^{-1})(\sigma\gamma_2\sigma^{-1})\cdots(\sigma\gamma_s\sigma^{-1}),$$

and we are done. The last statement of this lemma is obvious. $\square$

Proposition 6.2.4. *In S_n, two permutations are conjugate to each other if and only if they are of the same type.*

Proof. The "only if" part follows from Lemma 6.2.3. On the other hand, suppose given the disjoint cycle decompositions of two permutations of the same type

$$\pi = (i_1\ i_2\ \cdots\ i_r)(j_1\ j_2\ \cdots\ j_s)\cdots(k_1\ k_2\ \cdots\ k_t)$$

and

$$\pi' = (i_1'\ i_2'\ \cdots\ i_r')(j_1'\ j_2'\ \cdots\ j_s')\cdots(k_1'\ k_2'\ \cdots\ k_t')$$

where we include all the 1-cycles. Then $\{\,i_1, \ldots, i_r, j_1, \ldots, j_s, \ldots, k_1, \ldots, k_t\,\}$ and $\{\,i_1', \ldots, i_r', j_1', \ldots, j_s', \ldots, k_1', \ldots, k_t'\,\}$ are both $\{\,1, 2, \ldots, n\,\}$. If we define

$$\sigma = \begin{bmatrix} i_1 & \cdots & i_r & j_1 & \cdots & j_s & k_1 & \cdots & k_t \\ i_1' & \cdots & i_r' & j_1' & \cdots & j_s' & k_1' & \cdots & k_t' \end{bmatrix},$$

then $\pi' = \sigma\pi\sigma^{-1}$ is conjugate to π. $\square$

Proposition 6.2.4 makes it easy to find the class equations of S_n.

Example 6.2.5. We now determine the class equation of S_4. To do this, we find a representative for each conjugacy class in S_4. We will also find the centralizers in S_4 for these representatives.

Type	Representative	# of elements in this conjugacy class	$|C(\alpha)|$
(4)	(1 2 3 4)	$3! = 6$	4
(3, 1)	(1 2 3)	$\binom{4}{3}2! = 8$	3
(2, 2)	(1 2)(3 4)	$\binom{4}{2}\binom{2}{2}/2 = 3$	8
(2, 1, 1)	(1 2)	$\binom{4}{2} = 6$	4
(1, 1, 1, 1)	1	1	24

Clearly, $C(1) = S_4$. To compute $\|[(1\ 2\ 3\ 4)]\|$ is to find the number of 4-cycles in S_4. Fix 1 at the first position in a 4-cycle. Each distinct ordering of 2, 3 and 4 would give us a different 4-cycle. Hence there are exactly 3! distinct 4-cycles. Since clearly $C((1\ 2\ 3\ 4))$ contains $(1\ 2\ 3\ 4)$, it contains the subgroup $\langle (1\ 2\ 3\ 4) \rangle$. From the table above we know that $|C((1\ 2\ 3\ 4))| = 24/6 = 4$. It follows that

$$C((1\ 2\ 3\ 4)) = \langle (1\ 2\ 3\ 4) \rangle.$$

Similarly, we have that

$$C((1\ 2\ 3)) = \langle (1\ 2\ 3) \rangle.$$

By the table above, $|C((1\ 2))| = 24/6 = 4$. Since (1 2) and (3 4) commute with (1 2), we can see that

$$C((1\ 2)) = \langle (1\ 2),\ (3\ 4) \rangle = \{\, 1,\ (1\ 2),\ (3\ 4),\ (1\ 2)(3\ 4) \,\}.$$

The most tricky part is how to find $\|[(1\ 2)(3\ 4)]\|$ and the centralizer of (1 2)(3 4). To construct a permutation of type $(2, 2)$, we first pick 2 digits from $\{1, 2, 3, 4\}$ to make up the first 2-cycle and there are $\binom{4}{2}$ of these. Then we pick 2 digits from the rest of the 2 digits and there are $\binom{2}{2}$ of these. However, different orderings of the 2-cycles would give the same permutation and there are 2! different orderings of the two 2-cycles. Thus, altogether there are $\binom{4}{2}\binom{2}{2}/2! = 3$ elements in $[(1\ 2)(3\ 4)]$.

To find $C((1\ 2)(3\ 4))$, note that $(1\ 2)$ and $(3\ 4)$ both commute with $(1\ 2)(3\ 4)$. Thus, $C((1\ 2)(3\ 4))$ contains

$$\langle\,(1\ 2),\ (3\ 4)\,\rangle = \{\,1,\ (1\ 2),\ (3\ 4),\ (1\ 2)(3\ 4)\,\}.$$

From the table above, we see that there are 8 elements in the centralizer, and so we must find other permutations in it. We need to find β such that $\beta(1\ 2)(3\ 4)\beta^{-1} = (1\ 2)(3\ 4) = (3\ 4)(1\ 2)$. Use Lemma 6.2.3 we may choose

$$\beta = \begin{bmatrix} 1 & 2 & 3 & 4 \\ 3 & 4 & 1 & 2 \end{bmatrix} = (1\ 3)(2\ 4).$$

The order of $\langle\,(1\ 2),(3\ 4),(1\ 3)(2\ 4)\,\rangle$ is greater than 4 and is a divisor of 8. Hence it must be of order 8 and

$$C((1\ 2)(3\ 4)) = \langle\,(1\ 2),\ (3\ 4),\ (1\ 3)(2\ 4)\,\rangle.$$

From the table above we can see that $Z(S_4) = \{1\}$ and we conclude that the class equation for S_4 is

$$24 = 1 + 6 + 8 + 3 + 6.$$

Example 6.2.6. How many permutations of type $(3,3,2,2,2,1,1,1)$ are there in S_{15}?

Solution. There are $\binom{15}{3}2!\binom{12}{3}2!\binom{9}{2}\binom{7}{2}\binom{5}{2}/2!3!$ of them.   ◇

Example 6.2.7. We now determine the class equation of A_4. The difficulty here is that even though the conjugates are of the same type, permutations of the same type may not be conjugates. However, since for all $\alpha \in A_4$, we have that

$$|[\alpha]| = \frac{12}{|C_{A_4}(\alpha)|},$$

it suffices to know the order of the centralizers in A_4. It is clear that

$$C_{A_4}(\alpha) = C_{S_4}(\alpha) \cap A_4.$$

If $C_{S_4}(\alpha)$ contains only even permutations, then $C_{A_4}(\alpha) = C_{S_4}(\alpha)$. In this case, the conjugacy class of α in S_n will break into two conjugacy classes in A_4, each with half of the elements. On the other hand, if $C_{S_4}(\alpha)$ contains an odd permutation, then $C_{A_4}(\alpha)$ will contain only half of the elements in $C_{S_4}(\alpha)$ (see Exercise 8, §5.3). In this case, the conjugacy class of α in A_4 is the same as the conjugacy class of α in S_4.

In A_4, the even permutations are of types $(3,1)$, $(2,2)$ and $(1,1,1,1)$. In the previous example, we have that

$$C_{S_4}(1) = S_4 \quad \text{and} \quad C_{S_4}((1\ 2)(3\ 4)) = \langle\, (1\ 2),\ (3\ 4),\ (1\ 3)(2\ 4)\,\rangle,$$

both of which contain odd permutations. Hence, the conjugacy classes of types $(1,1,1,1)$ and $(2,2)$ remain intact in A_4. Here $|[1]_{A_4}| = 1$ and $|[(1\ 2)(3\ 4)]_{A_4}| = 3$. However,

$$C_{S_4}((1\ 2\ 3)) = \langle\, (1\ 2\ 3)\,\rangle.$$

Thus, the conjugacy class of type $(3,1)$ in S_4 splits into two conjugacy classes in A_4, each with 4 elements. Hence, the class equation of A_4 is

$$12 = 1 + 4 + 4 + 3.$$

Next we give an important consequence of the class equation.

Corollary 6.2.8. *If G is a finite group of order p^n where p is prime, then p is a divisor of $|Z(G)|$. In other words, $Z(G)$ is a nontrivial subgroup of G.*

Proof. For each $a \in G \setminus Z(G)$, $C(a)$ is a proper subgroup of G and its order is p^k, $k < n$, by Lagrange's theorem. It follows that p is a divisor of $|G|/|C(a)|$. Let (6.2.1) be the class equation of G. Thus we have that p divides $|Z(G)|$ since p both divides $|G|$ and $\sum_{i=1}^{s} |G|/|C(a_i)|$. We conclude that $Z(G)$ contains at least p elements. $\square$

Exercises 6.2

1. Let G be a group and $a \in G$. Show that $C_G(a)$ is a subgroup of G.

2. In a permutation group, show that the conjugate of any odd permutation is odd and the conjugate of any even permutation is even.

3. Let H be a subgroup of G and let $g \in G$. Show that

$$gHg^{-1} = \{\, ghg^{-1} \in G : h \in H \,\}$$

is also a subgroup of G. In fact, gHg^{-1} is called a **conjugation group** of H in G. Show that the orders of H and gHg^{-1} are the same when H is finite.

4. In a group G, we say that a subgroup H is **conjugate** to another subgroup K if there exists $g \in G$ such that $K = gHg^{-1}$. Show that the conjugation of subgroups of G is an equivalence relation.   •

5. Determine the class equation of D_4. Find a representative for each conjugacy class in D_4 and find the centralizer for each representative.

6. Determine the class equation of S_5. Find a representative for each conjugacy class in S_5 and find the centralizer for each representative.

7. Determine the class equation of A_5.

Review Exercises for Chapter 6

1. Suppose that a is conjugate to b in a group G. Show that $C_G(a)$ is also conjugate to $C_G(b)$ in G.

2. In a group, let $aba^{-1} = b^k$, $a^r = e$ and $b^p = e$. Show that $k^r \equiv 1 \pmod{p}$.

3. Let H be a subgroup of a group G in which $aH = bH$ implies $Ha = Hb$. Show that $aHa^{-1} = H$ for every $a \in G$.

4. Let H and K be subgroups of a finite abelian group G. Suppose that

$$HK = \{\, hk \in G : h \in H, \; k \in K \,\}.$$

   (a) Show that $|HK| = |H||K|/|H \cap K|$. (Hint: Let $\alpha \colon H \times K \to HK$ be the map which sends (h, k) to hk. Study the cardinality of $\alpha^{-1}(hk)$. *Cf.* Theorem 8.3.10.)

(b) If $|H|$ and $|K|$ are relatively prime, show that $|HK| = |H||K|$.

5. Let H and K be subgroups of a group G. Define $\sim$ on G by letting

$$a \sim b \iff a = hbk$$

for some $h \in H$ and $k \in K$.

(a) Show that $\sim$ is an equivalence relation.

(b) Show that the equivalence class of a is

$$HaK = \{\, hak \in G : h \in H,\ k \in K \,\}.$$

This is called the **double coset** of a relative to the pair (H, K).

(c) If G is finite, show that

$$|HaK| = |H||[K : a^{-1}Ha \cap K]|.$$

CHAPTER 7

Group Homomorphisms

The term *homomorphism* comes from the Greek words *homo* (like) and *morphe* (form). We will use this terminology to mean maps preserving form (the algebraic structures of groups, rings, etc.). The concept of group homomorphisms was introduced by Camille Jordan (1838–1922) in 1870 in his book *Traité des Substitutions*.

One cannot judge a group just by its appearance. For example, $\mathbb{Z}_2$ and S_2 are both cyclic groups of order 2. They are basically the same group except for the difference in notations. In this chapter we will establish what it means for two groups to be "the same". We will use this to show that all group are fundamentally permutation groups.

7.1   Examples and Basic Properties

A group homomorphism from a group G_1 to another group G_2 is a function from G_1 into G_2 which preserves the operations of groups. To be precise, we have the following definition.

Definition 7.1.1. Let $(G_1, \cdot)$ and $(G_2, *)$ be groups. We say a mapping $\varphi \colon G_1 \to G_2$ is a **group homomorphism** if

$$\varphi(a \cdot b) = \varphi(a) * \varphi(b)$$

for all a and $b \in G_1$.

In other words, a group homomorphism carries the operation of the domain into the operation of the codomain.

Example 7.1.2. Let $\mathbb{Z}$ be the additive group of integers and G be the multiplicative group of n-th roots of unity

$$G = \{1, \zeta, \ldots, \zeta^{n-1} : \zeta = \cos(2\pi/n) + i\sin(2\pi/n)\}.$$

Define $\varphi \colon \mathbb{Z} \to G$ by letting $\varphi(j) = \zeta^j$. Then we have

$$\varphi(i + j) = \zeta^{i+j} = \zeta^i \zeta^j = \varphi(i)\varphi(j).$$

The function φ is indeed a group homomorphism from $\mathbb{Z}$ into G.

Example 7.1.3. It is easy to check that both $(\mathbb{R}_+, \cdot)$ and $(\mathbb{R}, +)$ are groups. Define $\varphi \colon \mathbb{R}_+ \to \mathbb{R}$ by letting $\varphi(x) = \log x$. Then φ is a group homomorphism since

$$\varphi(xy) = \log xy = \log x + \log y = \varphi(x) + \varphi(y).$$

On the other hand, the function $\psi \colon \mathbb{R} \to \mathbb{R}^+$ sending x to e^x is also a group homomorphism from $(\mathbb{R}, +)$ into $(\mathbb{R}_+, \cdot)$ since

$$\psi(x + y) = e^{x+y} = e^x e^y = \psi(x)\psi(y).$$

Example 7.1.4. Define $\varphi \colon \mathbb{Z} \to \mathbb{Z}_n$ by letting $\varphi(m) = \overline{m}$. Since

$$\varphi(m + n) = \overline{m + n} = \overline{m} + \overline{n} = \varphi(m) + \varphi(n),$$

we conclude that φ is a group homomorphism.

Example 7.1.5. Let x be a variable and let $\mathbb{Q}[x]$ be the set of polynomial functions with rational coefficients. For fixed $a \in \mathbb{Q}$, define $\varphi \colon \mathbb{Q}[x] \to \mathbb{Q}$ by letting $\varphi[f(x)] = f(a)$. Then we have

$$\varphi[f(x) + g(x)] = f(a) + g(a) = \varphi[f(x)] + \varphi[g(x)].$$

Hence φ is a group homomorphism.

We list a few of the basic properties regarding group homomorphisms below.

Proposition 7.1.6. *Let $\varphi \colon G \to G'$ be a group homomorphism. Then φ has the following basic properties.*

(a) *Let e and e' be identities of G and G' respectively. Then $\varphi(e) = e'$.*

(b) *For $a \in G$ and for all $n \in \mathbb{Z}$, we have that $\varphi(a^{-1}) = \varphi(a)^{-1}$ and $\varphi(a^n) = \varphi(a)^n$.*

(c) *If $o(a) = n < \infty$, then $o(\varphi(a)) \mid n = o(a)$.*

(d) *For a and b in G, $\varphi(a)\varphi(b) = \varphi(b)\varphi(a)$ if $ab = ba$.*

(e) *Let $G = \langle S \rangle$. Then φ is totally determined by the images of the elements in S.*

(f) *Suppose φ is onto. If $G = \langle g \rangle$ then $G' = \langle \varphi(g) \rangle$.*

Proof. Since $\varphi(e)^2 = \varphi(e^2) = \varphi(e) = \varphi(e)e'$, part (a) follows by cancellation law. Part (b) is left as an exercise (see Exercise 1). Part (c) follows from the fact that $\varphi(a)^n = \varphi(a^n) = \varphi(e) = e'$. Part (d) is trivial and part (e) follows from Proposition 3.4.3. Part (f) is true since for any $a \in G'$, $a = \varphi(g^k) = \varphi(g)^k$ for some $k \in \mathbb{Z}$. $\qquad\square$

It is straightforward to understand a group homomorphism when its domain is the additive group $\mathbb{Z}$.

Proposition 7.1.7. *Let G be a group and let a be any element in G. Then there is a unique group homomorphism from $\mathbb{Z}$ to G sending 1 to a.*

Proof. Since $\mathbb{Z} = \langle 1 \rangle$, such homomorphism is unique if it exists by Proposition 7.1.6(e) above. It remains to show that the map φ from $\mathbb{Z}$ to G sending k to a^k is such a group homomorphism. This is true since for k and $\ell \in \mathbb{Z}$, we have $\varphi(k+\ell) = a^{k+\ell} = a^k a^\ell = \varphi(k)\varphi(\ell)$. $\qquad\square$

Proposition 7.1.8. *Let*

$$C_n = \langle a \rangle = \left\{ e, a, \ldots, a^{n-1} : a^n = e \right\}$$

be the cyclic group of order n and let G be a group. For $b \in G$, there is a group homomorphism from C_n to G sending a to b if and only if $b^n = e'$, the identity of G. Such a homomorphism is unique if it exists.

Proof. Let φ be a group homomorphism from C_n to G sending a to b. Then $b^n = \varphi(a)^n = \varphi(a^n) = \varphi(e) = e'$. This proves the "only if" part. Conversely, let $b \in G$ is such that $b^n = e'$. Define a map φ from C_n to G by sending a^i to b^i. First we need to check that φ is well defined. By Proposition 4.1.3, $a^i = a^j$ if and only if $n \mid i - j$. This implies that $b^{i-j} = e'$ and thus $b^i = b^j$. It then is easy to check that φ is a group homomorphism. Finally, the uniqueness part follows from Proposition 7.1.6(e) again. $\qquad\square$

Example 7.1.9. In this example we aim to find all the group homomorphisms from C_4 into C_6.

Let $C_4 = \left\{ e, a, a^2, a^3 \right\}$ and $C_6 = \left\{ e', b, b^2, b^3, b^4, b^5 \right\}$ with $o(a) = 4$ and $o(b) = 6$. Suppose that $\varphi \colon C_4 \to C_6$ is a group homomorphism. Then φ is wholly determined by $\varphi(a)$. Let $\varphi(a) = b^k$. By Proposition 7.1.8, we have that $b^{4k} = e$ and thus $4k \equiv 0 \pmod 6$. The solutions are $k = 0$ or $3 \pmod 6$. Thus, there are exactly two group homomorphism from C_4 to C_6. The first one is the trivial homomorphism since it sends a to e'. The second one is the homomorphism sending a to b^3.

Exercises 7.1

1. Prove Proposition 7.1.6(b).

2. Find all the group homomorphisms from G_1 into G_2.

(a) $G_1 = \{e, a, a^2 \mid a^3 = e\};$ $G_2 = \{e, b, b^2, b^3 \mid b^4 = e\}.$

(b) $G_1 = \mathbb{Z}_{12};$ $G_2 = \mathbb{Z}_8.$

(c) $G_1 = \mathbb{Z}_4;$ $G_2 = \{e, a, b, c\}$ is the Klein 4-group.

(d) $G_1 = \mathbb{Z};$ $G_2 = S_3.$

3. Suppose that m, n are relatively prime integers. Show that the only group homomorphism $\varphi \colon \mathbb{Z}_m \to \mathbb{Z}_n$ is the trivial homomorphism.

4. Let G, G' be groups and let a be any element in G'. If $G = \langle g \rangle$ is of infinite order, show that there is exactly one group homomorphism from G to G' sending g to a. (Note that this group homomorphism sends g^i to a^i.)

7.2   Isomorphisms

An injective group homomorphism is called a group **monomorphism**. A surjective group homomorphism is called a group **epimorphism**. At last, a bijective group homomorphism is called a group **isomorphism**. We say G_1 is **isomorphic to** G_2, denoted $G_1 \cong G_2$, if there is a group isomorphism from G_1 to G_2. The group homomorphisms in Example 7.1.3 are both isomorphisms. In fact, they are inverse to each other.

Example 7.2.1. Let $G = \{e, a, \ldots, a^{n-1} \mid a^n = e\}$ be a cyclic group of order n and $\mathbb{Z}_n = \{0, 1, \ldots, n-1\}$ be the additive group modulo n. Define $\varphi \colon G \to \mathbb{Z}_n$ by $\varphi(a^j) = j$. Then φ is a group isomorphism and hence $G \cong \mathbb{Z}_n$.

Example 7.2.2. Let

$$K = \left\{ \begin{bmatrix} a & b \\ -b & a \end{bmatrix} \;\middle|\; a, b \in \mathbb{R}, \ (a, b) \neq (0, 0) \right\}.$$

We leave it as an exercise to verify that K is a multiplicative group (see Exercise 2). Let $\mathbb{C}^*$ denote the set of all nonzero complex numbers. Define $\varphi \colon \mathbb{C}^* \to K$ by letting

$$\varphi(a + bi) = \begin{bmatrix} a & b \\ -b & a \end{bmatrix}, \qquad \text{where } a, b \in \mathbb{R}.$$

Then

$$\varphi((a+bi)(c+di)) = \varphi((ac-bd)+(bc+ad)i) = \begin{bmatrix} ac-bd & bc+ad \\ -(bc+ad) & ac-bd \end{bmatrix}$$

and

$$\varphi(a+bi)\varphi(c+di) = \begin{bmatrix} a & b \\ -b & a \end{bmatrix} \begin{bmatrix} c & d \\ -d & c \end{bmatrix} = \begin{bmatrix} ac-bd & bc+ad \\ -(bc+ad) & ac-bd \end{bmatrix}.$$

Hence we have that

$$\varphi((a+bi)(c+di)) = \varphi(a+bi)\varphi(c+di).$$

It is trivial to check that φ is one-to-one and onto, and so $\mathbb{C}^* \cong K$.

Proposition 7.2.3. *If $\varphi\colon G \to G'$ and $\psi\colon G' \to G''$ are group homomorphisms, so is $\psi\varphi$. Furthermore, if φ and ψ are isomorphisms, then so are $\psi\varphi$ and φ^{-1}.*

Proof. For a and $b \in G$, we have $(\psi\varphi)(ab) = \psi(\varphi(ab)) = \psi(\varphi(a)\varphi(b)) = \psi(\varphi(a))\psi(\varphi(b)) = (\psi\varphi)(a)(\psi\varphi)(b)$. Hence $\psi\varphi$ is a group homomorphism. When φ and ψ are both isomorphisms, so is $\psi\varphi$ by Corollary 1.2.3. At last, when the group homomorphism φ is bijective, it remains to show that φ^{-1} is a group homomorphism by Theorem 1.2.5.

Let $c, d \in G'$ and let a and b be the unique elements in G such that $\varphi(a) = c$ and $\varphi(b) = d$. Note that $\varphi(ab) = cd$. Thus $\varphi^{-1}(cd) = ab = \varphi^{-1}(c)\varphi^{-1}(d)$ and φ^{-1} is indeed a group homomorphism. $\square$

Since the identity map of any group is a group isomorphism, Proposition 7.2.3 above guarantees that $\cong$ is an equivalence relation. Thus, $\cong$ partitions all groups into classes of isomorphic groups. An isomorphism between two groups gives an one-to-one correspondence between the elements of the two groups. More than that, it also carries the binary operation of one group into the other. From Proposition 7.1.6(a) and (b) we see that the identity corresponds to the identity, the inverse of an element corresponds to the inverse of its image. We will show that isomorphisms preserve even more group properties below.

Lemma 7.2.4. *Let $\varphi\colon G \to G'$ be an isomorphism. Then the following statements are true:*

(a) $o(a) = o(\varphi(a))$ *for all* $a \in G$;

(b) G *is abelian if and only if* G' *is abelian*;

(c) G *is cyclic if and only if* G' *is cyclic*.

Proof. We can prove this lemma by using the properties in Proposition 7.1.6.

    (a) Using property (c), we have that

$$o(\varphi(a)) \mid o(a) \quad \text{and} \quad o(a) = o(\varphi^{-1}(\varphi(a))) \mid o(\varphi(a)).$$

    (b) This follows from property (d).

    (c) If $G = \langle g \rangle$, then property (f) says that $G' = \langle \varphi(g) \rangle$ is cyclic. On the other hand, if $G' = \langle g' \rangle$ is cyclic, then $G = \langle \varphi^{-1}(g') \rangle$ is cyclic. □

In fact, we anticipate that all "algebraic" properties (properties related to the group operation) of the two isomorphic groups should be the same! Boldly put, the two groups are exactly the same except for appearances. It is a job for algebraists (or all mathematicians) to be able to tell if two algebraic objects (or the objects in their respective disciplines) are the same *fundamentally*.

The job to find a representative for each isomorphic class and to determine whether a group belongs to a certain isomorphic class is called **classification of groups**. This is a most important job for group theorists. In this book we will introduce some basic tools to work towards the goal of classifying groups of small orders.

Theorem 7.2.5. *Two cyclic groups are isomorphic if and only if they are of the same order.*

Proof. The "only if" part is trivial. To prove the "if" part, let $G = \langle g \rangle$ and $G' = \langle g' \rangle$ be two cyclic groups of the same order. Remember that

$$o(g) = |G| = |G'| = o(g').$$

By Proposition 7.1.8 and Exercise 4, §7.1, we have a group homomorphism $\varphi\colon G \to G'$ sending g to g' and a group homomorphism $\psi\colon G' \to G$ sending g' to g. It is easy to check that φ and ψ are inverse to each other. Hence φ is a group isomorphism. □

The additive group $\mathbb{Z}$ can be chosen to be representative for the class of infinite cyclic group, and the additive group $\mathbb{Z}_n$ is a representative for the cyclic group of order n. Group theorists often use C_n as a representative for the isomorphic class of cyclic groups of order n.

Let $\varphi\colon G \to G'$ be a group homomorphism and let e and e' denote the identities of G and G' respectively. Remember that the **image** of φ is $\operatorname{Im}\varphi = \varphi(G) = \{\,\varphi(g) \in G' : g \in G\,\}$. We define the **kernel** of φ, denoted by $\ker\varphi$, to be the preimage of $e' = \{\,g \in G : \varphi(g) = e'\,\}$.

Proposition 7.2.6. *Let $\varphi\colon G \to G'$ be a group homomorphism. The following statements are true:*

(a) *$\varphi(G)$ is a subgroup of G';*

(b) *$\ker\varphi$ is a subgroup of G and $gkg^{-1} \in \ker\varphi$ for all $g \in G$ and $k \in \ker\varphi$.*

Proof. (a) First of all, $e' = \varphi(e) \in \varphi(G)$. Secondly, as $\varphi(a)\varphi(b) = \varphi(ab)$, $\varphi(G)$ is closed under the operation of G'. Finally, $\varphi(a)^{-1} = \varphi(a^{-1}) \in \varphi(G)$. We conclude that $\varphi(G)$ is a subgroup of G'.

(b) Obviously, $e \in \ker\varphi$. Suppose that $g, h \in \ker\varphi$. Then $\varphi(gh^{-1}) = \varphi(g)\varphi(h^{-1}) = e'\varphi(h)^{-1} = e'(e')^{-1} = e'$. This shows that $gh^{-1} \in \ker\varphi$. Hence $\ker\varphi$ is a subgroup of G. To prove the last statement, note that for all $g \in G$ and $k \in \ker\varphi$, we have that

$$\varphi(gkg^{-1}) = \varphi(g)\varphi(k)\varphi(g^{-1}) = \varphi(g)e'\varphi(g)^{-1} = e'.$$

Thus, $gkg^{-1} \in \ker\varphi$. $\qquad\qquad\qquad\qquad\qquad\qquad\qquad\qquad\square$

Example 7.2.7. Define the **sign function** from S_n into the multiplicative group $E = \{\,1, -1\,\}$ by

$$\operatorname{sgn}\sigma = \begin{cases} 1, & \text{if } \sigma \text{ is even;} \\ -1, & \text{if } \sigma \text{ is odd.} \end{cases}$$

The sign function is a group epimorphism from S_n onto E with A_n as the kernel.

The following proposition shows that the preimage of any element in the image is a coset of the kernel.

Proposition 7.2.8. *Let $\varphi\colon G \to G'$ be a group homomorphism. Then the following statements are true.*

(a) *For all $b \in \varphi(G)$, let a be an arbitrary element in G such that $\varphi(a) = b$. Then $\varphi^{-1}(b) = a \ker \varphi$.*

(b) *For all $b \in \varphi(G)$, $|\varphi^{-1}(b)| = |\ker \varphi|$.*

(c) *The group homomorphism φ is injective if and only if $\ker \varphi = \{e\}$.*

Obviously, if b is not in $\varphi(G)$, then $\varphi^{-1}(b)$ is empty.

Proof. (a) For any $c \in \ker \varphi$, $\varphi(ac) = \varphi(a)\varphi(c) = be' = b$. Hence $a \ker \varphi \subseteq \varphi^{-1}(b)$. On the other hand, let $c \in \varphi^{-1}(b)$. Then $\varphi(a^{-1}c) = \varphi(a)^{-1}\varphi(c) = b^{-1}b = e'$. Therefore, $a^{-1}c \in \ker \varphi$ and $c = a(a^{-1}c) \in a \ker \varphi$. Thus, $\varphi^{-1}(b) \subseteq a \ker \varphi$.

(b) Let a be an arbitrary element in $\varphi^{-1}(b)$. Then $|\varphi^{-1}(b)| = |a \ker \varphi| = |\ker \varphi|$.

(c) This follows from (b).   □

Exercises 7.2

1. Determine whether each of the following pairs of groups are isomorphic?

    (a) Let $\rho = (1\ 2\ \cdots\ n-1\ n)$. Is $\langle \rho \rangle \cong \mathbb{Z}_n$?

    (b) Is $D_5 \cong S_4$?

    (c) Is $\mathbb{Z}_4 \cong V$?

    (d) Is $S_3 \cong \mathbb{Z}_6$?

    (e) Is $D_6 \cong A_4$?

    (f) Is $D_n \cong \mathbb{Z}_{2n}$ for $n \geq 4$?

    (g) Is $(\mathbb{R}, +) \cong (\mathbb{Q}, +)$?

2. Show that the set K in Example 7.2.2 form a multiplicative group.

3. Let G and G' both be groups of order p where p is a positive prime integer. Show that $G \cong G'$.

4. Let $\varphi \colon G \to G'$ be a group homomorphism.

   (a) If G is cyclic, show that so is $\varphi(G)$.

   (b) If G is abelian, show that so is $\varphi(G)$.

5. Let φ be a group homomorphism from G to G'. Let H be a subgroup of G and H' be a subgroup of G'. Show that $\varphi(H)$ is a subgroup of G' and $\varphi^{-1}(H')$ is a subgroup of G.

6. Let $\varphi \colon G \to G'$ be a group homomorphism and let $G = \langle S \rangle$. Show that $\operatorname{Im} \varphi = \langle \varphi(S) \rangle$.

7. Let $\varphi \colon \mathbb{Z}_{100} \to \mathbb{Z}_{20}$ be the map sending $\overline{k}$ to $\overline{k}$.

   (a) Verify that φ is a well-defined map.

   (b) Verify that φ is a group homomorphism.

   (c) Find $\varphi(\mathbb{Z}_{100})$ and $\ker \varphi$. Is φ an isomorphism?

8. Recall the general linear group $GL_2(\mathbb{R})$ in Example 3.1.5. Let det be the map from $GL_2(\mathbb{R})$ to $\mathbb{R}^*$ by letting

$$\det \begin{bmatrix} a & b \\ c & d \end{bmatrix} = ad - bc.$$

Prove that det is a group homomorphism. Find the kernel of det.

9. Let G be a group. Define $\varphi \colon G \to \operatorname{Sym}(G)$ by $\varphi(g) = \sigma_g$ where $\sigma_g \colon G \to G$ is the function defined by $\sigma_g(a) = gag^{-1}$. Prove that φ is a group homomorphism and find the kernel of φ.

10. Let G be a group. A group isomorphism from G to G is called a group **automorphism** of G. Let $\operatorname{Aut}(G)$ be the set of all group automorphisms of G. Show that $\operatorname{Aut}(G)$ is a subgroup of $\operatorname{Sym}(G)$. The group $\operatorname{Aut}(G)$ is called the **automorphism group** of G.

11. Let $G = \{ 1, \zeta, \dots, \zeta^{n-1} : \zeta = \cos(2\pi/n) + i\sin(2\pi/n) \}$. Find $\operatorname{Aut}(G)$.

7.3 Cayley's Theorem

Suppose that $\varphi\colon G \to G'$ is a group monomorphism. Then it is clear that $G \cong \varphi(G)$. Since $\varphi(G)$ is a subgroup of G', G is identifiable as a subgroup of G via φ. In this case, we say that we **embed** G into G' via φ. The group monomorphism φ is also called an **embedding** of G into G'.

In this section, we shall embed any given group into a symmetric group. Thus any given group is isomorphic to a permutation group.

When G is a group and $a \in G$. The function $L_a\colon G \to G$ defined by $L_a(x) = ax$ is a bijective function from G into G since

(i) $L_a(x) = L_a(y) \implies ax = ay \implies x = y$ by cancellation law, and

(ii) $L_a(a^{-1}x) = a(a^{-1}x) = x$

for all $x, y \in G$. Thus, L_a is in $\mathrm{Sym}(G)$.

Theorem 7.3.1 (Cayley's Theorem). *Every group G is isomorphic to a subgroup of* $\mathrm{Sym}(G)$.

Proof. Define a map φ from G into $\mathrm{Sym}(G)$ by sending a to L_a. For all $x \in G$, we have that

$$(L_a \circ L_b)(x) = L_a(L_b(x)) = L_a(bx) = abx = L_{ab}(x).$$

Hence $\varphi(ab) = L_{ab} = L_a \circ L_b = \varphi(a)\varphi(b)$. We conclude that φ is a group homomorphism.

If $\varphi(a) = \varphi(b)$, then $a = L_a(e) = L_b(e) = b$. Thus, φ is a group monomorphism. We conclude that G is isomorphic to the subgroup $\varphi(G)$ of $\mathrm{Sym}(G)$. $\qquad\square$

Remark. Let G be a group and define $R_a\colon G \to G$ by $R_a(x) = xa^{-1}$. Then $R_a \in \mathrm{Sym}(G)$ and $R_a \circ R_b = R_{ab}$ since $(ab)^{-1} = b^{-1}a^{-1}$. Now the function $\psi\colon G \to \mathrm{Sym}(G)$ defined by $\psi(a) = R_a$ is also a group monomorphism from G onto the subgroup $\psi(G)$ of $\mathrm{Sym}(G)$. This is another embedding of G into $\mathrm{Sym}(G)$.

If G is a finite group of order n, then Cayley's theorem tells us that we can embed G into S_n. Here we shall give three examples to illustrate the embeddings of finite groups into symmetric groups through Cayley's theorem.

Example 7.3.2. Let $C_4 = \{ e, \alpha, \alpha^2, \alpha^3 : \alpha^4 = e \}$ be the cyclic group of order 4. Then

$$L_\alpha = \begin{bmatrix} e & \alpha & \alpha^2 & \alpha^3 \\ \alpha & \alpha^2 & \alpha^3 & e \end{bmatrix}.$$

If we index $e, \alpha, \alpha^2, \alpha^3$ by $1, 2, 3, 4$ respectively, L_α is identified as the permutation $\sigma = (1\ 2\ 3\ 4)$. Thus C_4 is isomorphic to the subgroup $H = \{ e, \sigma, \sigma^2, \sigma^3 \}$ of S_4.

Example 7.3.3. Let $V = \{ e, \alpha, \beta, \alpha\beta : \alpha^2 = \beta^2 = e, \alpha\beta = \beta\alpha \}$ be the Klein 4-group. Then

$$L_\alpha = \begin{bmatrix} e & \alpha & \beta & \alpha\beta \\ \alpha & e & \alpha\beta & \beta \end{bmatrix}, \quad L_\beta = \begin{bmatrix} e & \alpha & \beta & \alpha\beta \\ \beta & \alpha\beta & e & \alpha \end{bmatrix}$$

and

$$L_{\alpha\beta} = \begin{bmatrix} e & \alpha & \beta & \alpha\beta \\ \alpha\beta & \beta & \alpha & e \end{bmatrix}.$$

If we index $e, \alpha, \beta, \alpha\beta$ by $1, 2, 3, 4$ respectively, we can see that V is isomorphic to the subgroup $H = \{ e, (1\ 2)(3\ 4), (1\ 3)(2\ 4), (1\ 4)(2\ 3) \}$ of S_4.

Example 7.3.4. Let

$$D_3 = \{ e, \alpha, \alpha^2, \beta, \beta\alpha, \beta\alpha^2 : \alpha^3 = \beta^2 = e, \beta\alpha\beta = \alpha^2 \} = \langle \alpha, \beta \rangle$$

be the dihedral 3-group. Then

$$L_\alpha = \begin{bmatrix} e & \alpha & \alpha^2 & \beta & \beta\alpha & \beta\alpha^2 \\ \alpha & \alpha^2 & e & \beta\alpha^2 & \beta & \beta\alpha \end{bmatrix}$$

which can be identified as

$$\sigma = \begin{bmatrix} 1 & 2 & 3 & 4 & 5 & 6 \\ 2 & 3 & 1 & 6 & 4 & 5 \end{bmatrix} = (1\ 2\ 3)(4\ 6\ 5).$$

Also we have

$$L_\beta = \begin{bmatrix} e & \alpha & \alpha^2 & \beta & \beta\alpha & \beta\alpha^2 \\ \beta & \beta\alpha & \beta\alpha^2 & e & \alpha & \alpha^2 \end{bmatrix}$$

which can be identified as

$$\tau = \begin{bmatrix} 1 & 2 & 3 & 4 & 5 & 6 \\ 4 & 5 & 6 & 1 & 2 & 3 \end{bmatrix} = (1\ 4)(2\ 5)(3\ 6).$$

Thus D_3 is isomorphic to the subgroup $H = \{\, e, \sigma, \sigma^2, \tau, \tau\sigma, \tau\sigma^2 \,\}$ of S_6.

Cayley's Theorem tells us that to understand all groups, it is sufficient to understand all permutation groups, or to understand all symmetric groups. Especially, to understand all finite groups, it is sufficient to understand S_n for all n. Theoretically, this result has great significance. However in practice, S_n is simply too complicated to understand in general. In short, Cayley's Theorem is an extremely fundamental but totally impractical theorem. We need to find other tools to be able to move forward in group theory.

Exercises 7.3

1. Embed the dihedral group

$$D_4 = \{\, e,\ \alpha,\ \alpha^2,\ \alpha^3,\ \beta,\ \beta\alpha,\ \beta\alpha^2,\ \beta\alpha^3 : \alpha^4 = e = \beta^2,\ \beta\alpha\beta = \alpha^3 \,\}$$

   into the symmetric group S_8 through Cayley's Theorem.

2. Embed the symmetric group S_3 into the symmetric group S_6 using Cayley's Theorem. Could you find other embeddings?

3. Embed D_3 into S_6 using ψ in the remark to Cayley's Theorem. Do you get the same embedding as in Example 7.3.4?

4. Let G be a group.

   (a) Show that G contains an element of order 2 if $|G|$ is even. (See Exercise 4 in §3.2.)

   (b) Suppose $|G| = 2k$ where k is odd. Show that G contains a subgroup of index 2. (Hint: Let a be an element of order 2 and show that L_a is an odd permutation in $\mathrm{Sym}(G)$.)

Review Exercises for Chapter 7

1. Find all group homomorphisms from $\mathbb{Z}_{20}$ to $\mathbb{Z}_{25}$. Find the images and kernels for all these group homomorphisms.

2. In S_4, let $a = (1\ 2)$ and $b = (3\ 4)$. Show that a and b generate a subgroup of order 4 which is isomorphic to the Klein 4-group V.

3. Prove the following assertions.

   (a) If $|G| = 2$ then $G \cong C_2$.

   (b) If $|G| = 3$ then $G \cong C_3$.

   (c) If $|G| = 4$ then $G \cong C_4$ or V, the Klein 4-group.

4. Define the exponential map exp from $\mathbb{C}$ into $\mathbb{C}^*$ by

$$\exp(a + bi) = e^a(\cos b + i \sin b).$$

Show that exp is a group homomorphism. Find the kernel of exp.

5. Show that $\operatorname{Aut}(\mathbb{Z}) \cong U(\mathbb{Z})$ and $\operatorname{Aut}\mathbb{Z}_n \cong U(\mathbb{Z}_n)$ for all $n \geq 2$.

6. Let G be a group and H be a subgroup. Let

$$S = \{\, gH \mid g \in G \,\}$$

be the set of left cosets of H in G. For each $a \in G$, define $L_a : S \to S$ by $L_a(gH) = agH$. Now the function $\psi : G \to \operatorname{Sym}(S)$ is given by $\psi(a) = L_a$.

   (a) Show that ψ is a group homomorphism.

   (b) Describe the kernel of ψ.

CHAPTER 8

The Quotient Group

The collection of cosets of a *normal* subgroup H in a group G has a natural group structure and is used to form the *quotient group* G/H. The group G/H is a natural group homomorphic image of the group G and has a simpler structure than that of G. It preserves the subgroup lattice formed by the subgroups in G which contains H.

We will thus be able to study the *Fundamental Theorem of Group Homomorphisms* and its consequences. A great theorem often relies on whether one can construct the right homomorphism. These theorems will help us do just that.

8.1   Normal Subgroups

A subgroup H of a group G is a **normal** subgroup, denoted $H \lhd G$, if $ghg^{-1} \in H$ for $h \in H$ and $g \in G$. To check whether a subgroup K is normal is to check that the conjugate of every element in K still remains in K. In this case we say that K is closed under conjugation.

Example 8.1.1. Let G be a group.

(1) Both G and the trivial subgroup are normal in G.

(2) If G is abelian, then any subgroup H of G is normal in G since $ghg^{-1} = hgg^{-1} = he = h \in H$ for any $g \in G$ and $h \in H$.

(3) If K is a subgroup of $Z(G)$, then K is a normal subgroup in G. In particular, $Z(G) \lhd G$.

(4) Let $\varphi \colon G \to G'$ be a group homomorphism. Proposition 7.2.6(b) tells us that $\ker \varphi \lhd G$. In particular, Example 7.2.7 shows that $A_n \lhd S_n$.

In the following proposition, we give other tests to verify whether a subgroup is normal.

Proposition 8.1.2. *Let K be a subgroup of G. The following properties are equivalent:*

(i) $K \lhd G$, *that is,* $aka^{-1} \in K$ *for all* $a \in G$ *and* $k \in K$;

(ii) $aKa^{-1} \subseteq K$ *for all* $a \in G$;

(iii) $aKa^{-1} = K$ *for all* $a \in G$;

(iv) $aK = Ka$ *for all* $a \in G$.

Proof. (i) $\Rightarrow$ (ii): This is trivial.

(ii) $\Rightarrow$ (iii): This follows from the fact that $K = a(a^{-1}K(a^{-1})^{-1})a^{-1} \subseteq aKa^{-1}$.

(iii) $\Rightarrow$ (iv): Let $ak \in aK$ where $k \in K$. Since $aka^{-1} = k'$ for some $k' \in K$, we have $ak = k'a \in Ka$. Thus, $aK \subseteq Ka$. Similarly, $Ka \subseteq aK$.

(iv) $\Rightarrow$ (i): For any $a \in G$ and any $k \in K$, $ak \in aK = Ka$. Hence $ak = k'a$ for some $k' \in K$. This implies that $aka^{-1} = k' \in K$. $\square$

The product of two subgroups H and H' of a group G given by

$$HH' = \{\, ab \in G : a \in H,\ b \in H' \,\}$$

is not a subgroup of G in general (see Exercise 2). However, the situation is better when one of H or H' is a normal subgroup of G.

Proposition 8.1.3. *Suppose that H and K are subgroups of a group G. If K is a normal subgroup of G, then KH is a subgroup of G.*

Proof. First note that $e = ee \in KH$. For ab and $a_1 b_1 \in KH$, we have

$$(ab)(a_1 b_1) = [a(ba_1 b^{-1})](bb_1) \in KH$$

since $ba_1 b^{-1} \in K$. Moreover,

$$(ab)^{-1} = b^{-1}a^{-1} = (b^{-1}a^{-1}b)b^{-1} \in KH$$

since $b^{-1}a^{-1}b \in K$. $\qquad\qquad\qquad\qquad\qquad\qquad\qquad\qquad\square$

Here we give a few more examples of normal subgroups.

Example 8.1.4. The special linear group $SL_2(\mathbb{R})$ is a normal subgroup of the general linear group $GL_2(\mathbb{R})$. The determinant $\det\colon GL_2(\mathbb{R}) \to \mathbb{R}^*$ is a group homomorphism and its kernel is $SL_2(\mathbb{R})$. Indeed for $A \in SL_2(\mathbb{R})$ and $P \in GL_2(\mathbb{R})$, we have

$$\det(PAP^{-1}) = \det(P)\det(A)\det(P^{-1})$$
$$= \det(P)\cdot 1 \cdot \det(P)^{-1} = 1.$$

This proves again that $SL_2(\mathbb{R}) \lhd GL_n(\mathbb{R})$.

Example 8.1.5. Let G be the group of linear mappings from $\mathbb{R}$ into itself:

$$G = \{\, L_{a,b} : L_{a,b}(x) = ax + b,\ a,b \in \mathbb{R},\ a \neq 0 \,\}.$$

The operation of G is the composition of functions. Note that

$$L_{a,b} \circ L_{c,d} = L_{ac,ad+b}$$

since $L_{a,b} \circ L_{c,d}(x) = L_{a,b}(cx + d) = a(cx + d) + b = acx + (ad + b)$. We also have that

$$(L_{a,b})^{-1} = L_{a^{-1},-a^{-1}b}$$

since

$$L_{a,b} \circ L_{a^{-1},-a^{-1}b}(x) = L_{a,b}(a^{-1}x - a^{-1}b) = a(a^{-1}x - a^{-1}b) + b = x$$

and

$$L_{a^{-1},-a^{-1}b} \circ L_{a,b}(x) = L_{a^{-1},-a^{-1}b}(ax + b) = a^{-1}(ax + b) - a^{-1}b = x.$$

Let $H = \{ L_{1,k} \in G : k \in \mathbb{R} \}$. Then H is a normal subgroup of G since

$$L_{a,b} \circ L_{1,k} \circ (L_{a,b})^{-1} = L_{a,ak+b} \circ L_{a^{-1},-a^{-1}b} = L_{1,ak} \in H.$$

Proposition 8.1.6. *If H is a subgroup of index 2 in G, then H is normal in G.*

Proof. Let g be any element in $G \setminus H$. Then G has the right and left coset decompositions

$$G = H \, \dot\cup \, Hg = H \, \dot\cup \, gH.$$

It follows that $gH = G \setminus H = Hg$. Therefore, H is a normal subgroup of G by Proposition 8.1.2. $\qquad\qquad\square$

Definition 8.1.7. A group G is called a **simple** group if G has no proper nontrivial normal subgroups.

Example 8.1.8. Let G be a group of prime order p. Then G is simple by Corollary 6.1.6.

Example 8.1.9. Note that

$$H = \langle (1\ 2\ 3) \rangle = \{ e,\ (1\ 2\ 3),\ (1\ 3\ 2) \}$$

is not a normal subgroup of A_4 since $(2\ 3\ 4)(1\ 2\ 3)(2\ 3\ 4)^{-1} = (1\ 3\ 4) \notin H$. On the other hand, let

$$V = \{ e,\ (1\ 2)(3\ 4),\ (1\ 3)(2\ 4),\ (1\ 4)(2\ 3) \}.$$

Then V is a subgroup of A_4 (*Cf.* Example 7.3.3). Since V contains the identity element and all the permutations of type $(2,2)$ in A_4, all the conjugates of elements in V are still in V by Lemma 6.2.3. We conclude that V is normal in A_4 and so A_4 is not simple.

We would like to comment that A_n is simple for $n \geq 5$. We will not prove this fact for the moment.

Exercises 8.1

1. For each n, determine whether S_n and D_n are simple or not.

2. Find an example of (G, H, H') such that H and H' are subgroups of G but not HH'.

3. Let H be an arbitrary subgroup of G.

   (a) Show that $HH = H$.

   (b) Show that $H^{-1} = \{h^{-1} : h \in H\} = H$.

4. Suppose that H and K are subgroups of a group G and assume that $K \lhd G$.

   (a) Show that HK is a subgroup of G. In fact, show that $HK = KH$.

   (b) Show that in G, HK is the *smallest* subgroup containing both H and K. In other words, show that (i) H and $K \subseteq HK$ and (ii) if L is a subgroup of G such that H and $K \subseteq L$, then $HK \subseteq L$.

   (c) Suppose that H and K are both normal in G. Show that HK is also a normal subgroup of G.

5. Let G be a group. Let H and K be subgroups of G such that K is normal in G.

   (a) Show that $K \cap H$ is a normal subgroup of H.

   (b) If H is also normal in G, show that $H \cap K$ is a normal subgroup of G.

6. Suppose that H is a normal subgroup of a group G. Show that for all a and $b \in G$, we have

$$(aH)(bH) = (ab)H$$

as sets.

7. Let $\varphi\colon G \to G'$ be a group homomorphism.

    (a) If K is a normal subgroup of G, show that $\varphi(K)$ is normal in $\varphi(G)$. Give an example such that $\varphi(K)$ is not normal in G'.

    (b) If K' is a normal subgroup of G', show that $\varphi^{-1}(K')$ is normal in G.

8.2   Quotient Groups

Suppose that K is a normal subgroup of a group G. Let

$$\frac{G}{K} = \{\, gK : g \in G \,\}$$

be the set of left cosets of K in G. As K is a normal subgroup of G, each left coset gK is also a right coset Kg. Thus we also have

$$\frac{G}{K} = \{\, Kg : g \in G \,\}.$$

Define a product on the left cosets in G/K by

$$(aK)(bK) = abK.$$

This product is natural since by abuse of notation we have

$$(aK)(bK) = a(Kb)K = a(bK)K = (ab)(KK) = abK.$$

Next we prove formally that such an operation is well-defined.

Lemma 8.2.1. *Let K be a normal subgroup of a group G. If $aK = cK$ and $bK = dK$, then $abK = cdK$.*

Proof. There exist h and k in K such that $a = ch$ and $b = dk$. It follows that

$$ab = (ch)(dk) = cd(d^{-1}hd)k.$$

Since K is a normal subgroup of G, $d^{-1}hd \in K$. Hence $ab = cdp$ where $p = (d^{-1}hd)k \in K$. Thus, $abK = cdK$.   □

Now we verify that G/K is indeed a group under this operation.

(1) *Associativity.* For aK, bK and cK in G/K, we have

$$\begin{cases} [(aK)(bK)](cK) = (abK)(cK) = (ab)cK, \\ aK)[(bK)(cK)] = (aK)(bcK) = a(bc)K. \end{cases}$$

As $(ab)c = a(bc)$, we get $[(aK)(bK)](cK) = (aK)[(bK)(cK)]$.

In other words, the associativity of the operation in G implies the associativity of the operation in G/K.

(2) *The existence of the identity.* Let e be the identity of G. The coset $eK = K$ plays the role of the identity in G/K since

$$(K)(aK) = aK = (aK)(K).$$

(3) *The existence of inverse elements.* The inverse of aK is simply given by $a^{-1}K$ since

$$(aK)(a^{-1}K) = (aa^{-1})K = eK = K$$

and similarly $(a^{-1}K)(aK) = K$.

The previous discussions now give the following theorem.

Theorem 8.2.2. *If K is a normal subgroup of a group G, the set of left cosets of K in G given by*

$$\frac{G}{K} = \{\, gK : g \in G \,\}$$

forms a group under the operation $(aK)(bK) = abK$.

The group G/K is called the **quotient group** or the **factor group** of G modulo K. The element gK is often simply denoted as $\overline{g}$ when K is understood. Hence the operation in G/K can be denoted as $\overline{a}\,\overline{b} = \overline{ab}$.

When $K = \{\, e \,\}$, then

$$G/K = \{\, \{\, g \,\} : g \in G \,\} \cong G.$$

When $K = G$, G/K is the trivial group. The group $\mathbb{Z}_n$ consists of the cosets of $n\mathbb{Z}$ and thus it is the quotient group $\mathbb{Z}/n\mathbb{Z}$.

Remember that the index of K in G, $[G : K]$, is exactly the number of cosets of K in G.

Corollary 8.2.3. *Let K be a normal subgroup of G. Then*

$$|G/K| = [G : K].$$

In particular, if G and K are both finite, then $|G/K| = |G|/|K|$.

Example 8.2.4. Let

$$C_6 = \left\{ e, a, a^2, a^3, a^4, a^5 : a^6 = e \right\}$$

be the cyclic group of order 6 and $K = \left\{ e, a^2, a^4 \right\}$. Then $C_6/K = \left\{ K, aK \right\}$ is isomorphic to $C_2 = \left\{ e, b : b^2 = e \right\}$, the cyclic group of order 2.

Example 8.2.5. Let

$$D_3 = \left\{ e, a, a^2, b, ab, a^2b : a^3 = e, \ b^2 = e, \ bab = a^2 \right\}$$

be the dihedral group of order 6 and $K = \left\{ e, a, a^2 \right\}$. By Proposition 8.1.6, K is a normal subgroup of D_3. It is clear that $D_3/K = \left\{ K, bK \right\}$ is the cyclic group of order 2.

Example 8.2.6. Remember that $A_n \lhd S_n$ and $[S_n : A_n] = 2$ for $n \geq 2$. Thus $S_n/A_n = \left\{ A_n, (1\ 2)A_n \right\}$ is the cyclic group of order 2.

Example 8.2.7. Let's consider G, the group of linear mappings $L_{a,b}$ in Example 8.1.5 again. Let $K = \left\{ L_{1,k} \in G : k \in \mathbb{R} \right\}$. We have seen that K is a normal subgroup of G. Note that

$$L_{a,b} = L_{1,b} \circ L_{a,0}.$$

It follows that $K \circ L_{a,b} = K \circ L_{a,0}$. Moreover, if $KL_{a,0} = KL_{a',0}$ then for some $b \in \mathbb{R}$, one has

$$L_{a',0} = L_{1,b}L_{a,0} = L_{a,b}.$$

This implies that $a = a'$ and $b = 0$. We conclude that each coset in G/K is represented by $L_{a,0}$ for some unique a in $\mathbb{R}$. Since

$$(KL_{a,0})(KL_{a',0}) = KL_{aa',0},$$

we see that the mapping from G/K to $\mathbb{R}^*$ sending $KL_{a,0}$ to a is a group isomorphism.

We will use the arrow "↠" to denote a surjective map.

Theorem 8.2.8. *If K is a normal subgroup of a group G, then there is a canonical group homomorphism $\pi\colon G \twoheadrightarrow G/K$ defined by $\pi(g) = gK$. The kernel of π is K.*

This homomorphism is called the **canonical epimorphism** $G \twoheadrightarrow G/K$.

Proof. First we check that π is a group homomorphism:

$$\pi(ab) = abK = (aK)(bK) = \pi(a)\pi(b).$$

Clearly, π is onto. Note that $a \in \ker \pi$ if and only if $aK = K$ if and only if $a \in K$. We conclude that $\ker \pi = K$. □

When $H \supseteq K$, $\pi(H) = \{\bar{a} \in G/K : a \in H\}$ is often denoted $\overline{H}$ or H/K.

Theorem 8.2.9 (Correspondence Theorem). *Let K be a normal subgroup of G and let $\pi\colon G \twoheadrightarrow G/K$ be the canonical epimorphism. Let $\mathscr{A}$ be the family of subgroups of G containing K and let $\mathscr{B}$ be the family of subgroups of G/K. There is an* order-preserving *one-to-one correspondence*

$$\Phi\colon \quad \mathscr{A} \quad \longrightarrow \quad \mathscr{B}$$
$$H \quad \longmapsto \quad H/K.$$

The inverse of Φ is given by

$$\Psi\colon \quad \mathscr{B} \quad \longrightarrow \quad \mathscr{A}$$
$$\mathfrak{H} \quad \longmapsto \quad \pi^{-1}(\mathfrak{H}).$$

Furthermore, $H \in \mathscr{A}$ is normal in G if and only if H/K is normal in G/K.

Here the "order-preserving" means the mappings preserve containment of sets.

Proof. Since $\pi^{-1}(\mathfrak{H}) \supseteq \pi^{-1}(e) = K$ for all $\mathfrak{H} \in \mathscr{A}$, both Φ and Ψ are well-defined by Exercise 5, §7.2. It is clear that both Φ and Ψ are order-preserving. It remains to show that Φ and Ψ are inverse to each other.

First we show that $\Psi(\Phi(H)) = \pi^{-1}(H/K) = H$ for all $H \in \mathscr{A}$. The "⊇" part is obvious. Let $a \in \pi^{-1}(H/K)$. Then $aK \in H/K$. There exists

$h \in H$ and $k \in K$ such that $a = hk$. Since $K \subseteq H$, we have $a \in H$. We have proved the "$\subseteq$" part.

Next we show that $\Phi(\Psi(\mathfrak{H})) = \pi^{-1}(\mathfrak{H})/K = \mathfrak{H}$ for all $\mathfrak{h} \in \mathscr{B}$. For $aK \in \pi^{-1}(\mathfrak{H})/K$, there exists $b \in \pi^{-1}(\mathfrak{H})$ such that $aK = bK$. Hence $aK = \pi(b) \in \mathfrak{H}$. This proves "$\subseteq$". Conversely, let $aK \in \mathfrak{H}$. Then $a \in \pi^{-1}(\mathfrak{H})$. Thus, $aK \in \pi^{-1}(\mathfrak{H})/K$. This proves "$\supseteq$".

Finally, let $H \lhd G$. Then for all $g \in G$ and $h \in H$,

$$(gK)(hK)(g^{-1}K) = ghg^{-1}K \in H/K$$

since $ghg^{-1} \in H$. This shows that $H/K \lhd G/K$. On the other hand, assume $H/K \lhd G/K$. For all $g \in G$ and $h \in H$, we have

$$(gK)(hK)(g^{-1}K) = ghg^{-1}K$$

in H/K. Thus, $ghg^{-1} \in \pi^{-1}(H/K) = H$. This proves that $H \lhd G$. $\qquad\square$

Using the results in §4.2, we can construct and compare the subgroup lattice of $\mathbb{Z}_{12}$ and that of $\mathbb{Z}$ which contains $12\mathbb{Z}$.

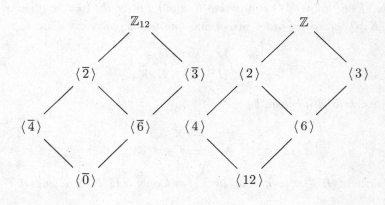

Exercises 8.2

1. Let $G = \mathbb{Q}[x]$ be the set of polynomials in the variable x with rational coefficients and H be the subgroup of polynomials which are multiples of $x^2 + 1$. Show that the cosets of H in G can be represented by

$$(a + bx) + H.$$

Exhibit the multiplication in G/H.

2. Let $\mathbb{R}^*$ be the set of nonzero real numbers and $H = \{1, -1\}$. Explain why H is normal in $\mathbb{R}^*$ with respect to multiplication. Describe $\mathbb{R}^*/H$.

3. Let w_1 and w_2 be nonzero real numbers such that 0, w_1, w_2 in the complex plane are *not* colinear in the complex plane. Consider

$$\mathscr{L} = \{aw_1 + bw_2 : a, b \in \mathbb{Z}\}$$

as an additive subgroup of the complex numbers $\mathbb{C}$. What does $\mathbb{C}/\mathscr{L}$ look like?

4. If G is a cyclic group and H is a subgroup of G, show that G/H is also a cyclic group.

5. If G is an abelian group and H is a subgroup of G, show that G/H is an abelian group.

6. Let G be a group and let K be a subgroup of G contained in $Z(G)$. Show that G is abelian if G/K is cyclic.

7. If H is a normal subgroup of a group G, show that G/H is abelian if and only if $aba^{-1}b^{-1} \in H$ for all $a, b \in G$.

8.3 Fundamental Theorem of Group Homomorphisms

In this section, we will go deeper into the theory of group homomorphisms. We want to discuss how to construct group homomorphisms and how to construct isomorphisms.

Theorem 8.3.1 (Fundamental Theorem of Group Homomorphisms). *Let* $\varphi\colon G \to G'$ *be a group homomorphism. Suppose K is a normal subgroup in G and let $K \subseteq \ker\varphi$. Then there is a unique group homomorphism* $\overline{\varphi}\colon G/K \to G'$ *such that* $\varphi = \overline{\varphi}\pi$ *where π is the canonical epimorphism*

$G \twoheadrightarrow G/K$.

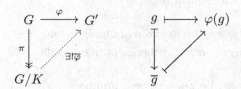

Moreover, if $K = \ker \varphi$, then $\overline{\varphi}$ is a monomorphism.

Proof. For any $\overline{g} \in G/K$, by the requirement of this theorem, we must have $\overline{\varphi}(\overline{g}) = \varphi \pi(g) = \varphi(g)$. Hence the choice of $\overline{\varphi}$ is unique if it exists. It remains to show that such a $\overline{\varphi}$ is indeed a group homomorphism.

First we need to check that $\overline{\varphi}$ is well-defined. Let $\overline{g} = \overline{g'}$. Then $g' \in gK$ and $g' = gk$ for some $k \in K$. We have that

$$\varphi(g') = \varphi(gk) = \varphi(g)\varphi(k) = \varphi(g)$$

since $k \in \ker \varphi$. Thus $\overline{\varphi}$ is well-defined. The map $\overline{\varphi}$ is a group homomorphism since

$$\overline{\varphi}[(aK)(bK)] = \overline{\varphi}(abK) = \varphi(ab) = \varphi(a)\varphi(b) = \overline{\varphi}(aK)\overline{\varphi}(bK).$$

At last, when $K = \ker \varphi$ we have

$$gK \in \ker \overline{\varphi} \iff \varphi(g) = \overline{\varphi}(gK) = e'$$
$$\iff g \in \ker \varphi = K \iff gK = K.$$

This shows that $\overline{\varphi}$ is one-to-one $\square$

Corollary 8.3.2. *Let* $\varphi \colon G \to G'$ *be a group homomorphism. Then*

$$G/\ker \varphi \cong \varphi(G).$$

If $G' = \varphi(G)$ for some group homomorphism φ, we say G' is a **homomorphic image** of G. In this case, we often simply assume that $G' = G/K$ for some normal subgroup K of G.

The first of the three isomorphism theorems is also an immediate result.

Theorem 8.3.3 (First Isomorphism Theorem). *Let* φ *be a group epimorphism from G onto G'. Then*

$$G/\ker \varphi \cong G'.$$

Example 8.3.4. Consider the integers $\mathbb{Z}$ as an additive subgroup of the real numbers $\mathbb{R}$. It is of course a normal subgroup since $\mathbb{R}$ is abelian. Then

$$\mathbb{R}/\mathbb{Z} = \{\, r + \mathbb{Z} : r \in \mathbb{R} \,\}.$$

Note that $r_1 + \mathbb{Z} = r_2 + \mathbb{Z}$ if and only if $r_1 - r_2 \in \mathbb{Z}$. Thus

$$\mathbb{R}/\mathbb{Z} = \{\, r + \mathbb{Z} : 0 \leq r < 1,\ r \in \mathbb{R} \,\}$$

where the elements of $\mathbb{R}/\mathbb{Z}$ bijectively correspond with the elements in the half open interval $[\,0, 1)$ where the endpoint 1 is identified with 0. Therefore, we can visualize $\mathbb{R}/\mathbb{Z}$ as the unit circle.

In fact, the exponential map $e^{2\pi i x} = \cos 2\pi x + i \sin 2\pi x$ maps the abelian group $\mathbb{R}$ onto the unit circle

$$S^1 = \left\{\, e^{2\pi i x} : 0 \leq x < 1 \,\right\}$$

as a multiplicative group. It is a group epimorphism and its kernel is precisely $\mathbb{Z}$. This leads to the isomorphism $\mathbb{R}/\mathbb{Z} \cong S^1$.

Example 8.3.5. The set of positive real numbers $\mathbb{R}_+ = \{\, x \in \mathbb{R} : x > 0 \,\}$ is a multiplicative subgroup of $\mathbb{C}^*$. Note that every nonzero complex number z has a unique polar form

$$z = r(\cos \theta + i \sin \theta), \qquad \text{where } r = |z| > 0 \text{ and } \theta \in [\,0, 2\pi).$$

Hence we have a group epimorphism from $\mathbb{C}^*$ to S^1 sending z to $\cos \theta + i \sin \theta$ with $\theta \in [\,0, 2\pi)$. Clearly, z is in the kernel if and only if $z = |z| \in \mathbb{R}_+$. Consequently, we have that

$$\mathbb{C}^*/\mathbb{R}_+ \cong \{\, \cos \theta + i \sin \theta : \theta \in [\,0, 2\pi) \,\} = S^1.$$

Note that the absolute value function $z \mapsto |z|$ is a group epimorphism from $\mathbb{C}^*$ onto $\mathbb{R}_+$ and its kernel is equal to S^1. We also have $\mathbb{C}^*/S^1 \cong \mathbb{R}_+$.

Example 8.3.6. Let $G = \mathbb{Q}[x]$ be the additive group of polynomials in the variable x with rational coefficients and

$$H = \{\, f(x) \in \mathbb{Q}[x] : f(0) = 0 \,\}.$$

Then H is the kernel of the group epimorphism φ from G onto $\mathbb{Q}$ defined by

$$\varphi[f(x)] = f(0).$$

Consequently, we have $G/H \cong \mathbb{Q}$.

Example 8.3.7. The determinant defined by

$$\det \begin{bmatrix} a & b \\ c & d \end{bmatrix} = ad - bc$$

is a group epimorphism from $GL_2(\mathbb{R})$ to $\mathbb{R}^* = \mathbb{R} \setminus \{\, 0 \,\}$. The kernel of det is $SL_2(\mathbb{R})$. Therefore

$$GL_2(\mathbb{R})/SL_2(\mathbb{R}) \cong \mathbb{R}^*.$$

Example 8.3.8. The exponential function defined by

$$e^z = 1 + z + \frac{z^2}{2!} + \cdots + \frac{z^n}{n!} + \cdots$$

is a group epimorphism from $(\mathbb{C}, +)$ onto $(\mathbb{C}^*, \cdot)$. The kernel of exponential function e^z is

$$K = \{\, 2n\pi i : n \in \mathbb{Z} \,\}.$$

Thus we have $\mathbb{C}/K \cong \mathbb{C}^*$. Note that K is also isomorphic to the additive group of integers $\mathbb{Z}$.

Example 8.3.9. Let $K \lhd G$. Define $\varphi \colon G \to \mathrm{Sym}(G/K)$ by $\varphi(g) = L_g$, where

$$L_g(aK) = gaK \qquad \text{for all } aK \in G/K.$$

The kernel of φ consists of those $g \in G$ such that $L_g = \mathbf{1}_{G/K}$, that is, those g such that $gaK = aK$ for all $a \in G$. This is equivalently to saying that $a^{-1}ga \in K$ for all $a \in G$. Thus $\ker \varphi \supseteq K$ and may contain more elements. We also conclude that $G/\ker \varphi$ is isomorphic to some subgroup of $\mathrm{Sym}(G/K)$.

We may use the first isomorphism theorem to derive two more useful tools regarding isomorphisms.

Theorem 8.3.10 (Second Isomorphism Theorem). *Let H be a subgroup of a group G and K a normal subgroup of G. Then*

$$HK = \{\, ab : a \in H,\ b \in K \,\}$$

is a subgroup of G, $H \cap K$ is a normal subgroup of H and

$$\frac{H}{H \cap K} \cong \frac{HK}{K}.$$

Proof. By Proposition 8.1.3 we have that HK is a subgroup of G. It is a straightforward verification that and $H \cap K$ is a normal subgroup of H and this was left as an exercise (see Exercise 5(a), §8.1).

Now define $\varphi \colon H \to (HK)/K$ by letting $\varphi(h) = hK$. Then φ is a group homomorphism since it is the composition of the inclusion homomorphism $H \hookrightarrow HK$ and the canonical epimorphism $HK \twoheadrightarrow (HK)/K$. Also we have

$$h \in \ker \varphi \iff hK = K \iff h \in H \cap K.$$

Thus $\ker \varphi = H \cap K$. An arbitrary element in $(HK)/K$ is of the form hkK for some $h \in H$ and $k \in K$. Since $\varphi(h) = hK = hkK$, φ is onto. The result now follows form the first isomorphism theorem. $\square$

Theorem 8.3.11 (Third Isomorphism Theorem). *If φ is a group epimorphism from a group G onto a group G'. Suppose that H' is a normal subgroup of G' and $H = \varphi^{-1}(H')$. Then $H \lhd G$ and*

$$\frac{G}{H} \cong \frac{G'}{H'}.$$

Since in the assumption of this theorem G' is a homomorphic image of G we may assume $G' = G/K$ for some $K \lhd G$. Let H be a normal subgroup of G containing K and let $H' = \varphi(H) = H/K$. Then the third isomorphism theorem can be restated as

$$\frac{G}{H} \cong \frac{(G/K)}{(H/K)}.$$

Proof. Define $\psi \colon G \to G'/H'$ by $\psi(g) = \varphi(g)H'$. Then ψ is a group epimorphism since it is the composition of $\varphi \colon G \to G'$ and the canonical epimorphism $G' \twoheadrightarrow G'/H'$, which are both group epimorphisms. It is straightforward to check that H is normal in G (see Exercise 7(b), §8.1).

Now we prove that the kernel of ψ is H. This is so since

$$h \in \ker \psi \iff \psi(h) = \varphi(h)H' = H'$$
$$\iff \varphi(h) \in H' \iff h \in \varphi^{-1}(H') = H.$$

The result now follows from the first isomorphism theorem.   □

Here we give another example concerning isomorphism theorems.

Example 8.3.12. The exponential function e^z is a group epimorphism from $\mathbb{C}$ onto $\mathbb{C}^*$. Let

$$S^1 = \{ z : |z| = 1 \}.$$

S^1 is a subgroup of $\mathbb{C}^*$ and the inverse image of S^1 is the set of purely imaginary numbers

$$P = \{ ib \in \mathbb{C} : b \in \mathbb{R} \}.$$

So we have $\mathbb{C}/P \cong \mathbb{C}^*/S^1 \cong (\mathbb{R}_+, \cdot)$. However, in $\mathbb{C}/P$ we have $\overline{z} = \overline{a}$ where a is the real part of z. It is easy to see that $\mathbb{C}/P \cong (\mathbb{R}, +)$. Thus the exponential function induces an isomorphism between $(\mathbb{R}, +)$ and $(\mathbb{R}_+, \cdot)$.

Exercises 8.3

1. Is the homomorphic image of a non-abelian group always non-abelian? Prove or disprove this assertion!

2. Find n such that $\mathbb{Z}_{36}/20\mathbb{Z}_{36} \cong \mathbb{Z}_n$.

3. Show that $\mathbb{Z}_n/k\mathbb{Z}_n \cong \mathbb{Z}_d$ where d is the positive g. c. d. of k and n.

4. Let G be a group and K be a normal subgroup. Suppose that $\{H_i\}_{i \in \Lambda}$ is a family of subgroups of G containing K.

   (a) Show that $\bigcap_{i \in \Lambda}(H_i/K) = (\bigcap_{i \in \Lambda} H_i)/K$.

   (b) For the rest of this exercise, assume that $\{H_i\}_{i \in \Lambda}$ is a family of normal subgroups of G containing K. Show that $\bigcap_{i \in \Lambda}(H_i/K)$ is normal in G/K.

(c) Is it true that $\dfrac{G}{\bigcap_{i\in\Lambda} H_i} \cong \dfrac{G/K}{\bigcap_{i\in\Lambda}(H_i/K)}$?

5. Let p be a positive prime integer. Let $GL_2(\mathbb{Z}_p)$ and $SL_2(\mathbb{Z}_p)$ be the general linear group and the special linear group with entries in $\mathbb{Z}_p$, respectively. Show that

$$|GL_2(\mathbb{Z}_p)| = (p^2 - 1)(p^2 - p)$$

and

$$|SL_2(\mathbb{Z}_p)| = p(p^2 - 1).$$

Review Exercises for Chapter 8

1. Let H be a subgroup of G. Suppose $H = \langle S \rangle$. Show that to prove $H \lhd G$ it suffices to check that $gsg^{-1} \in H$ for all $s \in S$ and $g \in G$.

2. Find all normal subgroups K in D_4. Compute the quotient group D_4/K for all such K.

3. Let G be a group. Let H be a subgroup of G and let K and K' be normal subgroups of G.

   (a) Is $K \cap K'$ normal in G?

   (b) Is KK' a subgroup of G? If yes, is KK' normal in G? Does $KK' = K'K$?

   (c) Is $H \cap K$ normal in H?

4. Let H and K be normal subgroups of G with $G = HK$ and $H \cap K = \{e\}$. Show that $G/H \cong K$ and $G/K \cong H$.

5. Let a, b be elements in a group G. Then the element $aba^{-1}b^{-1}$ is called a **commutator** in G.

   (a) Show that the subgroup generated by the commutators in G is a normal subgroup in G. This subgroup is denoted by $[G, G]$ and is called the **commutator subgroup** of G.

(b) Show that $G/[G, G]$ is abelian.

(c) Show that if K is a normal subgroup of G such that G/K is abelian, then $[G, G] \subseteq K$.

CHAPTER 9

Finite Abelian Groups

The direct product of n groups $G_1, G_2, \ldots, G_n$ is the set

$$G_1 \times G_2 \times \cdots \times G_n = \{(g_1, g_2, \ldots, g_n) : g_1 \in G_1, g_2 \in G_2, \ldots, g_n \in G_n\}$$

equipped with the binary operation of componentwise multiplication

$$(a_1, a_2, \ldots, a_n)(b_1, b_2, \ldots, b_n) = (a_1 b_1, a_2 b_2, \ldots, a_n b_n).$$

This is a quick way to form new groups.

We will develop a few elementary tools to classify finite abelian groups. The structure of a finite abelian group is almost determined by its order. Indeed we are able to classify all finite abelian groups up to isomorphism. For example, there are only two abelian groups of order 12, $C_4 \times C_3$ and $C_2 \times C_2 \times C_3$. Also there are only two abelian groups of order 18, $C_2 \times C_9$ and $C_2 \times C_3 \times C_3$.

9.1   Direct Products of Groups

There is an easy way to produce new groups from old groups. Let A and B be two groups. The Cartesian product of the sets A and B is given by

$$A \times B = \{ (a, b) : a \in A,\ b \in B \}.$$

We can define the componentwise operation

$$(a, b)(c, d) = (ac, bd)$$

on $A \times B$ to make it into a group for the following reasons.

(1) Associativity holds for the componentwise operation:

$$
\begin{aligned}
\big[(a_1, b_1)(a_2, b_2)\big](a_3, b_3) &= (a_1 a_2, b_1 b_2)(a_3, b_3) \\
&= ((a_1 a_2) a_3, (b_1 b_2) b_3) = (a_1 (a_2 a_3), b_1 (b_2 b_3)) \\
&= (a_1, b_1)(a_2 a_3, b_2 b_3) = (a_1, a_2)\big[(a_2, b_2)(a_3, b_3)\big]
\end{aligned}
$$

for all $a_1, a_2, a_3 \in A$ and $b_1, b_2, b_3 \in B$.

(2) The identity of $A \times B$ is (e_1, e_2), where e_1 and e_2 are identities of A and B, respectively.

(3) The inverse of the element (a, b) is (a^{-1}, b^{-1}) for $(a, b) \in A \times B$.

The group $A \times B$ is called the **direct product** of A and B.

The product $A \times B$ contains two particular subgroups

$$\overline{A} = \{ (a, e_2) : a \in A \} \quad \text{and} \quad \overline{B} = \{ (e_1, b) : b \in B \}.$$

For any $(c, d) \in A \times B$, we have

$$(c, d)(a, e_2)(c, d)^{-1} = (c, d)(a, e_2)(c^{-1}, d^{-1}) = (cac^{-1}, e_2) \in \overline{A}.$$

Thus $\overline{A}$ is normal in $A \times B$. It is also clear that the group homomorphism from A to $\overline{A}$ sending a to (a, e_2) is an isomorphism. Hence $\overline{A} \cong A$. Similarly $\overline{B}$ is also normal in $A \times B$ and $\overline{B} \cong B$.

We can extend the direct product of two groups to the direct product of a finite number of groups. Let $G_1, G_2, \ldots, G_n$ be n groups. Let

$$G = G_1 \times G_2 \times \cdots \times G_n = \{ (a_1, a_2, \ldots, a_n) : a_j \in G_j,\ j = 1, 2, \ldots, n \}.$$

Then G is a group called the (**external**) **direct product** of $G_1, G_2, \ldots, G_n$ under the componentwise operation

$$(a_1, a_2, \ldots, a_n)(b_1, b_2, \ldots, b_n) = (a_1 b_1, a_2 b_2, \ldots, a_n b_n).$$

This operation is easily seen to be associative. If $e_1, e_2, \ldots, e_n$ are identities of $G_1, G_2, \ldots, G_n$, respectively, then $e = (e_1, e_2, \ldots, e_n)$ is the identity of G and

$$(a_1, a_2, \ldots, a_n)^{-1} = (a_1^{-1}, a_2^{-1}, \ldots, a_n^{-1}).$$

The direct product of n groups $G_1, G_2, \ldots, G_n$ is denoted by

$$G_1 \times G_2 \times \cdots \times G_n \quad \text{or} \quad \prod_{i=1}^{n} G_i.$$

However, when $G_1, G_2, \ldots, G_n$ are additive groups, we can also use the **direct sum** notation

$$G_1 \oplus G_2 \oplus \cdots \oplus G_n \quad \text{or} \quad \bigoplus_{i=1}^{n} G_i.$$

Example 9.1.1. Let $C_2 = \{e, a : a^2 = e\}$ and $C_3 = \{e, b, b^2 : b^3 = e\}$. Then

$$C_2 \times C_3 = \{(e, e), (a, e), (e, b), (a, b), (e, b^2), (a, b^2)\}$$

is actually a cyclic group of order 6 since $o(c) = 6$ where $c = (a, b)$. Thus, c is a generator for $C_2 \times C_3$ and

$$C_2 \times C_3 \cong C_6 = \{e, c, c^2, c^3, c^4, c^5 : c^6 = e\}.$$

More generally, let m and n be relatively prime positive integers. Then $\mathbb{Z}_m \times \mathbb{Z}_n = \{(a, b) : 0 \leq a \leq m - 1, \ 0 \leq b \leq n - 1\}$ is a group of order mn. Since $(1, 1)$ is of order mn, we actually have that $\mathbb{Z}_m \times \mathbb{Z}_n$ is a cyclic group of order mn, that is, we have

$$\mathbb{Z}_m \times \mathbb{Z}_n \cong \mathbb{Z}_{mn} \quad \text{or} \quad C_m \times C_n \cong C_{mn}.$$

Example 9.1.2. Consider the group $G = \mathbb{Z}_{24} \times \mathbb{Z}_{36} \times \mathbb{Z}_{60}$. The maximal order of the elements in G is the l.c.m. of 24, 36 and 60 which is 360. In fact, the order of the element $(1, 1, 1)$ is 360. In G, $360a = 0$ for all $a \in G$. The order of every element in G is a divisor of 360.

Example 9.1.3. Let $\mathbb{Z}$ be the additive group of integers and $S = \{-1, 1\}$. The operation in $\mathbb{Z} \times S$ is given by

$$(i, a)(j, b) = (i + j, ab).$$

Observe that $\mathbb{Z} \times S$ is not cyclic. Suppose $\mathbb{Z} \times S$ is indeed cyclic. Let g be a generator. If $g = (k, 1)$ for some $k \in \mathbb{Z}$, then

$$\langle g \rangle = \{(ik, 1) : i \in \mathbb{Z}\} \neq \mathbb{Z} \times S,$$

a contradiction. If $g = (k, -1)$ for some $k \in \mathbb{Z}$, then

$$\langle g \rangle = \{(ik, 1) : i \text{ is even}\} \cup \{(ik, -1) : i \text{ is odd}\} \neq \mathbb{Z} \times S,$$

a contradiction again.

Let $G_1, \ldots, G_n$ be groups. Next, we make some observations on the direct product $G = G_1 \times \cdots \times G_n$.

(1) Let
$$\overline{G}_i = \{(e_1, \ldots, e_{i-1}, a_i, e_{i+1}, \ldots, e_n) \in G : a_i \in G_i\}.$$

Then $\overline{G}_i$ is a normal subgroup of G which is an isomorphic copy of G_i.

Let $(g_1, \ldots, g_n) \in G$. Then

$$\begin{aligned}
&(g_1, \ldots, g_n)(e_1, \ldots, e_{i-1}, a_i, e_{i+1}, \ldots, e_n)(g_1, \ldots, g_n)^{-1} \\
=&(g_1, \ldots, g_n)(e_1, \ldots, e_{i-1}, a_i, e_{i+1}, \ldots, e_n)(g_1^{-1}, \ldots, g_n^{-1}) \\
=&(e_1, \ldots, e_{i-1}, g_i a_i g_i^{-1}, e_{i+1}, \ldots, e_n) \in \overline{G}_i.
\end{aligned}$$

Hence $\overline{G}_i$ is normal in G.

Write
$$\overline{a}_i = (e_1, \ldots, e_{j-1}, a_j, e_{j+1}, \ldots, e_n) \in \overline{G}_j$$

for $a_i \in G_i$. It is also clear that

$$\begin{aligned}
\varphi: \quad G_i &\longrightarrow \overline{G}_i \\
a_i &\longmapsto \overline{a}_i
\end{aligned}$$

is an isomorphism.

(2) For each $i \neq j$, we have $\bar{a}_i \bar{a}_j = \bar{a}_j \bar{a}_i$. In other words, elements in $\overline{G}_i$ commute with elements in $\overline{G}_j$ for $i \neq j$.

(3) Given any element $g = (a_1, a_2, \ldots, a_n)$ in G, we have the expression

$$g = \bar{a}_1 \bar{a}_2 \cdots \bar{a}_n.$$

In other words,

$$G = \overline{G}_1 \overline{G}_2 \cdots \overline{G}_n.$$

(4) For each i, we have

$$G_i \cap G_1 \cdots G_{i-1} G_{i+1} \cdots G_n = \{e\}.$$

(5) Suppose that

$$g = g_1 g_2 \cdots g_n = h_1 h_2 \cdots h_n$$

where $g_i, h_i \in \overline{G}_i$ for $i = 1, \ldots, n$. Then

$$(h_1)^{-1} g_1 = (h_2 \cdots h_n)(g_2 \cdots g_n)^{-1} = (h_2 g_2^{-1}) \cdots (h_n h_n^{-1})$$

thanks to (2). It follows that

$$(h_1)^{-1} g_1 \in \overline{G}_1 \cap \overline{G}_2 \overline{G}_3 \cdots \overline{G}_n = \{e\}$$

where $e = (e_1, \ldots, e_n)$. Hence $g_1 = h_1$. Using the same argument, we can show that $g_2 = h_2$, $\ldots$, $g_n = h_n$.

To summarize, the group $G = G_1 \times \cdots \times G_n$ contains normal subgroups $\overline{G}_1, \overline{G}_2, \ldots, \overline{G}_n$ which are isomorphic of copies of its components $G_1, G_2, \ldots, G_n$. The elements of these subgroups commute with each other. (An element may not commute with its fellow subgroup members.) Moreover, any element g may be written as a product $g_1 g_2 \cdots g_n$, $g_i \in \overline{G}_i$, in a unique way.

What would happen if a group contains subgroups just as well-behaved? Let's look at the following result.

Proposition 9.1.4. *Suppose G contains two subgroups H and K such that $G = HK$ and $H \cap K = \{e\}$.*

(a) *The following two conditions are equivalent:*

(i) *Both H and K are normal in G;*

(ii) *Elements of H commute with elements of K.*

(b) *The group G is isomorphic to $H \times K$ if H and K satisfy one of the equivalent conditions in* (a).

Proof. (a) "(i)$\Rightarrow$(ii)": Let $h \in H$ and $k \in K$. Consider the element $g = hkh^{-1}k^{-1}$. Since K is normal in G, we have $g = (hkh^{-1})k^{-1} \in K$. On the other hand, we have $g = h(kh^{-1}k^{-1}) \in H$ since H is also normal in G. Thus $g \in H \cap K = \{e\}$. It follows that $hkh^{-1}k^{-1} = e$, or equivalently, $hk = kh$. We conclude that elements of H commute with elements of K.

"(ii)$\Rightarrow$(i)": Let $a \in H$. Take any $g \in G$. We may find $h \in H$ and $k \in K$ such that $g = hk$. Then

$$gag^{-1} = (hk)a(hk)^{-1} = hkak^{-1}h^{-1} = hah^{-1}kk^{-1} = hah^{-1} \in H.$$

Hence H is normal in G. Using a similar argument, we may also show that K is normal in G.

(b) We now construct a map $\varphi \colon H \times K \to G$ by sending (h, k) to hk. First, we check that this is indeed a group homomorphism. Let both (h, k) and (h', k') be elements in $H \times K$. Then

$$
\begin{aligned}
&\varphi((h, k)(h', k')) \\
&= \varphi(hh', kk') \\
&= hh'kk' = hkh'k', \qquad \text{by (ii) in (a),} \\
&= \varphi(h, k)\varphi(h', k').
\end{aligned}
$$

Next, we can see that φ is onto for we know that $G = HK$. Finally, we check that φ is injective. Let $\varphi(h, k) = hk = e$. Then $h = k^{-1} \in H \cap K = \{e\}$. We have just shown that $\ker \varphi = \{(e, e)\}$. Thus φ is indeed injective. To conclude, φ is indeed an isomorphism. $\qquad \square$

This proposition may be generalized to the case of n subgroups. See Exercises 7.

In Proposition 9.1.4, a group may be decomposed into a product from within itself when certain conditions are met. In this case we say that such a group is an (**internal**) **direct product** of its normal subgroups.

However, from the discussion above Proposition 9.1.4, we can see that $G = G_1 \times \cdots \times G_n$, as an *external* direct product, is also the *internal* direct product of its normal subgroups $\overline{G}_1, \overline{G}_2, \ldots, \overline{G}_n$. Since $\overline{G}_i \cong G_i$, we often don't distinguish between the external and the internal direct products. We will simply call either product the **direct product** of $G_1, \ldots, G_n$.

Example 9.1.5. The cyclic group $C_6 = \{ e, a, a^2, a^3, a^4, a^5 : a^6 = e \}$ contains two normal subgroups

$$H = \{ e, a^3 \} \quad \text{and} \quad K = \{ e, a^2, a^4 \}.$$

Observe that $H \cap K = \{e\}$. Since $a^j = a^{3j}a^{-2j} \in HK$, we have that $G = HK$. Thus, C_6 is the internal direct product of H and K. Since $H \cong C_2$, $K \cong C_3$ and $G \cong C_6$, this is a restatement of $C_6 \cong C_2 \times C_3$ (*Cf.* Example 9.1.1).

Example 9.1.6. The dihedral group

$$D_3 = \{ e, a, a^2, b, ab, a^2b : bab = a^2, \ a^3 = e = b^2 \}$$

contains two subgroups

$$H = \{ e, a, a^2 \} \quad \text{and} \quad K = \{ e, b \}.$$

The subgroup H is normal in D_3, but K is not. Thus, D_3 is not an internal direct product of H and K.

Exercises 9.1

1. Find the order of the element $(2, 3, 5)$ in $\mathbb{Z}_4 \times \mathbb{Z}_{12} \times \mathbb{Z}_{60}$.

2. Find the maximal order for the elements in the following groups.

   (a) $\mathbb{Z}_2 \times \mathbb{Z}_3$

   (b) $\mathbb{Z}_2 \times \mathbb{Z}_2 \times \mathbb{Z}_3$

   (c) $\mathbb{Z}_6 \times \mathbb{Z}_{12} \times \mathbb{Z}_{18}$

    (d) $\mathbb{Z}_{12} \times \mathbb{Z}_{30} \times \mathbb{Z}_{42}$

3. Let G, G', H and H' be groups.

    (a) Suppose $G \cong H$ and $G' \cong H'$. Show that $G \times G' \cong H \times H'$.

    (b) Show that $G \times G' \cong G' \times G$. Thus, when we decompose groups into direct products of subgroups, the ordering of the components in the product is irrelevant.

    (c) If G and G' are abelian, show that $G \times G'$ is abelian.

    (d) If $H \triangleleft G$ and $H' \triangleleft G'$, show that $H \times H' \triangleleft G \times G'$ and

$$\frac{G \times G'}{H \times H'} \cong \frac{G}{H} \times \frac{G'}{H'}.$$

4. Suppose given a finite group G. Let H and K be normal subgroups of G such that

    (i) $|G| = |H||K|$;

    (ii) $|H|$ and $|K|$ are relatively prime with each other.

Show that G is the internal direct product of H and K.

5. Let m and n be relatively prime positive integers. Show that

$$\mathbb{Z}_{mn} \cong \mathbb{Z}_m \times \mathbb{Z}_n$$

via the mapping $\varphi(\overline{x}) = (\overline{x}, \overline{x})$.

6. Are the groups $\mathbb{Z}_2 \times \mathbb{Z}_{12}$ and $\mathbb{Z}_4 \times \mathbb{Z}_6$ isomorphic to each other? Why or why not?

7. Let $H_1, \ldots, H_n$ be subgroups of G such that

    (i) the elements of H_i commutes with the elements of H_j for $i \neq j$,

  (ii) $G = H_1 H_2 \cdots H_n$, and

  (iii) the triangular condition holds, that is,

$$\begin{cases} H_1 \cap H_2 = \{e\}, \\ (H_1 H_2) \cap H_3 = \{e\}, \\ \cdots \\ (H_1 \cdots H_{n-1}) \cap H_n = \{e\}. \end{cases}$$

Show that G is the internal direct product of $H_1, \ldots, H_n$.

Show that this result still holds if condition (i) is replaced by

(i′) the subgroups $H_1, \ldots, H_n$ are normal in G.

9.2   Cauchy's Theorem

Let G be a finite group of order n. Lagrange's Theorem tells us that the order of a subgroup is a divisor of n. However, the converse of Lagrange's Theorem is not true in general (see Exercise 6). Nevertheless, Cauchy's Theorem guarantees that at least the converse of Lagrange's Theorem is true for prime divisors of n.

Theorem 9.2.1 (Cauchy's Theorem). *Let p be a prime divisor of the order of a finite group G. Then G contains an element of order p.*

Proof. First we prove the theorem when G is *abelian*. We proceed by induction on $|G|$. If $|G| = 1$, then $|G|$ has no prime divisor at all and we are done. Suppose that $|G| > 1$. Choose $a \in G \setminus \{e\}$ and let $H = \langle a \rangle$. We have the following possibilities.

(1) If $H = G$, then G is a cyclic group. Suppose that $|G| = kp$. Then $o(a) = kp$ and a^k is an element of order p.

(2) If H is a proper subgroup of G with p a divisor of $|H|$, then H contains an element of order p by the induction hypothesis.

(3) If H is a proper subgroup of G with p relatively prime to $|H|$. Since G is abelian, we have $H \lhd G$. Consider the quotient group G/H. Now p is a divisor of $|G/H| = |G|/|H| < |G|$. There exists an element gH in G/H of order p by the induction hypothesis. It follows that $H = (gH)^p = g^p H$ and hence $g^p = a^j$ for some integer j. Let k be the order of a^j. Then k is a divisor of $|H|$ and hence relatively prime to p. Note that $(g^k)^p = g^{kp} = e$. It follows that either $g^k = e$ or $o(g^k) = p$. Suppose that $g^k = e$. Then we have $(gH)^p = H = (gH)^k$. Note that p and k are relatively prime. There exist integers u and v such that $pu + kv = 1$. It follows that $gH = (gH)^{pu}(gH)^{kv} = H$,

a contradiction to the fact that gH is of order p. Thus $g^k \neq e$ and $o(g^k) = p$.

To prove the theorem for the general case, we need the class equation

$$|G| = |Z(G)| + \sum_{i=1}^{s} \frac{|G|}{|C(x_i)|},$$

for a choice of representatives of the x_i's with $C(x_i) \subsetneq G$. Again, we prove the assertion by induction on $|G|$. We have the following cases to consider.

(1) If p is a divisor of $|C(x_i)|$ for some i, then $C(x_i)$ contains an element of order p by the induction hypothesis.

(2) If p is not a divisor of $|C(x_i)|$ for any i, then p is a divisor of $|G|/|C(x_i)|$ for all i. This implies that p is a divisor of $|Z(G)|$. However, $Z(G)$ is an abelian group. So our assertion follows from the abelian case. $\square$

Cauchy's theorem has many implications in group theory. Here we use it to determine the structure of groups of order pq where p and q are distinct positive prime integers.

Lemma 9.2.2. *Let G be a group of order pq where p is a positive prime integers and $p > q$. Then G contains a normal subgroup of order p.*

Proof. By Cauchy's theorem, there exists an element a of order p. Take any $b \in G$. Let $A = \langle a \rangle$ and

$$B = bAb^{-1} = \left\{ e, bab^{-1}, ba^2b^{-1}, \ldots, ba^{p-1}b^{-1} \right\}.$$

Then A and B are both subgroups of order p. Consider the set

$$AB = \{ uv \in G : u \in A, \ v \in B \}.$$

We claim that AB has p^2 elements if $A \cap B = \{ e \}$ (*Cf.* Exercise 4, P. 93). Suppose that $uv = xy$ with $u, x \in A$ and $v, y \in B$. Then $x^{-1}u = yv^{-1} \in A \cap B = \{ e \}$. It follows that $u = x$ and $y = v$. Hence AB is a subset of G with p^2 elements, a contradiction since $pq < p^2$. It follows that $A \cap B \neq \{ e \}$. Since $A \cap B$ is a subgroup of A and $|A| = p$, we have that $A \cap B = A$. Thus $bAb^{-1} = A$ for all $b \in G$. This shows that A is a normal subgroup of G. $\square$

Example 9.2.3. Let G be a group of order 15. Then G must be cyclic.

Solution. The group G contains an element a of order 5 and another element b of order 3. Clearly $\langle a \rangle \cap \langle b \rangle = \{ e \}$. Since

$$|\langle a \rangle \langle b \rangle| = |\langle a \rangle||\langle b \rangle|/|\langle a \rangle \cap \langle b \rangle| = 15,$$

it follows that $G = \langle a \rangle \langle b \rangle$. We have $\langle a \rangle \lhd G$ by Lemma 9.2.2. Therefore $bab^{-1} = a^j$ for some integer $0 \le j \le 4$. Now we have

$$a = b^3 a b^{-3} = b(b(bab^{-1})b^{-1})b^{-1} = b(ba^j b^{-1})b^{-1} = b(bab^{-1})^j b^{-1}$$
$$= ba^{j^2}b^{-1} = (bab^{-1})^{j^2} = a^{j^3}.$$

Hence $j^3 \equiv 1 \pmod 5$. The only possibility is $j = 1$. Therefore we have that $bab^{-1} = a$, that is, $ab = ba$. It follows that the elements of $\langle a \rangle$ commutes with elements of $\langle b \rangle$. Hence $G = \langle a \rangle \times \langle b \rangle$ is abelian by Proposition 9.1.4. We conclude that ab is an element of order 15 (see Exercise 4, P. 63). $\diamond$

We may generalize the previous result to the following proposition.

Proposition 9.2.4. *Let G be a group of order pq where p and q are positive prime integers and $p > q$. If q is not a divisor of $p - 1$, then G must be cyclic.*

Proof. By Cauchy's theorem, G contains an element a of order p and another element b of order q. Also by Lemma 9.2.2, we have $bab^{-1} = a^j$ for some integer $0 \le j \le p - 1$. By Exercise 1 we have

$$a = b^q a b^{-q} = a^{j^q}.$$

It follows that

$$j^q \equiv 1 \pmod p.$$

However, by Fermat's Little Theorem, we have

$$j^{p-1} \equiv 1 \pmod p.$$

If q is not a divisor of $p - 1$, then q and $p - 1$ are relatively prime. There exist integers u and v such that

$$qu + (p-1)v = 1.$$

If follows that

$$j \equiv j^{qu+(p-1)v} = (j^q)^u (j^{p-1})^v \equiv 1^u 1^v = 1 \pmod{p}.$$

Thus $j = 1$. Consequently, $ab = ba$ and ab is an element of order pq. ☐

Let p be a positive prime integer. We call a group G a p-**group** if every element in G is of order p^m for some m. If a subgroup H of a group G is a p-group, we say H is a p-**subgroup** of G.

Cauchy's Theorem gives us another classification of p-groups.

Proposition 9.2.5. *Let p be a positive prime integer. A group G is a finite p-group if and only if it is of order p^n for some positive integer n.*

Proof. The "if" part follows directly from Lagrange's Theorem. To prove the "only if" part, suppose $|G| = mq$ where q is some positive prime integer $\neq p$. Then Cauchy's Theorem tells us that G contains an element of order q, a contradiction. ☐

Corollary 6.2.8 tells us that the center of any p-group is nontrivial. In the following section we will look more closely at the concept of p-groups.

Exercises 9.2

1. Let a and b be elements in a group G such that $bab^{-1} = a^j$. Show that $b^r ab^{-r} = a^{j^r}$.

2. Prove that a group of order 35 or 77 is cyclic.

3. Prove that a group of order 99 has a nontrivial normal subgroup.

4. Prove that a group of order 42 has a normal subgroup of order 21.

5. Find all the 2-subgroups and 3-subgroups of S_3.

6. Show that A_4 contains no subgroups of order 6 in the following steps.

   (a) Find all 3-cycles in A_4. There are altogether 8 of them.

   (b) Let H be a subgroup of order 6 in A_4 and α be any element in A_4. Show that at least two of H, αH and $\alpha^2 H$ coincide.

   (c) Show that all the 3-cycles are in H. This is a contradiction.

9.3   Structure Theorem of Finite Abelian Groups

For a finite abelian group G, we are going to show how to decompose G into a direct sum of p-groups.

Lemma 9.3.1. *Let G be a finite abelian group and p be a positive prime divisor of $|G|$. Consider the subset of G defined by*

$$G_p = \left\{ g \in G \mid g^{p^k} = e \text{ for some integer } k \geq 0 \right\}.$$

Then G_p is a subgroup of G of order p^m for the positive integer m such that $p^m \big| |G|$ but $p^{m+1} \nmid |G|$.

The subgroup G_p is called the p-**component** of G.

Proof. Clearly $e \in G_p$. Next, let a and $b \in G_p$. Suppose that $a^{p^i} = e$ and $b^{p^j} = e$. Then we have

$$(ab)^{p^{i+j}} = a^{p^{i+j}} b^{p^{i+j}} = (a^{p^i})^{p^j} (b^{p^j})^{p^i} = e.$$

This implies that $ab \in G_p$. Furthermore, $(a^{-1})^{p^i} = (a^{p^i})^{-1} = e$ and hence $a^{-1} \in G_p$. We have shown that G_p is a subgroup of G.

Note that G_p is a p-subgroup. By Proposition 9.2.5, we have that $|G_p| = p^m$ for some positive integer m. Finally, we show that $p^{m+1} \nmid |G|$. If not, p is still a divisor of $|G|/p^m = |G/G_p|$. By Cauchy's Theorem, there is a gG_p whose order is p. This implies that $g^p \in G_p$. This further implies that $g^{p^{m+1}} = (g^p)^{p^m} = e$. Hence $g \in G_p$ and the order of gG_p is 1, a contradiction. $\square$

Proposition 9.3.2. *Let G be a finite abelian group of order n. Suppose that $n = p_1^{\alpha_1} p_2^{\alpha_2} \cdots p_k^{\alpha_k}$ where the p_i's are positive prime integers such that $p_1 < p_2 < \cdots < p_k$. Then*

$$G \cong G_{p_1} \times G_{p_2} \times \cdots \times G_{p_k},$$

where $|G_{p_j}| = p_j^{\alpha_j}$ for each j.

Note that the recovering of the p-components of G is independent of the decompositions above. Hence this decomposition into p-components is unique.

Proof. Let $q_j = n/p_j^{\alpha_j}$, $j = 1, 2, \ldots, k$. Then the greatest common divisor of $q_1, q_2, \ldots, q_k$ is 1. There exist integers $m_1, m_2, \ldots, m_k$ such that

$$m_1 q_1 + m_2 q_2 + \cdots + m_k q_k = 1.$$

It follows that

$$g = g^{m_1 q_1} g^{m_2 q_2} \cdots g^{m_k q_k}$$

where $g^{(m_j q_j)} \in G_{p_j}$ since $\left(g^{m_j q_j}\right)^{p_j^{\alpha_j}} = (g^n)^{m_j} = e$. Consequently, we have $G = G_{p_1} G_{p_2} \cdots G_{p_k}$. Next we prove that the product is a direct product. Suppose that

$$g = a_1 a_2 \cdots a_k = b_1 b_2 \cdots b_k$$

where a_j and $b_j \in G_{p_j}$ for each j. Then

$$a_1 b_1^{-1} = (b_2 a_2^{-1}) \cdots (b_k a_k^{-1}) \in G_{p_1} \cap (G_{p_2} \cdots G_{p_k}).$$

As the order of G_{p_1} and the order of $G_{p_2} \cdots G_{p_k}$ are relatively prime, it follows that

$$G_{p_1} \cap (G_{p_2} \cdots G_{p_k}) = \{e\}.$$

Hence $a_1 = b_1$ and $a_2 \cdots a_k = b_2 \cdots b_k$. Inductively, we can also show that $a_2 = b_2, \ldots, a_k = b_k$.

The fact that the order of G_{p_j} is $p_j^{\alpha_j}$ is a result of Lemma 9.3.1. □

The elements of G whose orders are powers of a prime p are collected into the p-components. Once we decompose a finite abelian group into a direct sum of its p-components, we can focus our attention on abelian p-groups.

Lemma 9.3.3. *Let G be a finite abelian p-group, p a positive prime integer, and let a be an element of maximal order in G. Then $G \cong \langle a \rangle \times K$ for some subgroup K of G.*

Proof. We will prove this lemma by induction on n where $|G| = p^n$. If $n = 1$ then $G = \langle a \rangle$ and we simply choose K to be the trivial subgroup. Without loss of generality we may assume $n > 1$.

If $G = \langle a \rangle$, there is nothing to prove. We assume otherwise. Since $G/\langle a \rangle$ is a nontrivial p-group, find $b\langle a \rangle$ in $G/\langle a \rangle$ with $o(b\langle a \rangle) = p$. Then $b^p = a^i$ for some i. If $(i, p) \sim 1$ then $o(b^p) = o(a^i) = o(a)$. It follows that $o(b) = po(a) > o(a)$, a contradiction. Hence $i = pk$ for some k. Thus $(ba^{-k})^p = e$. We have that $ba^{-k} \notin \langle a \rangle$ and $o(ba^{-k}) = p$. Replace b by ba^{-k}. We may assume that $b \notin \langle a \rangle$ and $o(b) = p$. It follows that $\langle a \rangle \cap \langle b \rangle = \{e\}$. Consider the quotient group $\overline{G} = G/\langle b \rangle$. Note that $o(\overline{a}) = o(a)$ and $\overline{a}$ is of maximal order in $\overline{G}$. By the induction hypothesis, $\overline{G} = \langle \overline{a} \rangle \times \overline{K}$ for some subgroup K containing $\langle b \rangle$. We claim that $G \cong \langle a \rangle \times K$.

For all $g \in G$, $\overline{g} = \overline{a}^i \overline{x}$ for some integer i and some $x \in K$. It follows that for some j, $g = a^i x b^j \in \langle a \rangle K$. On the other hand, if $g \in \langle a \rangle \cap K$, then $\overline{g} \in \langle \overline{a} \rangle \cap \overline{K} = \{\overline{e}\}$. Thus $g = b^j$ for some j and $g \in \langle a \rangle \cap \langle b \rangle = \{e\}$. We have proved that $\langle a \rangle \cap K = \{e\}$. It follows that G is the internal direct product of $\langle a \rangle$ and K. □

Proposition 9.3.4. *Let G be a finite abelian p-group where p is a positive prime integer. Then*

$$G \cong C_{p^{j_1}} \times C_{p^{j_2}} \times \cdots \times C_{p^{j_s}}$$

where $C_{p^{j_i}}$ is the cyclic group of order p^{j_i}. The decomposition is unique up to different orderings of the components. More precisely, if we arrange it so that $j_1 \geq j_2 \geq \cdots \geq j_s$ and there is another decomposition of G into a direct product of cyclic subgroups

$$G \cong C_{p^{k_1}} \times C_{p^{k_2}} \times \cdots \times C_{p^{k_t}}$$

so that $k_1 \geq k_2 \geq \cdots \geq k_t$, then $s = t$ and $j_i = k_i$ for all i.

Proof. Let $|G| = p^n$ where $n \geq 1$. We shall prove the "existence" part by induction on n. There is nothing to prove if G is cyclic. We assume otherwise. Use Lemma 9.3.3 to find $a \in G$ such that $G \cong \langle a \rangle \times K$. Use the induction hypothesis to further decompose K into a product of cyclic groups. We have the "existence" part.

For the uniqueness of the decomposition, we leave it as an exercise (see Exercise 7). □

We sum up all results into the following theorem.

Theorem 9.3.5 (Structure Theorem of Finite Abelian Groups). *Let G be a finite abelian group. Then*

$$G \cong C_{p_1^{j_1}} \times C_{p_2^{j_2}} \cdots \times C_{p_s^{j_s}}$$

where $C_{p_i^{j_i}}$ is the cyclic group of order $p_i^{j_i}$ and $p_1, \ldots, p_s$ are (possibly repeated) prime integers such that $|G| = p_1^{j_1} p_2^{j_2} \cdots p_s^{j_s}$. The decomposition is unique except for different orderings of the summands.

Proof. The result follows from Proposition 9.3.2 and Proposition 9.3.4. $\square$

A decomposition of the form for a finite abelian group as in Theorem 9.3.5 is called a **primary decomposition**.

Example 9.3.6. Let C_n be the cyclic group of order $360 = 2^3 \times 3^2 \times 5$. There are exactly six different non-isomorphic abelian groups of order 360 as listed below:

$$C_8 \times C_9 \times C_5, \quad C_8 \times C_3 \times C_3 \times C_5,$$
$$C_2 \times C_4 \times C_9 \times C_5,$$
$$C_2 \times C_4 \times C_3 \times C_3 \times C_5,$$
$$C_2 \times C_2 \times C_2 \times C_9 \times C_5, \quad \text{and}$$
$$C_2 \times C_2 \times C_2 \times C_3 \times C_3 \times C_5.$$

Exercises 9.3

1. Determine the structure of abelian groups of order 540 up to isomorphism.

2. How many non-isomorphic finite abelian groups of order 3600 are there?

3. How many non-isomorphic finite abelian groups of order 5400 are there?

4. Find all the p-components in $U(\mathbb{Z}_{36})$. Find the primary decomposition for $U(\mathbb{Z}_{36})$.

5. Show that $C_8 \times C_9 \times C_5 \cong C_{360}$.

6. Let G be a finite group of order mn where m, n are relatively prime positive integers. Let

$$M = \{\, g \in G : g^m = e \,\} \quad \text{and} \quad N = \{\, g \in G : g^n = e \,\}.$$

   (a) Give an example to show that M or N might not be subgroups of G.

   (b) Show that both M and N are subgroups of G if G is assumed to be abelian. Also show that $G \cong M \times N$ in this case.

7. Let G be an abelian group and p be a positive prime integer. For each i, define

$$G^{p^i} = \left\{\, g^{p^i} : g \in G \,\right\}.$$

Of course, when G is an additive group, we will use the notation

$$p^i G = \left\{\, p^i g : g \in G \,\right\}.$$

   (a) Show that $G^{p^{i+1}}$ is a subgroup of G^{p^i} for each i.

   (b) Show that $\left| p^i \mathbb{Z}_{p^m} / p^{i+1} \mathbb{Z}_{p^m} \right| = \begin{cases} p, & \text{if } i < m, \\ 1, & \text{if } i \geq m. \end{cases}$

   (c) Suppose $G \cong C_{p^{j_1}} \times C_{p^{j_2}} \times \cdots \times C_{p^{j_s}}$. Show that $\left| G^{p^i} / G^{p^{i+1}} \right| = p^{n_i}$, where $n_i = \#\{\, k : 1 \leq k \leq s, \ j_k > i \,\}$.

   (d) Show that the j_k's can be recovered from the n_i's. Hence the j_k's are independent of the decomposition of G.

8. Let G be a finite abelian group of order n. Show that G is cyclic if and only if n is the least positive integer such that $a^n = e$ for all $a \in G$.

Review Exercises for Chapter 9

1. Find all the proper subgroups of $\mathbb{Z}_2 \oplus \mathbb{Z}_2$ and $\mathbb{Z}_2 \oplus \mathbb{Z}_3 \oplus \mathbb{Z}_3$.

2. Show that the infinite cyclic group cannot be the direct product of two nontrivial groups.

3. Find all the 2- and 3-subgroups of A_4.

4. Let G be an abelian group of order n. Let d be any positive divisor of n. Show that G contains a subgroup of order d. Hence the converse of Lagrange's Theorem holds for finite abelian groups.

CHAPTER 10

Group Actions

After having studied the basic definitions and properties regarding groups in general and permutation groups in particular, we are now ready to delve into the more technical aspect of the theory of groups. In this chapter we will study *group actions*. The notion of group actions is also useful in other disciplines of mathematics. As an application we will introduce Burnside's formula, which is a very nice little formula for combinatorial problems.

10.1   Definition and Basic Properties

Definition 10.1.1. Let S be a set and G be a group. A **group action** of G on the set S is a map from $G \times S$ into S, the image of (g, x) being denoted by $g \cdot x$, such that

  (i) $e \cdot x = x$ for all $x \in S$, and

  (ii) $(g_1 g_2) \cdot x = g_1 \cdot (g_2 \cdot x)$ for all $g_1, g_2 \in G$ and for all $x \in S$.

In this situation, we also say that G **acts** on S or that S is a G-**set**.

Theorem 10.1.2. *Let S be a G-set. For each $g \in G$, the function $L_g \colon S \to S$ defined by $L_g(x) = g \cdot x$ is an element in $\mathrm{Sym}(S)$. Moreover, the map $\varphi \colon G \to \mathrm{Sym}(S)$ defined by $\varphi(g) = L_g$ is a group homomorphism.*

Conversely, a group homomorphism $\varphi \colon G \to \mathrm{Sym}\, S$ also determines a group action of G on S by letting $g \cdot x = \varphi(g)(x)$.

Proof. To show that L_g is an element in $\mathrm{Sym}(S)$, we must show that it is one-to-one and onto. Suppose that $L_g(x_1) = L_g(x_2)$. Then

$$x_1 = e \cdot x_1 = (g^{-1}g) \cdot x_1 = g^{-1} \cdot (g \cdot x_1) = g^{-1} \cdot (g \cdot x_2) = (g^{-1}g) \cdot x_2 = e \cdot x_2 = x_2.$$

It follows that L_g is one-to-one. For any $x \in S$, we have

$$L_g(g^{-1} \cdot x) = g \cdot (g^{-1} \cdot x) = (gg^{-1}) \cdot x = e \cdot x = x.$$

Hence L_g is onto. Thus L_g is indeed an element in $\mathrm{Sym}(S)$.

To show that φ is a group homomorphism from G into $\mathrm{Sym}(S)$, we must show that $\varphi(g_1 g_2) = \varphi(g_1)\varphi(g_2)$. It is equivalent to showing that $L_{g_1 g_2} = L_{g_1} \circ L_{g_2}$. This follows from

$$L_{g_1 g_2}(x) = (g_1 g_2) \cdot x = g_1 \cdot (g_2 \cdot x) = g_1 \cdot L_{g_2}(x) = L_{g_1} \circ L_{g_2}(x).$$

Conversely, we need to show the mapping induced by the group homomorphism is a group action. Since $\varphi(e) = \mathbf{1}_S$, we have

$$e \cdot x = \varphi(e)(x) = \mathbf{1}_S(x) = x$$

for all $x \in S$. Furthermore, we have

$$(g_1 g_2) \cdot x = \varphi(g_1 g_2)(x) = (\varphi(g_1) \circ \varphi(g_2))(x)$$

$$= \varphi(g_1)(\varphi(g_2)(x)) = \varphi(g_1)(g_2 \cdot x) = g_1 \cdot (g_2 \cdot x)$$

for all $g_1, g_2 \in G$ and all $x \in G$. $\qquad\qquad \square$

We say a group action is **effective** if the group homomorphism φ it induces is injective.

Example 10.1.3. The most trivial example of group actions is the trivial action. Let G be an arbitrary group and let S be an arbitrary set. Let G act on S by letting $g \cdot x = x$ for all $g \in G$ and $x \in S$. This action is the *least* "effective" group action.

Example 10.1.4. Let S be any set and let $G = \mathrm{Sym}(S)$. Take $f \in \mathrm{Sym}(S)$ and $x \in S$. Then $\mathrm{Sym}(S)$ acts naturally on S by letting $f \cdot x = f(x)$. In fact, the group homomorphism induced by this action is the identity map on $\mathrm{Sym}\, S$. This is the most natural group action and it is extremely "effective".

Example 10.1.5. Let $G = \{\, e, a, a^2 \mid a^3 = e \,\}$ be a cyclic group of order 3. Then G can act on itself by left multiplication. This group action induces a group homomorphism $\varphi \colon G \to \mathrm{Sym}\, G$. We have

$$\varphi(e) = \begin{bmatrix} e & a & a^2 \\ e & a & a^2 \end{bmatrix}, \ \varphi(a) = \begin{bmatrix} e & a & a^2 \\ a & a^2 & e \end{bmatrix} \text{ and } \varphi(a^2) = \begin{bmatrix} e & a & a^2 \\ a^2 & e & a \end{bmatrix}.$$

If we index e, a, a^2 by $1, 2, 3$ respectively, we can view $\varphi(a)$ as the 3-cycle $(1\ 2\ 3)$ and the image of φ as $\langle (1\ 2\ 3) \rangle$. This group action is effective.

Example 10.1.6. Let H be a subgroup of the group G and $S = \{\, aH : a \in G \,\}$ be the set of left cosets of H in G. A very important group action is that of G acting on S by left translation, that is, the one defined by letting $g \cdot aH = gaH$. This action induces a group homomorphism $\varphi \colon G \to \mathrm{Sym}(S)$. If g is in the kernel of φ, then $gaH = aH$ for all $a \in G$. In particular when $a = e$, we have $gH = H$ and hence $g \in H$. Thus the kernel of φ is contained in H.

Example 10.1.7. Let

$$G = SL_2(\mathbb{R}) = \left\{ \begin{bmatrix} a & b \\ c & d \end{bmatrix} : a, b, c, d \in \mathbb{R}, \ ad - bc = 1 \right\}$$

be the special linear group and $\mathscr{H}$ be the upper half plane

$$\mathscr{H} = \{\, z = x + iy : x, y \in \mathbb{R}, \ \mathrm{Im}\, z = y > 0 \,\}$$

of the complex plane $\mathbb{C}$. Let G act on $\mathscr{H}$ by the action

$$\begin{bmatrix} a & b \\ c & d \end{bmatrix} \cdot z = \frac{az + b}{cz + d}.$$

This is well-defined since

$$\mathrm{Im}\, \frac{az + b}{cz + d} = \frac{\mathrm{Im}\, z}{|cz + d|^2} > 0$$

when $z \in \mathscr{H}$. Clearly,

$$\begin{bmatrix} 1 & 0 \\ 0 & 1 \end{bmatrix} \cdot z = \frac{z}{1} = z$$

for all $z \in \mathscr{H}$. For $g_1 = \begin{bmatrix} a & b \\ c & d \end{bmatrix}$ and $g_2 = \begin{bmatrix} p & q \\ r & s \end{bmatrix} \in G$, we have

$$\left(\begin{bmatrix} a & b \\ c & d \end{bmatrix} \begin{bmatrix} p & q \\ r & s \end{bmatrix} \right) \cdot z = \begin{bmatrix} ap + br & aq + bs \\ cp + dr & cq + ds \end{bmatrix} \cdot z$$

$$= \frac{(ap + br)z + (aq + bs)}{(cp + dr)z + (cq + ds)}.$$

On the other hand, we also have

$$\begin{bmatrix} a & b \\ c & d \end{bmatrix} \cdot \left(\begin{bmatrix} p & q \\ r & s \end{bmatrix} \cdot z \right) = \frac{a\left(\dfrac{pz + q}{rz + s} \right) + b}{c\left(\dfrac{pz + q}{rz + s} \right) + d}$$

$$= \frac{a(pz + q) + b(rz + s)}{c(pz + q) + d(rz + s)} = \frac{(ap + br)z + (aq + bs)}{(cp + dr)z + (cq + ds)}.$$

This is indeed a group action.

Example 10.1.8. Let S be a G-set and let H be a subgroup of G. Then H inherits an induced action on S from G. Similarly, if T is a subset of S stabilized by G, that is, $G \cdot T \subseteq T$, the restriction of the action to T is an action of G on T.

Lemma 10.1.9. *Let S be a G-set. Suppose $g \cdot a = b$ for some $g \in G$ and $a, b \in S$. Then $g^{-1} \cdot b = a$.*

Proof. It is clear that $g^{-1} \cdot b = g^{-1} \cdot (g \cdot a) = (g^{-1}g) \cdot a = 1 \cdot a = a$. $\qquad\square$

Proposition 10.1.10. *Let S be a G-set. Define a relation $\sim$ on S by letting*

$$a \sim b \text{ iff } g \cdot a = b \text{ for some } g \in G.$$

Then $\sim$ is an equivalence relation on S.

We leave the proof of this theorem as an exercise (see Exercise 1).

Let S be a G-set. The equivalence classes determined by the equivalence relation $\sim$ in the previous theorem are called the **orbits** of G (or G-**orbits**) on S. The equivalence class of $x \in S$ is also called the **orbit of x under G** (or the G-**orbit of** x), denoted $G \cdot x$. In fact,

$$G \cdot x = \{g \cdot x : g \in G\}$$

for $x \in S$. The G-orbits form a partition of S.

If the action of G on S has exactly one orbit, we say G acts **transitively** on S. In this case, $S = G \cdot x$ for some $x \in S$.

Proposition 10.1.11. *Let S be a G-set. For $x \in S$, let*

$$\mathrm{Stab}_G\, x = \{\, g \in G : g \cdot x = x \,\}.$$

Then $\mathrm{Stab}_G\, x$ is a subgroup of G for each x in S.

Proof. Clearly $e \in \mathrm{Stab}_G\, x$. Suppose that $g_1, g_2 \in \mathrm{Stab}_G\, x$. Then $g_1 \cdot x = x$ and $g_2 \cdot x = x$. Consequently, $(g_1 g_2) \cdot x = g_1 \cdot (g_2 \cdot x) = g_1 \cdot x = x$ and so $g_1 g_2 \in \mathrm{Stab}_G\, x$. If $g \in \mathrm{Stab}_G\, x$, then $g^{-1} \cdot x = g^{-1} \cdot (g \cdot x) = e \cdot x = x$. It follows that $g^{-1} \in \mathrm{Stab}_G\, x$. Thus $\mathrm{Stab}_G\, x$ is a subgroup of G. $\qquad\square$

For each $x \in S$, we call the subgroup $\mathrm{Stab}_G\, x$ the **stabilizer** or the **isotropy group** of x in G. Sometimes the stabilizer group $\mathrm{Stab}_G\, x$ is denoted as $\mathrm{Stab}\, x$ when G is understood.

We are now ready to count the number of elements in an orbit.

Proposition 10.1.12. *Let S be a G-set and let $x \in S$. If G is a finite group, then*

$$|G \cdot x| = [G : \operatorname{Stab}_G x] = \frac{|G|}{|\operatorname{Stab}_G x|}.$$

Proof. Let

$$T = \{\, g \operatorname{Stab}_G x : g \in G \,\}$$

be the set of left cosets of $\operatorname{Stab}_G x$ in G. Define $\varphi \colon T \to G \cdot x$ by letting $\varphi(g \operatorname{Stab}_G x) = g \cdot x$. First we need to check that φ is well-defined. Suppose $g \operatorname{Stab}_G x = g' \operatorname{Stab}_G x$. Then $g' = gh$ for some $h \in \operatorname{Stab}_G x$. It follows that $g' \cdot x = (gh) \cdot x = g \cdot (h \cdot x) = g \cdot x$.

Next we shall show that φ is bijective. Since φ is clearly onto, it remains to show that φ is one-to-one. Suppose that $\varphi(g_1 \operatorname{Stab}_G x) = \varphi(g_2 \operatorname{Stab}_G x)$. Then $g_1 \cdot x = g_2 \cdot x$. It follows that $(g_2^{-1} g_1) \cdot x = x$ and $g_2^{-1} g_1 \in \operatorname{Stab}_G x$. Consequently, $g_1 \operatorname{Stab}_G x = g_2 \operatorname{Stab}_G x$. Hence φ is bijective and so

$$|G \cdot x| = |T| = \frac{|G|}{|\operatorname{Stab}_G x|}. \qquad \square$$

Exercises 10.1

1. Prove Proposition 10.1.10.

2. Let the group G act on the set S. Suppose $G = \langle T \rangle$. Show that $x \in S$ is fixed by G if it is fixed by g for all $g \in T$.

3. Find the number of orbits in $\{\, 1, 2, \ldots, 8 \,\}$ under the subgroup of S_8 generated by (1 3) and (2 4 8).

4. Let the group G act transitively on a set S. Show that the isotropy subgroups of two different elements x and y of S are conjugate to each other in G.

5. Let G act on S and let $\mathscr{O}$ be one of the orbits under this action.

   (a) Show that there is an induced transitive group action of G on $\mathscr{O}$.

(b) Let $\eta \in G$. Show that $\{\eta \cdot x \in S : x \in \mathcal{O}\} = \mathcal{O}$.

6. Let S be a set and let $\mathrm{Sym}(S)$ act naturally on S. Find the orbits of this group action. Is this action transitive? Find the stabilizer of $x \in S$ in G.

7. Let G act on S as in Example 10.1.6. Show that the kernel of the group homomorphism induced by this action is H if $H \triangleleft G$.

8. Let $\mathbb{R}$ be the additive group of real numbers. Then $\mathbb{R}$ acts on the real plane $\mathbb{R}^2$ by rotating the plane counterclockwise about the origin by the angle θ, *i.e.*,

$$\theta \cdot (x, y) = (x \cos\theta - y \sin\theta, x \sin\theta + y \cos\theta).$$

Let P be a point other than the origin in the plane.

(a) Describe the orbit containing P.

(b) Find the isotropy subgroup of P in $\mathbb{R}$.

10.2 Orbits and Stabilizers

Let S be a G-set. For $g \in G$ we call the set

$$S_g = \{x \in S : gx = x\}$$

the **fixed set of g in** S. We call the set

$$S_G = \{x \in S : g \cdot x = x \text{ for all } g \in G\} = \{x \in S : \mathrm{Stab}_G\, x = G\}$$

the **fixed set** of G in S. In other words, S_G is the subset of S of elements with one-element orbit.

Remember that the orbits of G form a partition of S. Hence we may find a set of representatives $x_1, \ldots, x_t$ in S such that

$$S = G \cdot x_1 \,\dot{\cup}\, G \cdot x_2 \,\dot{\cup} \cdots \dot{\cup}\, G \cdot x_t.$$

We may arrange it so that the orbit of x_i contains more than one element for $i = 1, \ldots, s$ while $G \cdot x_i$ is a singleton orbit for $i = s + 1, \ldots, t$. Thus, $S_G = \{x_{s+1}, \ldots, x_t\}$ and

(10.2.1) $$S = S_G \,\dot{\cup}\, G \cdot x_1 \,\dot{\cup} \cdots \dot{\cup}\, G \cdot x_s.$$

Now we can make the following conclusion.

Theorem 10.2.1. *Let G be a group and let S be a finite G-set. Then there exist $x_1, \ldots, x_s \in G$ such that*

$$(10.2.2) \qquad |S| = |S_G| + \sum_{i=1}^{s} \frac{|G|}{|\operatorname{Stab}_G x_i|}$$

where $\operatorname{Stab}_G x_i \subsetneq G$ *for all i.*

Proof. The result follows from (10.2.1) and Proposition 10.1.12. $\square$

Corollary 10.2.2. *Let p be a positive prime integer and let a p-group G act on a finite set S. Then*

$$|S| \equiv |S_G| \pmod{p}.$$

Proof. If $G \cdot x$ contains more than one element, p must divide $|G|/|\operatorname{Stab}_G x_i|$ for G is a p-group. Hence the result follows from Theorem 10.2.1. $\square$

Let G be a group. A very important group action of G on $S = G$ is the one of conjugation, that is, the one defined by $g \cdot x = gxg^{-1}$. In this case, the orbit containing x is the conjugacy class of x.

The fixed set of G in S is

$$(10.2.3) \qquad S_G = \{\, x \in G : gxg^{-1} = x \text{ for all } g \in G \,\}$$
$$= \{\, x \in G : gx = xg \text{ for all } g \in G \,\} = Z(G),$$

the *center* of G.

The stabilizer of x in G is

$$\operatorname{Stab}_G x = \{\, g \in G : gxg^{-1} = x \,\} = \{\, g \in G : gx = xg \,\} = C_G(x),$$

the centralizer of x in G. Hence the relation (10.2.2) for the group action of conjugation is simply the *class equation* for G (*Cf.* the class equation in (6.2.1)). We restate the class equation of G here for easy reference.

Corollary 10.2.3. *Let G be a finite group. We may find $x_1, \ldots, x_s$ in G such that*

$$|G| = Z(G) + \sum_{i=1}^{s} \frac{|G|}{|C_G(x_i)|}$$

where $C_G(X_i)$ *is a proper subgroup of G for each i.*

Corollary 10.2.4. *Let p be a positive prime integer and let G be a p-group, then the center of G is non-trivial.*

Proof. The result follows from (10.2.3) and Corollary 10.2.2. *Cf.* Corollary 6.2.8. □

Proposition 10.2.5. *For any prime integer p, every group of order p^2 is abelian.*

Proof. Let G be a group of order p^2. Then $Z(G)$ is a nontrivial normal subgroup of G. We have the following cases to be considered.

(1) If $|Z(G)| = p^2$ then $G = Z(G)$ is an abelian group.

(2) If $|Z(G)| = p$, G is non-abelian. In this case $G/Z(G)$ is a cyclic group of order p. Let a be a generator of $Z(G)$ and $bZ(G)$ be a generator of $G/Z(G)$. Then elements in G can be expressed as $b^j a^k$. Also we have

$$(b^\ell a^m)(b^j a^k) = b^{\ell+j} a^{m+k} = (b^j a^k)(b^\ell a^m)$$

since $ab = ba$. Hence G is abelian, a contradiction. □

We can make some deductions about a group by observing its class equation.

Example 10.2.6. In this example, we explain how to find the normal subgroups in S_4 and in A_4.

Note that a normal subgroup must contain the identity element whose conjugacy class is a singleton set. A normal subgroup is closed under conjugation. Thus it must be a union of some conjugacy classes.

In Example 6.2.5 we calculated the class equation for S_4. Let's reproduce the table here for easy reference:

Type	Representative	# of elements in this conjugacy class
(4)	(1 2 3 4)	$3! = 6$
(3, 1)	(1 2 3)	$\binom{4}{3}2! = 8$
(2, 2)	(1 2)(3 4)	$\binom{4}{2}\binom{2}{2}/2 = 3$
(2, 1, 1)	(1 2)	$\binom{4}{2} = 6$
(1, 1, 1, 1)	1	1

If K is a normal subgroup of S_4, we know the following properties must be satisfied.

(1) The order $|K|$ is a divisor of $|S_4| = 24$.

(2) The order $|K|$ is the sum of 1 and some of the numbers among 6, 8, 3, and 6.

After examining all combinations, we may find all the possibilities listed below:

- $|K| = 1$;

- $|K| = 1 + 6 + 8 + 3 + 4$;

- $|K| = 1 + 8 + 3$;

- $|K| = 1 + 3$.

In the first three cases, K is clearly seen to be the trivially subgroup, the group S_4 and the group A_4 respectively. For the last case,

$$(10.2.4) \qquad K = \{1,\ (1\ 2)(3\ 4),\ (1\ 3)(2\ 4),\ (1\ 4)(2\ 3)\}$$

if it indeed is a subgroup of S_4. This can be easily checked to be true by using Theorem 3.3.6. We conclude that there are altogether 4 normal subgroups in S_4.

Next, we proceed to find the normal subgroups of A_4. Again, we make use of the class equation of A_4 which we found in Example 6.2.7:

$$12 = 1 + 4 + 4 + 3.$$

Suppose K is a normal subgroup of A_4. After examining all combinations, we may find all the possibilities listed below:

- $|K| = 1$;

- $|K| = 1 + 4 + 4 + 3$;

- $|K| = 1 + 3$.

The first two cases give the two obvious normal subgroups: the trivial subgroup and A_4. The last case again gives the subgroup in (10.2.4). There are altogether three normal subgroups in A_4.

Finally, we give another application of group actions.

Theorem 10.2.7 (Cauchy's Theorem revisited). *Let p be a prime divisor of the order of a finite group G. Then G contains an element of order p.*

McKay's proof of Cauchy's Theorem. Let

$$S = \{ (a_1, a_2, \ldots, a_p) \in G^p : a_j \in G \text{ and } a_1 a_2 \cdots a_p = e \}.$$

It is easy to see that $|S| = |G|^{p-1}$ since a_p must be equal to $(a_1 a_2 \cdots a_{p-1})^{-1}$ while $a_1, a_2, \ldots, a_{p-1}$ are freely chosen from G. Let $\sigma = (1\ 2\ 3\ \cdots\ p) \in S_p$ and let $K = \langle \sigma \rangle$ act on S by the action

$$\sigma \cdot (a_1, a_2, \ldots, a_p) = (a_2, a_3, \ldots, a_p, a_1).$$

Note that $(a_2, a_3, \ldots, a_1) \in S$ if $(a_1, a_2, \ldots, a_p) \in S$ since

$$a_2 a_3 \cdots a_p a_1 = a_1^{-1}(a_1 a_2 \cdots a_p)a_1 = a_1^{-1} e a_1 = e.$$

It is routine to check that $S_K = \{ (a, a, \ldots, a) \in G^p : a^p = e \}$. By Corollary 10.2.2, we have

$$|S_K| \equiv |S| \equiv 0 \pmod{p}.$$

Since S_K contains $(e, e, \ldots, e)$, S_K contains at least p elements. Hence there exists $a \in G \setminus \{ e \}$ such that $(a, a, \ldots, a) \in S_K$. We thus obtain an element a of order p. $\qquad\square$

Exercises 10.2

1. How many non-isomorphic groups of order p^2, p a prime integer, are there?

2. Let p be a positive prime integer. If G is a group of order p^n and $|Z(G)| \geq p^{n-1}$, show that G must be abelian.

3. Use the class equation of S_n to deduce that $Z(S_n)$ is trivial for $n \geq 3$.

4. Find all subgroups of order 4 in S_4. Show that they are all isomorphic to the Klein 4-group V. Explain why only one of them is normal in S_4.

5. Find all the normal subgroups on S_5 and in A_5. Is S_5 simple? Is A_5 simple?

10.3 Burnside's Formula

As an application of group actions, we will derive a useful and interesting formula in combinatorics.

We will start with one combinatorial problem than can be solved by a high school student.

Example 10.3.1. How many distinguishable ways can the six faces of a cube be marked with from one to six dots to form a dice?

Solution. The reasoning of a high school student would probably go like this. We may roll the dice so that the face up is marked with one dot. There are 5 possibilities of marking the dice at the face down. For argument's sake, suppose we mark the face down by two dots. Next we may roll the dice sideways so that the face towards us is marked by three dots. There are $3! = 6$ ways to mark the remaining three faces. Thus we conclude that there are altogether $5 \cdot 3! = 30$ ways to mark a dice.   ◇

We will review this problem using group actions. If we are not allowed to move the dice, there are $6! = 720$ different ways to mark it. However, one certain marking can be carried into another by a rolling of the dice. These two different markings make indistinguishable dices. The actions of rolling form a group of actions. The most significant trait is that you may reverse any rolling back to the original position for the dice. The rolling of the dice may be considered as a group action on the set of all markings of this dice.

The group in question may be viewed as a subgroup of S_6. What is the order of this group? Any one of the six faces can be placed up, and then any one of the remaining four faces can be placed towards the front. Hence there are $6 \cdot 4 = 24$ possible positions of a cube on a table. We now have a group G of order 24 acting on the set of $6! = 720$ different markings (before rolling). Two markings are indistinguishable if one can be obtained by a rolling (a permutation) in G, that is, if they are in the same orbits under the group action in question. Hence, Example 10.3.1 is asking us to count the number of orbits of this action.

Below we give a formula for counting the number of orbits under a group action.

Theorem 10.3.2 (Burnside's formula). *Let G be a finite group and let S be a finite G-set. Then*

$$the \ number \ of \ orbits \ of \ G = \frac{1}{|G|} \sum_{g \in G} |S_g|.$$

Proof. Let $T = \{(g, x) \in G \times S : g \cdot x = x\}$. We now count the number of elements in T in two different ways. First, we have

$$|T| = \sum_{g \in G} \left| \{x \in S : g \cdot x = x\} \right| = \sum_{g \in G} |S_g|.$$

On the other hand,

$$(10.3.1) \qquad |T| = \sum_{x \in S} \left| \{g : g \cdot x = x\} \right| = \sum_{x \in S} |\operatorname{Stab}_G x| = \sum_{x \in S} \frac{|G|}{|G \cdot x|}$$

by Proposition 10.1.12. Suppose $\mathscr{O}_1, \ldots, \mathscr{O}_s$ are all the orbits of S under G. Note that $\mathscr{O}_i = G \cdot x$ if $x \in \mathscr{O}_i$. Thus

$$(10.3.2) \qquad |T| = \sum_{i=1}^{s} \sum_{x \in \mathscr{O}_i} \frac{|G|}{|G \cdot x|} = \sum_{i=1}^{s} \sum_{x \in \mathscr{O}_i} \frac{|G|}{|\mathscr{O}_i|} = \sum_{i=1}^{s} |G| = s|G|.$$

Comparing (10.3.1) and (10.1), we obtain that

$$s = \frac{1}{|G|} \sum_{g \in G} |S_g|. \qquad \square$$

We may now use this formula to revisit Example 10.3.1.

Solution 2 to Example 10.3.1. The problem may be reinterpreted as a group of order 24 acting on

$$S = \{(x_1, x_2, \ldots, x_6) : x_i \in \{1, 2, \ldots, 6\} \ and \ the \ x_i\text{'s are distinct}\},$$

a set of 6! = 720 elements. It is easy to see

$$S_g = \begin{cases} S, & \text{if } g = e; \\ \varnothing, & \text{otherwise.} \end{cases}$$

Hence by Burnside's formula the answer is $720/24 = 30$. ◇

Example 10.3.3. How many distinguishable ways can eight people be seated at a round table?

Solution. Let S be the set of all seating of the eight guests. Then $|S| = 8!$. If the guests make a rotation, the position-relation among the guests remains the same. Hence this problem is asking us to find the number of orbits under the group action $G = \langle (1\ 2\ \cdots\ 8) \rangle$ on S. It is clear

$$S_g = \begin{cases} S, & \text{if } g = e; \\ \varnothing, & \text{otherwise.} \end{cases}$$

Hence the number of orbits of this action is $8!/8 = 7!$. ◇

Next, we try a more complicated problem.

Example 10.3.4. If you want to paint a Ferris wheel of 20 cars choosing from 5 different colors of paint, how many indistinguishable ways are there to do it?

Solution. Let S be the set of all possible coloring of the cars before they are allowed to move. Then $|S| = 5^{20}$. In this problem, the group is $G = \langle \sigma \rangle$ where $\sigma = (1\ 2\ \cdots\ 20)$ since the cars of the Ferris wheel may be rotated. We will use Burnside's formula to solve this problem. For this we need to find S_g for all $g \in G$.

It is obvious $S_e = S$ and $|S_e| = 5^{20}$.

What is S_σ? In this case, car 1 is moved to the original position of car 2, and car 2 is moved to that of car 3 and so on. If the coloring remain the same after this rotation, it can only mean that the color of car 1 equals the color of car 2, the color of car 2 equals the color of car 3, and so on. Thus, all cars of the Ferris wheel are painted the same color. We conclude that $|S_\sigma| = 5$.

Before we continue, we would like to point out that σ is not the only permutation that would achieve this. The permutation σ is a 20-cycle in which appearing numbers indicate which cars must be painted the same color as car 1. Similarly, when $\tau = \sigma^3$, σ^7, σ^9, σ^{11}, σ^{13}, σ^{17} or σ^{19}, it is also a 20-cycle. Hence

$$|S_\sigma| = |S_{\sigma^3}| = |S_{\sigma^7}| = |S_{\sigma^9}| = |S_{\sigma^{11}}| = |S_{\sigma^{13}}| = |S_{\sigma^{17}}| = |S_{\sigma^{19}}| = 5.$$

We know that

$$\sigma^2 = (1\ 3\ 5\ \cdots\ 17\ 19)(2\ 4\ 6\ \cdots\ 18\ 20).$$

This says that under this rotation, car 1 is moved to where car 3 was, car 3 to car 5 and so on. Hence car 1, car 3, ..., car 17 and car 19 are painted with the same color of paint, while car 2, car 4, ..., car 18 and car 20 are painted with the same color of paint. Hence $|S_{\sigma^2}| = 5^2$. Similarly,

$$|S_{\sigma^2}| = |S_{\sigma^6}| = |S_{\sigma^{14}}| = |S_{\sigma^{18}}| = 5^2.$$

The permutation

$$\sigma^4 = (1\ 5\ 9\ 13\ 17)(2\ 6\ 10\ 14\ 18)(3\ 7\ 11\ 15\ 19)(4\ 8\ 12\ 16\ 20).$$

Again, this tells us which cars should be painted with the same color. Thus, $|S_{\sigma^4}| = 5^5$. Similarly

$$|S_{\sigma^4}| = |S_{\sigma^8}| = |S_{\sigma^{12}}| = |S_{\sigma^{16}}| = 5^4.$$

The permutation

$$\sigma^5 = (1\ 6\ 11\ 16)(2\ 7\ 12 17)(3\ 8\ 13\ 18)(4\ 9\ 14\ 19)(5\ 10\ 15\ 20).$$

We have that

$$|S_{\sigma^5}| = |S_{\sigma^{15}}| = 5^5.$$

Finally,

$$\sigma^{10} = (1\ 11)(2\ 12)(3\ 13)\cdots(9\ 19)(10\ 20)$$

and we have $|S_{\sigma^{10}}| = 5^{10}$. Using Burnside's formula, we know that there are

$$\frac{5^{20} + 8 \cdot 5 + 4 \cdot 5^2 + 4 \cdot 5^4 + 2 \cdot 5^5 + 5^{10}}{20}$$

distinguishable ways to paint the Ferris wheel in question.   ◇

Exercises 10.3

1. How many distinguishable ways are there to a string necklace of 12 beads choosing from 6 different colors of beads?

2. Here is a wooden block in the shape of a regular tetrahedron as shown in Figure 10.1. You are asked to paint a color on each face. If you have 4 colors of paint to choose from, how many distinguishable ways are there to do this?

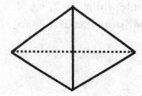

Figure 10.1: A wooden block in the shape of a regular tetrahedron

3. A child is given a card as shown in Figure 10.2. He has 6 crayons of different colors. If he is to fill each box with a color, how many distinguishable ways are there to color this card?

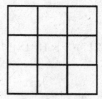

Figure 10.2: A card waiting to be colored

4. A new merry-go-round is to be installed in an amusement park. It will contain 12 seats, 8 in the outer circle and 4 in the inner circle as shown in Figure 10.3. The form of the seats in the outer circle are to be chosen among a horse, a unicorn or a dragon. The form

of the seats in the inner circle are to be chosen between a swan or a phoenix. How many distinguishable ways are there to arrange seats for this merry-go-round?

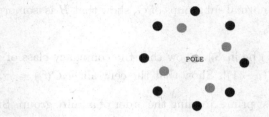

Figure 10.3: Designing a merry-go-round

Review Exercises for Chapter 10

1. If G is a non-abelian group, show that $G/Z(G)$ is never cyclic.

2. Let G be a group. Show that the "action" $g \cdot x = xg$ (the right translation) is not necessarily a group action! How do you adjust this false action to make it a valid action?

3. Let G be a group of order 77 acting on a set S of 20 elements. Show that G must have a fixed point (an x in S such that $gx = x$ for all $g \in G$).

4. Let $G = SL_2(\mathbb{R})$ act on $\mathscr{H}$ as in Example 10.1.7. Show that the isotropy subgroup of $z = i$ in G is given by

$$K = \left\{ \begin{bmatrix} a & b \\ -b & a \end{bmatrix} \middle| \ a^2 + b^2 = 1, \ a, b \in \mathbb{R} \right\}.$$

5. Show that if a finite group G contains a subgroup H of index $n > 1$ then H contains a normal subgroup of G whose index is a divisor of $n!$.

Furthermore, if $|G| \nmid n!$, then G contains a proper nontrivial normal subgroup.

6. Let H be a subgroup of a group G of index $n > 1$. If H does not contain any nontrivial normal subgroups of G, show that H is isomorphic to a subgroup of S_n.

7. Let $\gamma = (1, 2, \ldots, n)$ be in S_n. Show that the conjugacy class of γ in S_n has cardinality $(n-1)!$. Show that the centralizer $C(\gamma) = \langle \gamma \rangle$.

8. Let p be the smallest prime dividing the order of a finite group. Show that any subgroup H of G of index p is normal.

CHAPTER 11

Sylow Theorems and Applications

The well-known Lagrange's theorem tells us that the order of a subgroup of a group is a divisor of the order of the group. Suppose given a divisor d of the order n of a group G. Is there a subgroup H of G of order d? When G is abelian, this is indeed true. However, it is not true in general for non-abelian groups. Sylow's theorems provide a weaker version: the assertion holds when the divisor d is a power of a prime integer. Sylow's theorems also provide some information about the number of maximal p-subgroups in a finite group and the relationship between these maximal p-subgroups.

Sylow's theorems are due to the Norwegian mathematician Peter Ludvig Mejdell Sylow (1832–1918) who stated the theorems in terms of permutation groups in a short paper in 1872. Ferdinand Georg Frobenius (1849–1917) reproved these theorems for abstract groups in 1887 even though he knew that every finite group is a permutation group.

11.1   The Three Sylow Theorems

An important problem in group theory is to classify all finite order groups. The fundamental theorem for finitely generated abelian groups gives us complete information on finite abelian groups. However, the non-abelian groups are much harder to decipher. In order to do this, it is often helpful to study subgroups (or better yet, normal subgroups) in a group of a given order.

Remember that Lagrange's theorem (Theorem 6.1.1) tells us that the order of a subgroup must divide the order of its mother group. Unfortunately, the converse is not true (see Exercise 6, §9.2). However, the Sylow theorems give a weak converse: the existence of subgroups of certain orders. The Sylow theorems also give us some information concerning the number of subgroups of certain orders and their relation to each other. These results are very useful in studying finite groups.

Before we state and prove the Sylow theorems, we first introduce the following concept.

For each subgroup H of a group G, define

$$N_G(H) = \{\, a \in G : aHa^{-1} = H \,\}.$$

It is called the **normalizer** of H in G. We will sometimes write $N(H)$ for $N_G(H)$ if G is understood.

Proposition 11.1.1. *Let H be a subgroup of the group G. The following statements are true.*

(a) *Show that $N_G(H)$ is a subgroup of G and $H \subseteq N_G(H)$.*

(b) *Show that H is a normal subgroup of $N_G(H)$.*

(c) *If K is a subgroup of G such that H is normal in K, show that $K \subseteq N_G(H)$.*

(d) *The normalizer $N_G(H)$ is the largest subgroup in G which contains H as a normal subgroup.*

Proof. (a) We have $e \in N_G(H)$ since $eHe^{-1} = H$. Let $a,\, a' \in N_G(H)$. Then aHa^{-1} and $a'H(a')^{-1}$ are both contained in H. It follows that

$$(aa')H(aa')^{-1} = aa'H(a')^{-1}a^{-1} = a(a'H(a')^{-1})a^{-1} = aHa^{-1} = H.$$

This implies that $aa' \in N_G(H)$. Moreover,

$$aHa^{-1} = H \implies H = a^{-1}(aHa^{-1})a = a^{-1}H(a^{-1})^{-1}.$$

This implies that a^{-1} is also in $N_G(H)$. To conclude, $N_G(H)$ is a subgroup of G.

(b) This follows from definition.

(c) Since H is normal in K, this says that $kHk^{-1} = H$ for all $k \in K$. By definition, we have that $k \in N_G(H)$ for all $k \in K$. Hence $K \subseteq N_G(H)$.

(d) This follows from (b) and (c). $\square$

Lemma 11.1.2. *Let H be a p-subgroup of a finite group G. Then*

$$[N_G(H) : H] \equiv [G : H] \pmod{p}.$$

Proof. Let $S = \{ xH : x \in G \}$. Let H act on S by left translation. In other words, let $h \cdot xH = hxH$ for $h \in H$. We verify that

$$\begin{aligned}
S_H &= \{ xH : hxH = xH \text{ for all } h \in H \} \\
&= \{ xH : x^{-1}Hx = H \text{ for all } h \in H \} \\
&= \{ xH : x \in N_G(H) \}.
\end{aligned}$$

Since $H \subseteq N_G(H)$, S_H is the set of left cosets of H in $N_G(H)$. By Corollary 10.2.2, we have that

$$[N_G(H) : H] = |S_H| \equiv |S| = [G : H] \pmod{p}.$$ $\square$

Theorem 11.1.3 (Sylow's First Theorem). *Let G be a finite group of order $p^r m$ where p is a positive prime integer and m is an integer with $p \nmid m$. The following statements are true.*

(a) *The group G contains a subgroup of order p^i for each i with $1 \leq i \leq r$.*

(b) *In G every subgroup H of order p^i is normal in a subgroup K of order p^{i+1} for $1 \leq i < r$.*

When G is a group of order $p^r m$ where $p \nmid m$, a subgroup H of order p^r in G is called a **Sylow p-subgroup** of G.

Proof. (a) We will prove this theorem by induction on $|G|$. When $r = 1$, from Cauchy's Theorem (Theorem 10.2.7) there is an element of order p which generates a subgroup of order p in G. We now assume $r \geq 2$.

If p^r is a divisor of the order of some proper subgroup H of G, then by induction H contains a subgroup of order p^i for $1 \leq i \leq r$. We are done since these are also subgroups of G.

Now suppose that p^r is not a divisor of the order of any proper subgroup of G. Find $a_1, \ldots, a_s \notin Z(G)$ such that

$$p^r m = |Z(G)| + \sum_{i=1}^{s} \frac{|G|}{|C(a_i)|}$$

is the class equation for G. Under our assumption, p^r is not a divisor of $|C(a_i)|$ for all i. This implies that p divides $|G|/|C(a_i)|$. Hence we obtain that p is a divisor of $|Z(G)|$. By Cauchy's theorem, there exists an element a of order p in $Z(G)$. Then $\langle a \rangle$ is a normal subgroup of order p in G and $G/\langle a \rangle$ is a group of order $p^{r-1}m$. By induction, $G/\langle a \rangle$ contains a subgroup K_i of order p^i for $1 \leq i \leq r - 1$. By correspondence theorem (Theorem 8.2.9) for groups, for $1 \leq i \leq r - 1$, there is a subgroup L_i of G with $\langle a \rangle \subseteq L_i$ such that $K_i = L_i/\langle a \rangle$. Therefore, L_i is a subgroup of G of order p^{i+1}. This completes the induction step.

To prove part (b), let H be a subgroup of order p^i where $1 \leq i < r$. By Lemma 11.1.2, we have

$$[N_G(H) : H] \equiv [G : H] \equiv 0 \pmod{p}.$$

Thus p divides $[N_G(H) : H]$. By Cauchy's Theorem we may find a subgroup L in $N_G(H)$ containing H such that L/H is a group of order p. It follows that L is a subgroup of order p^{i+1} and H is normal in L. $\qquad\square$

If H is a Sylow p-subgroup of G, then any of its conjugate gHg^{-1} is also a Sylow p-subgroup since these two subgroups are of the same order.

Theorem 11.1.4 (Sylow's Second Theorem). *Let P and P' be Sylow p-subgroups of a finite group G. Then there exists $g \in G$ such that $P' = gPg^{-1}$. In other words, any two Sylow p-subgroups of G are conjugate to each other in G.*

Proof. Let P and P' be two Sylow p-subgroups of G. Let

$$S = \{\, gP \mid g \in G \,\}$$

be the set of left cosets of P in G. Let P' act on S by left translation. Since P' is a p-group, we have that

$$|S_{P'}| \equiv |S| = [G : P] = m \not\equiv 0 \pmod{p}$$

by Corollary 10.2.2. It follows that $S_{P'}$ is nonempty. Find $gP \in S$ such that $h \cdot gP = gP$ for all $h \in P'$. Then $hgP = gP$, or equivalently $g^{-1}hg \in P$, for all $h \in P'$. This implies that $g^{-1}P'g \subseteq P$. Since $g^{-1}P'g$ and P are of the same order, it follows that $g^{-1}P'g = P$. $\qquad\square$

Corollary 11.1.5. *Let p be a positive prime integer. There is only one Sylow p-subgroup in a finite group if and only if it is a normal subgroup.*

Theorem 11.1.6 (Sylow's Third Theorem). *Let G be a group of order $p^r m$ where p is a positive prime integer, $r > 0$ and m is an integer with $p \nmid m$. Let n_p be the number of Sylow p-subgroups of G. Then $n_p = kp + 1$ for some nonnegative integer k and $n_p | m$.*

Proof. Let P be a Sylow p-subgroup of G. By Sylow's Second Theorem

$$S = \{\, gPg^{-1} : g \in G \,\}$$

is the set of all Sylow p-subgroups of G. We want to derive numerical results on $|S|$.

Let P act on S by conjugation. To be precise, for each $x \in P$ and each $T \in S$, let $x \cdot T = xTx^{-1}$. If $T \in S_P$, then $xTx^{-1} = T$ for all $x \in P$. Thus $P \subseteq N_G(T)$. Note that both P and T are Sylow p-subgroups of $N_G(T)$. Sylow's second theorem implies that P and T must be conjugate to each other in $N_G(T)$. However, T is a normal subgroup of $N_G(T)$. It follows that $P = T$. This shows that $S_P = \{\, P \,\}$. Hence

$$|S| \equiv |S_P| = 1 \pmod{p}.$$

Now consider the group action of G on S by conjugation. This is a transitive group action by Sylow's second theorem. Hence $|S| = |G|/|\operatorname{Stab}_G P|$ is a divisor of $|G|$. Since $|S|$ is relatively prime to p, we have that $|S|$ is a divisor of m. $\qquad\square$

In the proof above we have observed that for a Sylow p-subgroup P of G, the only Sylow p-subgroup contained in $N_G(P)$ is P. We can slightly generalize this result.

Corollary 11.1.7. *Let P be a Sylow p-subgroup of a group G and let $a \in G$ be an element whose order is a power of p. Then*

$$a \in N_G(P) \quad \Longleftrightarrow \quad a \in P.$$

Proof. The "$\Leftarrow$" part is trivial. To show the "$\Rightarrow$" part, note that $\langle a \rangle \subseteq N_G(P)$. It follows that $\langle a \rangle P$ is a subgroup of G (see Exercise 3). Note that

$$|\langle a \rangle P| = \frac{|\langle a \rangle||P|}{|\langle a \rangle \cap P|}$$

is a power of p (see Exercise 4 in Review Exercises for Chapter 6). The subgroup $\langle a \rangle P$ is a p-group containing the Sylow p-subgroup P. Hence $\langle a \rangle P = P$ and $a \in P$. $\qquad \square$

Exercises 11.1

1. Find the Sylow 2-subgroups and Sylow 3-subgroups of S_3.

2. If G is a group of order p^n with p a positive prime integer and n a positive integer, show that G contains a normal subgroup of order p^{n-1}.

3. Let H and K be subgroups of G. If $K \subseteq N(H)$, show that $HK = KH$ is a subgroup of G.

4. Let H be a p-subgroup of G.

   (a) Show that H is contained in a Sylow p-subgroup of G.

   (b) Let P be a given Sylow p-subgroup of G. Show that there exists $g \in G$ such that $gHg^{-1} \subseteq P$.

5. If P is a Sylow p-subgroup of a finite group G, show that $N(N(P)) = N(P)$.

   Let H be a p-subgroup in the group G. Show that $p \nmid [N[H] : H]$ iff $p \nmid [G : H]$.

6. If P and Q are two distinct Sylow p-subgroups of G, show that PQ is not a subgroup of G.

11.2   Applications of Sylow Theorems

First, we would like to review a couple of results in §9.2.

Let G be a finite group of order pq where p and q are prime integers such that $p > q$. Let n_p be the number of Sylow p-subgroups in G. By Sylow's third theorem, $n_p \mid q$ and $n_p \equiv 1 \pmod{q}$. This says that n_p must be 1 and hence the unique Sylow p-subgroup P of G is normal in G. This gives a straightforward argument to Lemma 9.2.2.

Furthermore, let n_q be the number of Sylow q-subgroups on G. By Sylow's third theorem, $n_q \mid p$, and so $n_q = 1$ or p. If we add one more condition that $p \not\equiv 1 \pmod{q}$, we may conclude that $n_q = 1$ and hence the unique Sylow q-subgroup Q is also normal in G. We thus have found two normal subgroups in G such that (i) $|G| = |P||Q|$, and (ii) $|P|$ and $|Q|$ are relatively prime with each other. We now conclude that G is the direct product $P \times Q$ by Proposition 9.1.4 and Exercise 4, §9.1. Hence

$$G \simeq \mathbb{Z}_p \times \mathbb{Z}_q \simeq \mathbb{Z}_{pq}$$

is cyclic. This also gives a straightforward argument to Proposition 9.2.4.

From the discussion above, we may see that the Sylow theorems are useful in uncovering normal subgroups in a finite group. Let's look at more examples.

Example 11.2.1. Let G be a group of order $45 = 3^2 \cdot 5$. The Sylow 3-subgroups of G are of order 9. Let n_3 be the number of Sylow 3-subgroups in G. Then $n_3 | 5$ and $n_3 \equiv 1 \pmod{3}$. The only possibility is that $n_3 = 1$. Hence the unique Sylow 3-subgroup P is a normal subgroup of order 9 in G.

On the other hand, Let n_5 be the number of Sylow 5-subgroups of G. Then $n_5 | 9$ and $n_5 \equiv 1 \pmod 5$. The only possibility is still $n_5 = 1$. Hence the unique Sylow 5-subgroup Q is also a normal subgroup in G.

We have found two normal subgroups P and Q in G. Note that $|G| = |P||Q| = 45$ and the orders $|P|$ and $|Q|$ are relatively prime with each other. Thus

$$G \simeq P \times Q \simeq \mathbb{Z}_9 \times \mathbb{Z}_5 \text{ or } \mathbb{Z}_3 \times \mathbb{Z}_3 \times \mathbb{Z}_5.$$

We have successfully classified groups of order 45.

Example 11.2.2. Let G be a group of order $5 \cdot 7 \cdot 19$. Let n_5, n_7 and n_{19} be the numbers of Sylow 5-, Sylow 7- and Sylow 19-subgroups of G respectively. Using Sylow's third theorem, we have the following possibilities:

$$\begin{cases} n_5 = 1, 7, 19, \text{ or } 7 \cdot 19; \\ n_7 = 1, 5, 19, \text{ or } 5 \cdot 19; \\ n_{19} = 1, 5, 7, \text{ or } 5 \cdot 7. \end{cases}$$

Since none of 7, 19, $7 \cdot 19$ is congruent to 1 modulo 5, we have that $n_5 = 1$ by Sylow's third theorem again. Similarly, we have that $n_7 = n_{19} = 1$. This implies that the Sylow 5-subgroup P, the Sylow 7-subgroup Q and the Sylow 19-subgroup R are all normal in G. We leave it to the Reader to verify that

$$G \simeq P \times Q \times R \simeq \mathbb{Z}_5 \times \mathbb{Z}_7 \times \mathbb{Z}_{19} \simeq \mathbb{Z}_{665}$$

is cyclic. (See Exercise 2.)

Example 11.2.3. Let G be a group of order $50 = 2 \cdot 5^2$. Let n_5 be the number of Sylow 5-subgroups in G. Then $n_5 | 2$ and $n_5 \equiv 1 \pmod 5$. The only possibility is $n_5 = 1$. The Sylow 5-subgroup is unique and normal in G.

On the other hand, let n_2 be the number of Sylow 2-subgroups in G. Then $n_2 | 25$ and $n_5 \equiv 1 \pmod 5$. At this stage, we can only deduce that $n_2 = 1, 5$ or 25. It is not clear whether a Sylow 2-subgroup is normal in G.

Before we look at more examples, we first give a useful formula.

Lemma 11.2.4. *Let g and a be elements in a group. Suppose $gag^{-1} = a^j$. Then $g^k a g^{-k} = a^{j^k}$.*

Proof. We prove this lemma by induction on k. There is nothing to prove when $k = 1$. We now assume the induction hypothesis for k. Then

$$g^{k+1}ag^{-(k+1)} = g(g^k ag^{-k})g^{-1} = g(a^{j^k})g^{-1}$$
$$= (gag^{-1})^{j^k} = (a^j)^{j^k} = a^{j^{k+1}}.$$

This completes the induction steps. $\qquad\qquad\qquad\qquad\qquad\qquad$ □

Example 11.2.5. Let G be a group of order $255 = 3 \cdot 5 \cdot 17$. By Sylow's third theorem, the number of 17-Sylow subgroup of G is a divisor of 15 and is congruent to 1 modulo 17. There is only one Sylow 17-subgroup in G. Find subgroups H, K, L of order 3, 5, 17, respectively. Then L is a normal subgroup of G. Now that G/L has order 15, it is a cyclic group by Proposition 9.2.3. Let gL be a generator of G/L and b a generator of L. Note that $o(g) \geq 15$.

If $o(g) > 15$, then $o(g) = 15 \times 17 = 225$ by Exercise 4, P. 63. In this case $G = \langle g \rangle$ is cyclic.

Now suppose that $o(g) = 15$. Since L is a normal subgroup of G, $gbg^{-1} = b^j$ for some $1 \leq j \leq 16$. It follows that $b = g^{15}bg^{-15} = b^{j^{15}}$ by Lemma 11.2.4. Hence

$$j^{15} \equiv 1 \pmod{17}.$$

In other words, $\overline{j}^{15} = \overline{1}$ in $\mathbb{Z}_{17}$. Since $\overline{j^{16}} = \overline{1}$ in the group $U(\mathbb{Z}_{17})$, it follows that $\overline{j} = \overline{1}$. We now know that $j = 1$. Thus $gb = bg$ such that every element of $\langle g \rangle$ commutes with every element of $\langle b \rangle$. Since $\langle g \rangle \cap \langle b \rangle = \{ e \}$, we have that

$$G \cong \langle g \rangle \times \langle b \rangle \cong C_{15} \times C_{17} \cong C_{225}$$

is cyclic (see Exercise 4, §9.1).

Example 11.2.6. A group of order 30 is never simple.

Solution. Let G be a group of order $30 = 2 \times 3 \times 5$. If any of the Sylow subgroups of G is normal then we are done. We assume otherwise. Let n_p be the number of Sylow p-subgroups in G. By the Sylow's third theorem we have that

$$n_2 = 3, \ 5 \text{ or } 15, \quad n_3 = 10 \quad \text{and} \quad n_5 = 6.$$

Note that any nontrivial element in a Sylow 3-subgroup is a generator, and so it cannot be contained in two distinct Sylow 3-subgroups. Hence there are 20 distinct elements of order 3 in G (each of them a generator of the Sylow 3-subgroup containing it). Similarly, there are also 24 elements of order 5 and at least 3 elements of order 2. This says that G contains at least $1 + 20 + 24 + 3 = 48$ elements, a contradiction.   ◇

Example 11.2.7. No groups of order 48 are simple.

Solution. Suppose G is a group of order $48 = 2^4 \times 3$. Assume that the Sylow 2-subgroups of G are not normal in G. By the Sylow's third theorem, there are 3 distinct Sylow 2-subgroups of order 16. Let H_1 and H_2 be two of them. Since the subgroup $H_1 \cap H_2 \subsetneq H_1$, its order is 1, 2, 4 or 8. If $|H_1 \cap H_2| \leq 4$, then

$$|H_1 H_2| = \frac{|H_1||H_2|}{|H_1 \cap H_2|} \geq \frac{16 \cdot 16}{4} = 64,$$

a contradiction. Hence $|H_1 \cap H_2| = 8$. Since $[H_1 : H_1 \cap H_2] = [H_2 : H_1 \cap H_2] = 2$, we have that $H_1 \cap H_2$ is normal both in H_1 and in H_2. In other words, H_1, H_2 are both subgroups of $N(H_1 \cap H_2)$. It follows that $H_1 H_2 \subseteq N(H_1 \cap H_2)$. Since

$$|N(H_1 \cap H_2)| \geq |H_1 H_2| = \frac{16 \cdot 16}{8} = 32$$

and is a divisor of 48, we have that $|N(H_1 \cap H_2)| = 48$. Thus, we can conclude that $N(H_1 \cap H_2) = G$. We have shown that either the Sylow 2-subgroup or the intersection of two Sylow 2-subgroups is normal in G. Thus, G contains a normal subgroup of order 16 or 8.   ◇

Exercises 11.2

1. Let P_1, P_2, ..., P_s be the Sylow subgroups in the group G. Suppose P_i is the Sylow p_i-subgroup of G for each i and assume the p_i's are distinct. Show that $G \simeq P_1 \times P_2 \times \ldots P_s$.

2. Suppose G is a group of order $p_1 p_2 \ldots p_s$ where the p_i's are distinct prime integers. Suppose that the Sylow p_i-subgroups are normal in G for all i. Show that G is cyclic.

3. Classify groups of order 99.

4. Show that every group of order 30 contains a subgroup of order 15.

5. Prove that no group of order 160 is simple.

6. If G is a group of order p^n with p a positive prime integer and n a positive integer, show that G contains a normal subgroup of order p^{n-1}.

7. Show that a group of order 36 has a normal subgroup of order 3 or 9.

8. Show that a group of order 108 has a normal subgroup of order 9 or 27. (Hint: Let P be a Sylow 3-subgroup of G and let S be the set of left cosets of P in G. Let G act on S by left translation. Use the result of Exercise 6 in Review Exercises for Chapter 7.)

Review Exercises for Chapter 11

1. Find all Sylow 2-subgroups of S_5.

2. If G is a non-abelian group, show that $G/Z(G)$ is never cyclic.

3. Let p and q be positive prime integers such that $p > q$. Suppose that G is a finite group of order $p^n q$. Show that G contains a normal subgroup of order p^n.

4. If H is a normal Sylow p-subgroup of a finite group G, show that $\varphi(H) = H$ for every automorphism φ of G.

5. Show that groups of order 12, 28, 56 or 200 must contain a normal Sylow subgroup. Hence groups of these orders are never simple.

6. Let G be a group of order 168. How many elements are of order 7?

7. Show that every group of order 567 has a normal subgroup of order 27.

8. Show that every group of order 352 has a normal subgroup of order 16.

9. Let G be a group and H be a subgroup of G. Define $C_G(H)$, the **centralizer** of H in G, to be

$$\{ g \in G : ghg^{-1} = h \text{ for all } h \in H \}.$$

   (a) Show that $C_G(H)$ is a subgroup of G.

   (b) Show that $H \subseteq C_G(C_G(H))$. Is it true that $H \subseteq C_G(H)$?

   (c) Show that $C_G(H)$ is a subgroup of $N_G(H)$.

   (d) Show that $C_G(H) \lhd N_G(H)$.

10. Let G be a group and H be a subgroup of G.

   (a) Let $g \in G$. Define a map $\phi_g : G \to G$ by sending $a \in G$ to gag^{-1}. Show that ϕ_g is a group automorphism of G. The map ϕ_g is called the **inner automorphism** of G induced by g.

   (b) For $g \in N_G(H)$, show that ϕ_g induces an inner automorphism on H, which we will call ϕ_g^H for simplicity.

   (c) Let

$$\text{Inn}_G(H) = \{ \phi_g^H \in \text{Aut}(H) : g \in G \}.$$

   Show that $\text{Inn}_G(H)$ is a subgroup of $\text{Aut}(H)$.

   (d) Show that $N_G(H)/C_G(H) \cong \text{Inn}_G(H)$. (Hint: Consider the map $\Psi : N_G(H) \to \text{Inn}_G(H)$ sending g to ϕ_g^H.)

CHAPTER 12

Introduction to Group Presentations

How to construct group homomorphisms between two groups can be an extremely taxing problem if approached blindly. There are too many hidden relations between group elements which must be taken into consideration. This is where group presentations come in. It is a precise way to describe a group by describing its generators and the essential relations among these generators. Group presentations make it possible to construct group homomorphisms between two groups. This is also the language used in classification of finite groups.

12.1   Free Groups and Free Abelian Groups

Let X be a set. The group $F[X]$ is called the **free group** on the subset of **free generators** X (not necessarily finite) if it satisfies the following two conditions:

(i) $F[X]$ is generated by X, and

(ii) for any mapping φ from X into a group G, we can *uniquely* extend φ to a group homomorphism from $F[X]$ to G.

Now we shall take a close look at the elements in $F[X]$. Set

$$X^{-1} = \{\, x^{-1} : x \in X \,\}.$$

Here are a few facts about the free group $F[X]$.

(1) The set X is called the **alphabet** and elements in X and X^{-1} are called the **letters** of the free group $F[X]$. Elements in $F[X]$ are **words** which are finite strings of letters from X or X^{-1} written in juxtaposition. For example when $X = \{\, a, b, c \,\}$, $abbc$, $baa^{-1}a^{-1}b^{-1}ba^{-1}bc$ and $cb^{-1}b^{-1}b^{-1}ab$ are words or elements in the free group $F[X]$. However, we often denote a word by its **reduced word** for the sake of simplicity. For example, the aforementioned words are usually denoted by the simpler forms ab^2c, $ba^{-2}bc$ and $cb^{-3}ab$ respectively.

(2) For two words w_1 and w_2 in $F[X]$, $w_1 \cdot w_2$ is the word obtained by the juxtaposition $w_1 w_2$ of the two words w_1 and w_2. For example, $ab^2c \cdot ba^{-2} = ab^2cba^{-2}$. Sometimes, the juxtaposition of two words can be reduced to a shorter one. For example, $ab^2cb \cdot b^{-1}c^{-1} \cdot a^{-2} = ab^2a^{-2}$. Juxtaposition of words satisfies associativity.

(3) We introduce the **empty word** which is the word with no letters at all. The empty word is the identity of $F[X]$. The empty word is usually denoted by a blank space or by 1 when a blank space seems awkward. In particular, if $X = \varnothing$, then $F[X] = \{\, 1 \,\}$.

(4) If $w = a_1^{n_1} a_2^{n_2} \cdots a_k^{n_k}$ is a word in $F[X]$, then the inverse word of w is

$$w^{-1} = a_k^{-n_k} \cdots a_2^{-n_2} a_1^{-n_1}.$$

(5) Let G be any group and let $\varphi\colon X \to G$ be any function. We can extend φ uniquely to a group homomorphism from $F[X]$ to G by letting

$$a_{i_1}^{k_1} a_{i_2}^{k_2} \cdots a_{i_s}^{k_s} \longmapsto \varphi(a_{i_1})^{k_1} \varphi(a_{i_2})^{k_2} \cdots \varphi(a_{i_s})^{k_s},$$

where $a_{i_j} \in X$ and $k_j \in \mathbb{Z}$.

Proposition 12.1.1. *The free group $F = F[\{x\}]$ with one free generator is the infinite cyclic group.*

Proof. By definition, there is a group homomorphism $\varphi\colon F \to \mathbb{Z}$ sending x to 1. By Proposition 7.1.7, there is a group homomorphism $\psi\colon \mathbb{Z} \to F$ sending 1 to x. These two homomorphisms are inverse to each other.   □

Thus, the additive group $\mathbb{Z}$ is also the free group with one free generator.

Proposition 12.1.2. *Free groups with at least one generator is infinite. Free groups with more than one free generators are non-abelian.*

Proof. Let F be the free group on the set of free generators X. Suppose X contains at least one element x. Let $\varphi\colon X \to \mathbb{Z}$ be the function sending x to 1 and any other element in X to 0. Then φ extends uniquely to a group homomorphism from F to $\mathbb{Z}$. Since φ is onto an infinite set, F is infinite.

If X contains two distinct elements x and y. Define $\psi\colon X \to S_3$ by sending x to $(1\ 2\ 3)$, y to $(1\ 2)$ and any other element in X to 1. Then ψ can be extended to a group homomorphism from F into S_3. However, $\psi(x)\psi(y) \neq \psi(y)\psi(x)$. It follows that $xy \neq yx$ in F. Thus F is non-abelian. (In fact, since xy and yx are two distinct words in F, we know that $xy \neq yx$.)   □

The additive group $\mathbb{Z} \times \mathbb{Z}$ is an abelian but non-cyclic group. Hence $\mathbb{Z} \times \mathbb{Z}$ is not free. However, we will see that it is still "free" for abelian groups (*Cf.* Theorem 12.1.5). Observe that every element in $\mathbb{Z} \times \mathbb{Z}$ can be expressed as

$$m, n) = m(1,0) + n(0,1)$$

uniquely. We will say that $(1,0)$ and $(0,1)$ from a *basis* for the abelian group $\mathbb{Z} \times \mathbb{Z}$.

An abelian group G is a **free abelian group** if it contains a nonempty subset X such that each element g in G can be expressed uniquely in the form

$$g = \begin{cases} n_1 x_1 + n_2 x_2 + \cdots + n_r x_r, & \text{in additive form,} \\ x_1^{n_1} x_2^{n_2} \cdots x_r^{n_r}, & \text{in multiplicative form,} \end{cases}$$

where $n_j \in \mathbb{Z}$ and $x_1, x_2, \ldots, x_r$ are distinct elements in X. The set X is called a **basis** for the free abelian group.

For the following discussions we will assume the abelian groups are additive. The reader should modify all statements for multiplicative groups accordingly.

Example 12.1.3. Let r be a positive integer. In general, we use $\mathbb{Z}^r$ to denote the direct product of r copies of $\mathbb{Z}$. Observe that $\mathbb{Z}^r$ is a free abelian group with a basis

$$(12.1.1) \quad \{\, e_1 = (1, 0, \ldots, 0), \; e_2 = (0, 1, \ldots, 0), \; \ldots, \; e_r = (0, 0, \ldots, 1) \,\}$$

which consists of r elements.

Proposition 12.1.4. *Let X be a subset of an abelian group G. Then X is a basis if and only if the following conditions are satisfied:*

(i) *G is generated by X, and*

(ii) *for $n_j \in \mathbb{Z}$ and distinct elements $x_1, x_2, \ldots, x_r$ in X, we have*

$$n_1 x_1 + n_2 x_2 + \cdots + n_r x_r = 0 \quad \Longrightarrow \quad n_1 = n_2 = \cdots = n_r = 0.$$

Proof. "Only if": Clearly X generates G. Suppose that $n_1 x_1 + n_2 x_2 + \cdots + n_r x_r = 0$. Since $0 \cdot x_1 + 0 \cdot x_2 + \cdots + 0 \cdot x_r = 0$, we have that $n_1 = n_2 = \cdots = n_r = 0$ by definition.

"If": By assumption X generates G. Suppose that

$$a = n_1 x_1 + n_2 x_2 + \cdots + n_r x_r = m_1 x_1 + m_2 x_2 + \cdots + m_r x_r$$

for $m_j, n_j \in \mathbb{Z}$ and distinct elements $x_1, x_2, \ldots, x_r$ in X. Then

$$(n_1 - m_1) x_1 + (n_2 - m_2) x_2 + \cdots + (n_r - m_r) x_r = 0.$$

Hence $n_1 = m_1, n_2 = m_2, \ldots, n_r = m_r$ by condition (ii). $\qquad \square$

Theorem 12.1.5. *Let* $X = \{x_1, x_2, \ldots, x_r\}$ *be a basis for a free abelian group* G. *Suppose given an abelian group* G'. *Then for any mapping* φ *from* X *into* G', *we can extend* φ *uniquely to a group homomorphism from* G *into* G'.

Proof. A group homomorphism is uniquely determined by the images of a generating set. It suffices to show the existence of the extension. Define $\varphi \colon G \to G'$ by

$$\varphi \left(\sum_{j=1}^{r} n_j x_j \right) = \sum_{j=1}^{r} n_j \varphi(x_j).$$

This is well-defined since the expression of elements in G is unique. Since

$$\varphi \left(\sum_{j=1}^{r} n_j x_j + \sum_{j=1}^{r} m_j x_j \right) = \varphi \left(\sum_{j=1}^{r} (n_j + m_j) x_j \right)$$

$$= \sum_{j=1}^{r} (n_j + m_j) \varphi(x_j) = \sum_{j=1}^{r} n_j \varphi(x_j) + \sum_{j=1}^{r} m_j \varphi(x_j)$$

$$= \varphi \left(\sum_{j=1}^{r} n_j x_j \right) + \varphi \left(\sum_{j=1}^{r} m_j x_j \right),$$

φ is a group homomorphism. □

Theorem 12.1.6. *If* G *is a free abelian group with a finite basis* $X = \{x_1, x_2, \ldots, x_r\}$, *then* $G \cong \mathbb{Z}^r$.

Proof. Let e_i be as in (12.1.1). Remember that $\{e_1, \ldots, e_r\}$ forms a basis for $\mathbb{Z}^r$. There is a group homomorphism $\varphi \colon \mathbb{Z}^r \to G$ sending e_i to x_i and there is a group homomorphism $\psi \colon G \to \mathbb{Z}^r$ sending x_i to e_i. These two homomorphisms are inverse to each other. Thus, φ is an isomorphism. □

Proposition 12.1.7. *Let* G *be a free abelian group with a finite basis* $X = \{x_1, x_2, \ldots, x_r\}$. *Then every basis for* G *is finite with* r *elements.*

Proof. By the previous theorem, $G \cong \mathbb{Z}^r$ and hence

$$\frac{G}{2G} \cong \frac{\mathbb{Z} \times \mathbb{Z} \times \cdots \times \mathbb{Z}}{2\mathbb{Z} \times 2\mathbb{Z} \times \cdots \times 2\mathbb{Z}} \cong \mathbb{Z}_2 \times \mathbb{Z}_2 \times \cdots \times \mathbb{Z}_2$$

(see Exercise 3(d), §9.1). Thus $|G/2G| = 2^r$. If G has another finite basis of s elements, then we also have $|G/2G| = 2^s$. Hence $r = s$.

Next, we show that G cannot have an infinite basis. Suppose otherwise. Let $S = \{y_1, y_2, \ldots, y_{r+1}\}$ be distinct elements in an infinite basis Y. Let H be the subgroup of G generated by S and K be the subgroup of G generated by $Y \setminus S$. Then $G \cong H \times K$ (see Exercise 5). Thus,

$$\frac{G}{2G} \cong \frac{H \times K}{2H \times 2K} \cong \frac{H}{2H} \times \frac{K}{2K}.$$

It follows that $|G/2G| \geq 2^{r+1} > 2^r$, a contradiction.   □

If G is an abelian group with a basis of r elements, we say that G is free abelian of **rank** r.

Example 12.1.8. $\mathbb{Z}[i] = \{a + bi : a, b \in \mathbb{Z}\}$ is a free abelian group of rank 2 with $\{1, i\}$ as a basis.

Exercises 12.1

1. Is the set $\{(2,1), (3,2)\}$ a basis for $\mathbb{Z}^2$? Find a condition on a, b, c, $d \in \mathbb{Z}$ for the set $\{(a,b), (c,d)\}$ to be a basis for $\mathbb{Z}^2$.

2. Show that in a free abelian group any element which is not the identity is of infinite order.

3. If $X = \{x_1, x_2, \ldots, x_r\}$ is a basis for a free abelian (additive) group G and $n \in \mathbb{Z}$, then the set

$$Y = \{x_1 + nx_2, x_2, \ldots, x_r\}$$

is also a basis for G.

4. Show that the additive group $\mathbb{Z}[\sqrt[3]{2}] = \{a + b\sqrt[3]{2} + c\sqrt[3]{4} : a, b, c \in \mathbb{Z}\}$ is a free abelian group of rank 3.

5. Let $\mathscr{B}$ be a basis for the free abelian group G. Let S and T be subsets of $\mathscr{B}$ such that $S \cup T = \mathscr{B}$ and $S \cap T = \varnothing$. Let $H = \langle S \rangle$ and $K = \langle T \rangle$. Show that $G = H \times K$.

6. Let $X = \{x, y\}$. Show that $\langle x \rangle$ is not a normal subgroup of $F[X]$.

12.2 Generators and Relations

Let G be any group. Find a generating set S for G. Pick an alphabet set X of your choice which is of the same cardinality as S. Construct a bijective map φ from X onto S. Then φ extends uniquely to a group homomorphism from $F[X]$ to G. Since the image of φ contains a generating set for G, we know that φ is onto. Thus we have the following result by the first isomorphism theorem.

Theorem 12.2.1. *Every group G is a homomorphic image of a free group. One can find a alphabet set X such that $G \cong F[X]/N$ for some normal subgroup N of the free group $F[X]$.*

It is important to know how to describe N.

Definition 12.2.2 (Group Presentation). Let $X = \{a_i\}_{i \in I}$ be a set and let $\{r_j\}_{j \in J} \subseteq F[X]$. Let N be the smallest normal subgroup containing $\{r_j\}_{j \in J}$ (*Cf.* Exercise 1). We will use the notation

$$(12.2.1) \qquad \langle a_i,\ i \in I : r_j,\ j \in J \rangle$$

to denote the group $F[X]/N$.

If $G \cong F[X]/N$, then we say that (12.2.1) is a **group presentation** of G. The x_i's are called the **generators** for the presentation, and the r_j's are called the **relators**. An element $r \in N$ is called a **consequence** of $\{r_j\}_{j \in J}$. An equation $r_i = 1$ (or any equivalent equation) is called a **relation**. A **finite presentation** is a group presentation in which both X and $\{r_j\}_{j \in J}$ are finite sets. Sometimes a group presentation is also denoted by $\langle a_i,\ i \in I : r_j = 1,\ j \in J \rangle$ using relations instead of relators.

For example, $\langle a : a^5 \rangle$, $\langle a : a^5 = 1 \rangle$ and $\langle a : a^7 = a^2 \rangle$ all stand for the same group presentation. This presentation says that this group is generated by one element whose order is 5. Customarily, one simply use a to denote the class of a in the presentation.

Example 12.2.3. Could you see that $\langle \varnothing : \varnothing \rangle$ as well $\langle a : a \rangle$ are both group presentations for the trivial group? In the first presentation, $F[\varnothing]$ is the trivial group, and obviously the empty set of relators generate the trivial subgroup. In the second presentation, the relator makes the sole generator into 1. Hence the given group is generated by 1 and is thus the trivial group.

Since $\mathbb{Z}$ is the free group on one generator, $\langle a : \varnothing \rangle$ is a group presentation for $\mathbb{Z}$. Similarly, $\langle a_1, a_2, \ldots, a_n : \varnothing \rangle$ is a group presentation for the free group on n generators. This free group is generated by n elements with no relations among themselves at all.

Example 12.2.4. Show that $\langle x, y : y^2 x = y, \ yx^2 y = x \rangle$ is the trivial group.

Solution. The relation $y^2 x = y$ implies that $yx = 1$. Then we have $x = yx^2 y = (yx)xy = xy$. This implies that $y = 1$. It in turns implies that $x = 1$. The group presented is generated by the identity element and is hence the trivial subgroup.   ◇

Next we describe how to construct group homomorphisms using group presentations.

Let G and G' be groups. Suppose G is presented by the set X of generators and the set R of relators. We may construct a group homomorphism from G to G' in two steps.

(1) *Construct a function $\widetilde{\varphi}$ from X to G'.* By the definition of free groups, $\widetilde{\varphi}$ extends uniquely to a group homomorphism from $F[X]$ to G'.

(2) *Verify if every relator in R is mapped to the identity of G'*, that is, check if $R \subseteq \ker \widetilde{\varphi}$. If this is so, then $\ker \widetilde{\varphi}$ also contains the smallest normal subgroup N containing R. The fundamental theorem of group homomorphisms guarantees that $\widetilde{\varphi}$ induces a group homomorphism from $G \cong F[X]/N$ to G' which sends $\overline{x}$ to $\widetilde{\varphi}(x)$ for all $x \in X$.

If the presentation is expressed in terms of relations, to show that the relators are contained in $\ker \widetilde{\varphi}$ is equivalent to showing that the relations are preserved under $\widetilde{\varphi}$.

Next, we shall find group presentations for some familiar groups.

Example 12.2.5. For any positive integer n, a reasonable guess is that $\mathbb{Z}_n$ is presented by $G = \langle a : a^n \rangle$ since $\mathbb{Z}_n$ is generated by an element of order n. To show that this is indeed so, map a to $\overline{1} \in \mathbb{Z}_n$. Then we check that a^n

is mapped to $n \cdot \overline{1} = \overline{0}$. Hence we have a group homomorphism $\varphi \colon G \to \mathbb{Z}_n$ sending a to $\overline{1}$.

Conversely, there is a group homomorphism from $\mathbb{Z}$ to G sending 1 to $\overline{a}$ by Proposition 7.1.7. Since $n = n \cdot 1$ is mapped to $a^n = \overline{1}$ under $\widetilde{\psi}$, it induces a group homomorphism ψ from $\mathbb{Z}_n = \mathbb{Z}/\langle n \rangle$ to G sending $\overline{1}$ to a.

Since φ and ψ are inverse to each other, we have that φ is an isomorphism and $\mathbb{Z}_n \cong \langle a : a^n \rangle$.

Example 12.2.6. Remember that the Klein 4-group $V = \{e, a, b, c\}$ is generated by a and b (or any two non-identity elements) in V. Since each generator is of order 2 and V is abelian, it is reasonable to guess that

$$V \cong \langle x, y : x^2 = y^2 = 1, \ xy = yx \rangle.$$

Let G be the group presented. Map x to a and y to b. It is easy to see that all relations in the presentation are preserved in V since that is how we arrange them. Hence there is a group homomorphism $\varphi \colon G \to V$ as described. Note that φ is onto since its image contains a generating set of V. On the other hand, every element in G can be written as $x^i y^j$ since $xy = yx$. Using the relation $x^2 = y^2 = 1$, we may assume $i = 0, 1$ and $j = 0, 1$. Thus, G contains at most 4 elements. However, since φ is onto, G contains exactly 4 elements. By the pigeonhole principle, φ is also one-to-one. We conclude that φ is an isomorphism.

Example 12.2.7. In the discussion on finding useful relations regarding the dihedral group D_n as presented in (5.2.2), we see that it is reasonable to conjecture that

$$D_n \cong \langle a, \ b : a^n = b^2 = 1, \ bab^{-1} = a^{-1} \rangle.$$

Let G be group presented above. Let α be a rotation of order n and β be a reflection in D_n. We have a group homomorphism φ from G to D_n sending a to α and b to β since α and β satisfy the relations of a and b. Remember that $D_n = \langle \alpha, \beta \rangle$. Thus φ is onto. Using the relation $ba = a^{-1}b$ we can rewrite any element in G as $a^i b^j$ for i and $j \in \mathbb{Z}$. Using the relations $a^n = b^2 = 1$, any element of G is of the form $a^i b^j$ where $i = 0, 1, \ldots, n-1$, and $j = 0, 1$. Thus, G contains at most $2n$ elements. Since φ is onto, G contains exactly $2n$ elements. By the pigeonhole principle, φ is also one-to-one. We conclude that φ is an isomorphism.

Example 12.2.8. Let's try to construct a group homomorphism from the Klein 4-group V to D_4. Remember that

$$V \cong \langle \alpha, \beta : \alpha^2 = \beta^2 = 1, \ \alpha\beta = \beta\alpha \rangle$$

and

$$D_4 \cong \langle a, b : a^4 = b^2 = 1, \ bab^{-1} = a^3 \rangle.$$

Try mapping α to a^2 and β to b. Then the relations $\alpha^2 = \beta^2 = 1$ is preserved. It remains to check that $\alpha\beta\alpha^{-1}\beta^{-1}$ is mapped to 1. We verify that this is indeed so since

$$a^2 b a^{-2} b^{-1} = a^2 (b a^2 b^{-1}) = a^2 (bab^{-1})^2 = a^2 a^6 = (a^4)^2 = 1.$$

Thus, we have a group homomorphism φ from V to D_4 sending α to a^2 and β to b. Note that the image contains $1, a^2, b$ and $a^2 b$. Thus this group homomorphism is one-to-one and

$$V \cong \operatorname{Im} \varphi = \left\{ 1, a^2, b, a^2 b \right\} \subseteq D_4.$$

If we choose $a = (1\ 2\ 3\ 4)$ and $b = (1\ 2)(3\ 4)$, then

$$\{\, 1, \ (1\ 3)(2\ 4), \ (1\ 2)(3\ 4), \ (1\ 4)(2\ 3) \,\}$$

is a subgroup of D_4 which is isomorphic to V.

It is convenient to use generators and relations to describe a group. However, it is often very difficult to understand the group described. The goal of *combinatorial group theory* is to study the structure of a presented group.

Let G and G' be groups given by generators and relations. German mathematician M. Dehn (1878–1952) had raised three basic problems in combinatorial group theory.

(1) *The word problem*: is there an algorithm to determine whether a word in a presented group is 1?

(2) *The transformation problem*: is there an algorithm to determine whether two words in a presented group are conjugate to each other?

(3) *The isomorphism problem*: is there an algorithm to determine whether two presented groups are isomorphic?

So far, none of these problems have been answered.

Exercises 12.2

1. Let G be a group.

   (a) Let $\{\,H_i\,\}_{i \in \Lambda}$ be a family of normal subgroups of G. Show that $\bigcap_{i \in \Lambda} H_i$ is also a normal subgroup of G.

   (b) Let S be a subset of G. Show that there is a unique smallest normal subgroup containing S in G.

2. Show that $\langle\, a_1, \ldots, a_r : a_i a_j = a_j a_i, \ i \neq j \,\rangle$ is a group presentation for the free abelian group of rank r. Basically speaking, the free abelian group is the group generated by r generators in which the only relation is commutativity.

3. Show that $\langle\, x, \ y : x^2, \ y^5, \ xyx^{-1}y^{-1} \,\rangle \cong \mathbb{Z}_{10}$.

4. Show that $D_n \cong \langle\, y, \ z : y^2 = z^2 = (yz)^n = 1 \,\rangle$.

5. Find a group presentation for S_4 using the generators (1 2), (1 3) and (1 4) (*Cf.* Exercise 4, §5.3).

6. Let $D_{12} = \langle\, a, \ b : a^{12} = b^2 = 1, \ bab^{-1} = a^{-1} \,\rangle$.

   (a) Show that the subgroup $H = \langle\, a^3, \ b \,\rangle$ is isomorphic to D_4.

   (b) Show that H is not a normal subgroup in D_{12}.

12.3 Classification of Finite Groups of Small Orders

In this section, we shall classify all groups of orders up to 15. We first summarize what we know so far.

(1) $|G| = 1$: G is the trivial group

(2) $|G| = p$, p a positive prime integer: $G \cong C_p$. This follows from Corollary 6.1.6.

(3) $|G| = p^2$, p a positive prime integer: $G \cong C_{p^2}$ or $C_p \times C_p$. This follows from Proposition 10.2.5 and Theorem 9.3.5.

(4) $|G| = pq$, where $p > q$ are positive primes: if $q \nmid p - 1$ then $G \cong C_{pq}$. This follows from Proposition 9.2.4.

With these results we can classify the groups of order 1, 2, 3, 4, 5, 7, 9, 11, 13 and 15. We are yet to determine groups of order 6, 8, 10, 12 and 14.

Proposition 12.3.1. *Let p be a prime integer ≥ 3. If G is a group of order $2p$, then $G \cong C_{2p}$ or D_p.*

Proof. By Cauchy's theorem, there exists an element a of order p and an element b of order 2 in G. Then $H = \langle a \rangle$ is a normal subgroup of G since it is of index 2 in G. Remember that the conjugate of a is also of order p. Hence $bab^{-1} = a^j$ for some j with $1 \leq j \leq p - 1$. It follows that

$$a = b^2 a b^{-2} = b(bab^{-1})b^{-1} = a^{j^2}.$$

Thus, $j^2 \equiv 1 \pmod{p}$. It follows from Lemma 2.2.5 that $j = 1$ or $p - 1$. Hence

$$bab^{-1} = a \qquad \text{or} \qquad bab^{-1} = a^{-1}.$$

(1) If $bab^{-1} = a$, then $ab = ba$. It follows that G is abelian. Thus ab is an element of order $2p$ and $G \cong C_{2p}$.

(2) If $bab^{-1} = a^{-1}$, we claim that $G \cong D_p$.

    Recall that

$$D_p = \langle x, y : x^p = y^2 = 1, \ yxy^{-1} = x^{-1} \rangle.$$

There is a group homomorphism φ from D_p to G sending x to a and y to b. Note that $\langle a \rangle \cap \langle b \rangle = \{e\}$ by Lagrange's Theorem. It follows that $|\langle a \rangle \langle b \rangle| = |\langle a \rangle||\langle b \rangle|/|\langle a \rangle \cap \langle b \rangle| = 2p$. So $G = \langle a \rangle \langle b \rangle$ is generated by a and b. This implies that φ is onto. By the pigeonhole principle, φ is bijective since $|G| = 2p = |D_p|$. Thus, $G \cong D_p$. $\qquad \square$

Using this result, we can classify groups of order 6, 10 and 14. It remains to classify groups of order 8 and 12. Since finite abelian groups can be classified using Theorem 9.3.5. Our focus will be on the non-abelian groups.

The only non-abelian group of order 8 that we have discussed so far is the dihedral group D_4. In fact, there is another well-known group of order 8: the **quaternion group** Q_8. Let

$$Q_8 = \{\pm 1, \pm i, \pm j, \pm k\}$$

be a set of 8 elements. The product in Q_8 is an associative operation which satisfies the following conditions:

(i) 1 is the identity and $(-1)^2 = 1$;

(ii) $\pm 1 \in Z(Q_8)$;

(iii) $(-1)i = -i$, $(-1)j = -j$ and $(-1)k = -k$;

(iv) $i^2 = j^2 = k^2 = -1$;

(v) $ij = -ji = k$, $jk = -kj = i$ and $ki = -ik = j$.

We leave it as an exercise to verify that Q_8 is indeed a non-abelian group of order 8 (see Exercise 1). In Q_8 there are exactly 6 elements ($\pm i$, $\pm j$ and $\pm k$) of order 4, while in D_4 there are only two elements of order 4. Hence, Q_8 and D_4 are not isomorphic.

Proposition 12.3.2. *Up to isomorphism, there are exactly two distinct non-abelian groups of order 8: the quaternion group Q_8 and the dihedral group D_4.*

Proof. Let G be a non-abelian group of order 8. If all elements in G are of orders ≤ 2, then G is abelian (see Exercise 6, §3.2). If in G there is an element of order 8, G is abelian. Hence G must contain an element a of order 4.

Let $H = \langle a \rangle$. Then H is normal in G. Now choose $b \in G \setminus H$. Then $bab^{-1} = a^j$ for some j. Since $o(a) = o(bab^{-1})$, we have that $j = 1$ or 3. If $bab^{-1} = a$, by Exercises 7, §3.4, we have that G is abelian, a contradiction. Hence $bab^{-1} = b^3 = b^{-1}$. Moreover, $G = \langle a, b \rangle$ since it properly contains H. Note that $b^2 \in H$. We have the following possibilities.

(1) If $b^2 = a$ or a^3, then b is an element of order 8. In this case G is cyclic, a contradiction.

(2) If $b^2 = e$ and $bab^{-1} = a^{-1}$, then we have a group homomorphism φ from

$$D_4 = \langle\, x,\ y : x^4 = x^2 = 1,\ yxy^{-1} = x^{-1}\,\rangle$$

to G sending x to a and y to b. Since φ is onto and $|D_4| = 8$, it follows that φ is one-to-one and hence an isomorphism. We conclude that $G \cong D_4$.

(3) If $b^2 = a^2$ and $bab^{-1} = a^{-1}$, we conjecture that G is presented by

$$\langle\, a,\ b : a^4 = b^4 = 1,\ a^2 = b^2,\ bab^{-1} = a^{-1}\,\rangle.$$

Let G' be the presented group. Then G is a homomorphic image of G'. If we can prove that $|G'| = 8$, then $G \cong G'$ and we are done. Using the relation $ba = a^{-1}b$, we may express any element in G' as $a^i b^j$ where $i,\, j \in \mathbb{Z}$. Using the relation $a^4 = b^4 = 1$, we may further assume $0 \leq i,\, j \leq 3$. However, since $b^2 = a^2$ and $b^3 = a^2 b$, every element in G' is of the form $a^i b^j$ where $0 \leq i \leq 3$ and $j = 0, 1$. Thus, $|G'| \leq 8$.

On the other hand, there is a group homomorphic φ from G' to Q_8 sending a to i and b to j. It is easy to see that $Q_8 = \langle i, j \rangle$. Thus φ is onto and we have $|G'| \geq 8$. We conclude that $|G'| = 8$ and φ is an isomorphism. Thus we have found a group presentation for G and in the process we also showed that $G \cong Q_8$. $\qquad\square$

Now we are ready to classify non-abelian groups of order 12.

Proposition 12.3.3. *Up to isomorphism, there are exactly three distinct non-abelian groups of order* 12: *the dihedral group* D_6, *the alternating group* A_4 *and the group*

$$T = \langle\, a,\ b : a^6 = 1,\ b^2 = a^3,\ bab^{-1} = a^{-1}\,\rangle.$$

Proof. We assume that G is a non-abelian group of order 12. Let H be a Sylow 3-subgroup of G. Then $|H| = 3$ and the index of H in G is 4. There is a group homomorphism f from G into S_4 with kernel K contained in H (see Example 10.1.6).

If $K = \{e\}$, then f is one-to-one. We have that $f(G) \cong G$ is a subgroup of order 12. If $f(G)$ contains an odd permutation, $f(G) \cap A_4$ is a group of order 6 in A_4, a contradiction (see Exercise 8, §5.3 and Exercise 6, §9.2). Hence $f(G)$ contains 12 even permutations in S_4. Thus $G \cong f(G) = A_4$.

We now assume that $K \neq \{e\}$. We then have $K = H$ is a normal subgroup of G. Hence G contains exactly two elements of order 3. Let c be one of them. The number of elements in the conjugacy class of c is 1 or 2. Thus, $C_G(c)$ is of index 1 or 2 in G. It follows that $C_G(c)$ is of order 6 or 12. In either case, there exists an element $d \in C_G(c)$ such that d is of order 2. Let $a = cd$. Then a is an element of order 6 by Exercise 4, P.63.

Now $L = \langle a \rangle$ is a normal subgroup of G of order 6. Take any $b \in G \setminus L$. Then $G = \langle a, b \rangle$, $b^2 \in L$ and $bab^{-1} \in L$. Remember that $o(bab^{-1}) = 6$. Since G is non-abelian,

$$(12.3.1) \qquad\qquad bab^{-1} = a^5 = a^{-1}$$

is the only possibility. If $b^2 = a$ or a^5 then $o(b^2) = 6$ is a factor of $o(b)$. In this case $o(b) = 12$ and G is a cyclic group of order 12, a contradiction. If $b^2 = a^2$ then $ba^2b^{-1} = bb^2b^{-1} = b^2 = a^2$. On the other hand, we get $ba^2b^{-1} = (bab^{-1})^2 = a^{-2}$ from (12.3.1). It follows that $a^4 = e$, a contradiction to the fact that $o(a) = 6$. Similarly, if $b^2 = a^4$, we will also reach a similar contradiction. Therefore, the only possibilities are $b^2 = e$ or $b^2 = a^3$.

(1) If $b^2 = e$, then G is a homomorphic image of

$$D_6 = \langle a, b : a^6 = b^2 = 1, \ bab^{-1} = a^{-1} \rangle.$$

Since $|D_6| = 12$, we have $G \cong D_6$.

(2) If $b^2 = a^3$, then G is a homomorphic image of

$$(12.3.2) \qquad T = \langle a, b : a^6 = 1, \ b^2 = a^3, \ bab^{-1} = a^{-1} \rangle.$$

We leave it as an exercise to show that T is a group of order 12 (see Exercise 5). Thus, we have that $G \cong T$. $\qquad\qquad\qquad\square$

At this point, we have thus fulfilled our promises to classify all groups of orders up to 15. Below is a complete listing of all isomorphic classes of groups of orders ≤ 15 (see Exercises 5 and 6).

Order	No. of isom. types	Abelian groups	Non-abelian groups
1	1	$\mathbb{Z}_1$	
2	1	$\mathbb{Z}_2$	
3	1	$\mathbb{Z}_3$	
4	2	$\mathbb{Z}_4, \mathbb{Z}_2 \times \mathbb{Z}_2$	
5	1	$\mathbb{Z}_5$	
6	2	$\mathbb{Z}_6$	S_3
7	1	$\mathbb{Z}_7$	
8	5	$\mathbb{Z}_8, \mathbb{Z}_4 \times \mathbb{Z}_2$ $\mathbb{Z}_2 \times \mathbb{Z}_2 \times \mathbb{Z}_2$	D_4 Q_8
9	2	$\mathbb{Z}_9, \mathbb{Z}_3 \times \mathbb{Z}_3$	
10	2	$\mathbb{Z}_{10}$	D_5
11	1	$\mathbb{Z}_{11}$	
12	5	$\mathbb{Z}_{12}, \mathbb{Z}_2 \times \mathbb{Z}_2 \times \mathbb{Z}_3$	$D_6, \mathbb{Z}_3 \rtimes \mathbb{Z}_4, A_4$
13	1	$\mathbb{Z}_{13}$	
14	2	$\mathbb{Z}_{14}$	D_7
15	1	$\mathbb{Z}_{15}$	

Table 12.1: Classification of groups of orders up to 15

Exercises 12.3

1. Verify that the quaternion group Q_8 is indeed a non-abelian group of order 8.

2. Compare the centers of the groups Q_8 and D_4, and determine their normal subgroups.

3. Let N and H be two groups. Let $\varphi \colon H \to \text{Aut}(N)$ be a group homomorphism. We can define a group $N \rtimes_\varphi H$, called the (**external**) **semi-direct product** of N and H with respect to φ, as follows.

   As a set, $N \rtimes_\varphi H = N \times H$. We define the product $*$ in $N \rtimes_\varphi H$ by letting
   $$(n_1,\ h_1) * (n_2,\ h_2) = (n_1 \varphi(h_1)(n_2),\ h_1 h_2).$$

   (a) Show that $N \rtimes_\varphi H$ is indeed a group where $e = (e_N,\ e_H)$ is the identity element and $(n,\ h)^{-1} = (\varphi(h^{-1})(n^{-1}),\ h^{-1})$.

   (b) Let $\overline{N} = \{\,(n,\ e_H) : n \in N\,\}$. Show that $\overline{N}$ is a normal subgroup of $N \rtimes H$ and $\overline{N} \cong N$.

   (c) Let $\overline{H} = \{\,(e_N,\ h) : h \in H\,\}$. Show that $\overline{H}$ is a subgroup of $N \rtimes H$ and $\overline{H} \cong H$.

   (d) Show that $N \rtimes H = \overline{N}\,\overline{H}$ and $\overline{N} \cap \overline{H} = \{\,e\,\}$.

   (e) Show that $(N \rtimes H)/\overline{N} \cong H$.

   (f) If φ is the trivial homomorphism, show that $N \rtimes_\varphi H = N \times H$.

4. Let G be a group. Let N be a normal subgroup and H be a subgroup of G.

   (a) For $h \in H$, show that $\varphi_h \colon N \to N$ sending $n \in N$ to hnh^{-1} is a group automorphism of N.

   (b) Define a map $\varphi \colon H \to \text{Aut}\,N$ by sending $h \in H$ to φ_h. Show that φ is a group homomorphism.

   (c) Suppose $G = NH$ and $N \cap H = \{\,e\,\}$. Show that $G \cong N \rtimes_\varphi H$. Thus G is called the (**internal**) **semi-direct product** of N and H.

   (d) Show that $G/N \cong H$.

5. (a) Show that there is exactly one non-trivial group homomorphism φ from $\mathbb{Z}_4$ to $\text{Aut}(\mathbb{Z}_3)$.

   (b) Show that $\mathbb{Z}_3 \rtimes_\varphi \mathbb{Z}_4$ is a non-abelian group of order 12.

   (c) Show that $T \cong \mathbb{Z}_3 \rtimes_\varphi \mathbb{Z}_4$ where T is the group presentation in (12.3.2).

6. Show that D_6, A_4 and T as in (12.3.2) are non-isomorphic to each other.

7. Let p be a prime integer ≥ 3. Prove that there are at most two non-abelian groups of order p^3 up to isomorphism.

8. Classify the groups of order 20 up to isomorphism.

9. Classify the groups of order 21 up to isomorphism.

10. Classify the groups of order 30 up to isomorphism.

Review Exercises for Chapter 12

1. Let G be a free abelian group of rank r. Show that a nonzero subgroup H of G is free abelian of rank $\leq r$.

2. Show that $\langle x, y : x^2, y^2 \rangle$ is an infinite non-abelian group. (Hint: show that there is a surjective group homomorphism from this group onto D_n for all n.)

3. Let $G = \langle a, b : a^4 = b^4 = 1, abab^{-1} = 1 \rangle$.

   (a) Show that $|G| \leq 16$.

   (b) Find $Z(G)$.

   (c) Show that $G/\langle b^2 \rangle \cong D_4$.

4. Construct a non-abelian group of order 55.

5. Construct a non-abelian group of order 252.

6. Classify all groups of order 18 up to isomorphism.

CHAPTER 13

Types of Rings

The set of integers, the set of polynomials over a field and the set of matrices over a field are three classical examples of rings. It was David Hilbert (1862–1943) who first introduced the term *ring* in connection with polynomial rings and rings of integers of algebraic number fields. However, a fully abstract definition of rings does not appear until the second decade of the twentieth century. The theory of commutative rings was given a formal axiomatic foundation by Emmy Noether (1882–1935) in her paper *Ideal Theory in Rings* which appeared in 1921.

13.1   Definitions and Examples

Definition 13.1.1. A **ring** is a triple $(R, +, \cdot)$ such that R is a nonempty set and $+$ together with $\cdot$ are two binary operations in R satisfying the following properties:

(R1) $(R, +)$ is an abelian group;

(R2) $(R, \cdot)$ is associative;

(R3) the distributive laws hold, *i.e.*, for all a, b, $c \in R$ we have

$$a(b + c) = ab + ac \quad \text{and} \quad (a + b)c = ac + bc.$$

The identity element in $(R, +)$ is denoted by 0 and is called the **zero element** of the ring R.

Below are some special types of rings which are endowed with additional properties besides those mentioned in Definition 13.1.1.

(1) A **ring with identity** is a ring R in which $(R, \cdot)$ contains an identity 1 such that $1 \neq 0$. The identity 1 is also called the **unity** of R. A ring with identity is also called a **ring with unity**. A ring with identity contains at least two distinct elements, 0 and 1. Only in a ring with identity, it starts to make sense to talk about elements with multiplicative inverses. In such a ring, those elements with multiplicative inverses are called **invertible elements** or **units**.

(2) A **commutative ring** is a ring in which the multiplication is commutative. Otherwise, it is called a **non-commutative ring**.

(3) A **division ring** is also called a **skew field**. A division ring is a ring with identity in which every nonzero element is a unit.

(4) A **field** is a commutative division ring. If a field contains only finitely many elements, it is, not surprisingly, called a **finite field**.

(5) In general, not all nonzero elements are units. Worse, some are *zero divisors*. A nonzero element a in a ring R is called a **left** (or **right**) **zero divisor** if there exists $b \in R$, $b \neq 0$ such that $ab = 0$ (or $ba = 0$,

respectively). A **non-zero-divisor** or a **NZD** is a nonzero element which is not a zero divisor. A **domain** is a ring with identity which is without any zero divisors. An **integral domain** is a commutative domain.

Since units are never zero divisors, division rings are domains and fields are integral domains (see Exercise 1). It is often very important to determine which elements are units in a ring. We use $U(R)$ to denote the set of all units in the ring R. The set $U(R)$ is a group and is called the **group of units** of the ring R.

Now we give a few examples of rings.

Example 13.1.2. The triple $(\mathbb{Z}, +, \cdot)$ is an integral domain and is called the ring of integers.

Example 13.1.3. Clearly, $(\mathbb{Q}, +, \cdot)$, $(\mathbb{R}, +, \cdot)$ and $(\mathbb{C}, +, \cdot)$ are all examples of fields. They are called the field of rational numbers, the field of real numbers and the field of complex numbers respectively.

Example 13.1.4. Recall that in $\mathbb{Z}_n = \{\, \overline{0}, \overline{1}, \overline{2}, \ldots, \overline{n-1} \,\}$, we have defined $+$ and $\cdot$ in (2.2.1) and (2.2.2) as

$$\overline{a} + \overline{b} = \overline{a + b} \quad \text{and} \quad \overline{a} \cdot \overline{b} = \overline{ab}.$$

It is easy to check that $\mathbb{Z}_n$ is a commutative ring with identity. If n is a composite number then $\mathbb{Z}_n$ contains zero divisors. For example, in $\mathbb{Z}_6$, $\overline{2}$ and $\overline{3}$ are zero divisors since $\overline{2} \cdot \overline{3} = \overline{0}$. Thus, $\mathbb{Z}_n$ is not an integral domain or a field if n is a composite number. However, if $n = p$ is a prime integer, then $\mathbb{Z}_p$ is a field. By Proposition 2.2.2 we have that $U(\mathbb{Z}_p) = \mathbb{Z}_p^*$, that is, all the nonzero elements of $\mathbb{Z}_p$ are units.

Since fields are also integral domains, we sum up the results in the following proposition.

Proposition 13.1.5. *Let n be a positive integer. Then $\mathbb{Z}_n$ is a finite field if and only if it is an integral domain if and only if n is prime.*

Let R' be a nonempty subset of a ring R. We say that R' is a **subring** of R if $a+b$, $-a$ and $ab \in R'$ for all $a, b \in R'$. The conditions guarantee that

$(R', +)$ is a subgroup of $(R, +)$ and there is an inherited multiplication in R'. Associativity and distributivity are inherited from R. Thus, R' is a ring itself.

Example 13.1.6. The set $\mathbb{Z}[x]$ of polynomials in the variable x with integer coefficients is an integral domain under the usual addition and multiplication of polynomials. The same is true for $\mathbb{Q}[x]$, $\mathbb{R}[x]$ and $\mathbb{C}[x]$. Observe that $\mathbb{Z}[x]$ is a subring of $\mathbb{Q}[x]$, $\mathbb{Q}[x]$ is a subring of $\mathbb{R}[x]$, etc.

Example 13.1.7. Let

$$\mathbb{Z}[i] = \{\, a + bi \in \mathbb{C} : a, b \in \mathbb{Z} \,\}.$$

The addition and multiplication is given by

$$\begin{cases} (a + bi) + (c + di) = (a + c) + (b + d)i; \\ (a + bi)(c + di) = (ac - bd) + (ad + bc)i \end{cases}$$

for all $a, b, c, d \in \mathbb{Z}$. Under these operations $\mathbb{Z}[i]$ becomes a commutative ring with identity called the **ring of Gaussian integers**.

Since $\mathbb{C}$ contains no zero divisors, neither does $\mathbb{Z}[i]$. Hence $\mathbb{Z}[i]$ is an integral domain. However, the multiplicative inverse of $1 + i$ is $(1 - i)/2$ in $\mathbb{C}$. It follows that $1 + i$ is not a unit in $\mathbb{Z}[i]$. Thus, $\mathbb{Z}[i]$ is not a field.

Example 13.1.8. Let F be a field. Denote by $M_n(F)$ the set of $n \times n$ matrices with entries in F. In $M_n(F)$, we have the usual addition and multiplication of matrices. It is easy to see that $(M_n(F), +, \cdot)$ is a non-commutative ring with identity. An element B in $M_n(F)$ is invertible or a unit if and only if $\det(B) \neq 0$. Also note that $M_n(F)$ is not a domain as shown by an example such as

$$\begin{bmatrix} 1 & 0 \\ 0 & 0 \end{bmatrix} \begin{bmatrix} 0 & 0 \\ 0 & 1 \end{bmatrix} = \begin{bmatrix} 0 & 0 \\ 0 & 0 \end{bmatrix}.$$

Example 13.1.9. Denote by $C[0, 1]$ the set of real-valued continuous functions defined on $[0, 1]$. With the operations

$$(f + g)(x) = f(x) + g(x) \quad \text{and} \quad (f \cdot g)(x) = f(x)g(x),$$

$C[0,1]$ becomes a commutative ring with the identity function $1(x) = x$ as the unity and $0(x) = 0$ as the zero element. Note that $fg = 0$ if

$$f(x) = \begin{cases} -x + \frac{1}{2}, & \text{if } 0 \le x \le \frac{1}{2}; \\ 0, & \text{if } \frac{1}{2} \le x \le 1; \end{cases} \quad \text{and} \quad g(x) = \begin{cases} 0, & \text{if } 0 \le x \le \frac{1}{2}; \\ x - \frac{1}{2}, & \text{if } \frac{1}{2} \le x \le 1. \end{cases}$$

Thus f and g are zero divisors in $C[0,1]$ and the ring $C[0,1]$ is not an integral domain.

Even though we will observe more strange phenomena in other rings to come, there are properties that behave according to our intuitions. Before stating some of these basic properties, we would like to stress again that in a ring, 0 is the additive identity and 1 is the multiplicative identity. For a ring element a, $-a$ stands for the additive inverse and a^{-1} stands for the multiplicative inverse of a. For a and b in a ring, we write $a - b$ for $a + (-b)$.

Proposition 13.1.10. *Let R be a ring and a, b be elements in R. Then*

(a) $a0 = 0a = 0$;

(b) $a(-b) = (-a)b = -(ab)$;

(c) $(-a)(-b) = ab$;

(d) *if R contains a unity 1, then $(-1)a = -a = a(-1)$*;

(e) $(a - b)c = ac - bc$ *and* $c(a - b) = ca - cb$.

Proof. (a) As $0 = 0 + 0$, it follows that $a0 = a(0 + 0) = a0 + a0$. Cancel $a0$ from both sides to yield $a0 = 0$. Similarly, we have that $0a = 0$.

(b) Note that $ab + a(-b) = a[b + (-b)] = a0 = 0$. It follows $a(-b) = -(ab)$. We can similarly show that $(-a)b = -(ab)$.

Clearly, (c) holds since $(-a)(-b) = -(a(-b)) = -(-ab) = ab$. It is also clear that (d) follows from (b), and (e) follows from (b) and the distributive laws. $\qquad\square$

Using these basic properties and the definition of rings, we may see that some of the formulas that we know so well now get a new twist. Let R be a ring and let $a, b \in R$. Then

$$\begin{cases} (a + b)^2 = (a + b)(a + b) = a(a + b) + b(a + b) = a^2 + ab + ba + b^2; \\ (a + b)^3 = a^3 + a^2b + aba + ab^2 + ba^2 + bab + b^2a + b^3. \end{cases}$$

Only when R is commutative do we have the familiar formulae

$$(a+b)^2 = a^2 + 2ab + b^2 \quad \text{and} \quad (a+b)^3 = a^3 + 3a^2b + 3ab^2 + b^3$$

for all a and b in R (*Cf.* Exercise 13).

Let R be a ring with identity and let k be a positive integer. It is customary to use k to denote the element

$$\underbrace{1 + 1 + \cdots + 1}_{k \text{ times}}$$

in R. The element $-k$ stands for the additive inverse of k and is actually

$$\underbrace{(-1) + (-1) + \cdots + (-1)}_{k \text{ times}}.$$

For $a \in R$ and a positive integer k, ka now is an ambiguous notation. Does it stand for the sum of k copies of a, or the product of k and a in R? Fortunately, distributive laws solve this problem since

$$\underbrace{(1 + 1 + \cdots + 1)}_{k \text{ times}}a = \underbrace{a + a + \cdots + a}_{k \text{ times}}.$$

Similarly, we have

$$-ka = \underbrace{((-1) + (-1) + \cdots + (-1))}_{k \text{ times}}a = \underbrace{(-a) + (-a) + \cdots + (-a)}_{k \text{ times}} = -(ka).$$

We can thus use integers to denote elements in any ring without causing ambiguity. However, note that two different integers may represent the same element in a ring. For example, $2 = 6$ in $\mathbb{Z}_4$.

An element a is called an **idempotent** element, or simply an **idempotent**, if $a^2 = a$. The zero element and the unity are both idempotents in a ring.

A ring R is called a **Boolean ring** if every element in R is idempotent. For example, $\mathbb{Z}_2 = \{0, 1\}$ is a commutative Boolean ring. Next we give an example of a different type of Boolean rings.

Example 13.1.11. Let S be a nonempty set and 2^S be the set of subsets of S. Define the operations in 2^S as

$$A + B = (A \setminus B) \cup (B \setminus A) \quad \text{and} \quad A \cdot B = A \cap B.$$

Note that

$$
\begin{aligned}
A \cdot (B + C) &= A \cap [(B \setminus C) \cup (C \setminus B)] \\
&= [A \cap (B \setminus C)] \cup [A \cap (C \setminus B)] \\
&= [(A \cap B) \setminus (A \cap C)] \cup [(A \cap C) \setminus (A \cap B)] \\
&= A \cap B + A \cap C = A \cdot B + A \cdot C.
\end{aligned}
$$

We leave it to the reader to verify $(2^S, +, \cdot)$ satisfies all the other requirements for rings. Hence 2^S is a Boolean ring since $A \cdot A = A \cap A = A$.

Here we shall prove a nice result on Boolean rings.

Proposition 13.1.12. *Every Boolean ring is commutative.*

Proof. Let R be a Boolean ring and $a, b \in R$. Then

$$
a + b = (a + b)^2 = a^2 + ab + ba + b^2 = a + ab + ba + b.
$$

It follows that $ab + ba = 0$. Thus $ab = -ba$. Note that for all $x \in R$, $x + x = (x + x)^2 = x + x + x + x$. It follows that $x + x = 0$. Hence $2ba = ba + ba = 0$. We conclude that $ba = -ba = ab$. Therefore, R is commutative. $\qquad\square$

An element a in a ring R is called a **nilpotent** element or simply a **nilpotent** if there is a positive integer n such that $a^n = 0$.

Example 13.1.13. Inside $\mathbb{Z}_6$ the only nilpotent element is $\overline{0}$. Inside $\mathbb{Z}_8$, the elements $\overline{0}, \overline{4}, \overline{6}$ are all nilpotent.

Example 13.1.14. In this example, we introduce the notion of **group rings**. Let $G = \{ g_i : i \in I \}$ be a group and R be a commutative ring with identity. Let $R(G)$ be the set of all formal sums $\sum_{i \in I} a_i g_i$ where $a_i \in R$ are such that all but a finite number of the a_i's are 0 and $0g = 0$ for all $g \in G$. Define the addition of two elements in $R(G)$ by

$$
\left(\sum_{i \in I} a_i g_i \right) + \left(\sum_{i \in I} b_i g_i \right) = \sum_{i \in I} (a_i + b_i) g_i.
$$

Observe that the sum is also an element in $R(G)$ since $a_i + b_i = 0$ except for a finite number of indices i. It is easy to see that $(R(G), +)$ is an abelian group.

The product of two elements in $R(G)$ is defined so that the distributive laws and

$$(a_i g_i)(a_j g_j) = (a_i a_j)(g_i g_j).$$

are satisfied. It is natural to define

$$(13.1.1) \qquad \left(\sum_{i \in I} a_i g_i \right) \left(\sum_{j \in I} b_j g_j \right) = \sum_{k \in I} \left(\sum_{g_i g_j = g_k} a_i b_j \right) g_k.$$

Since a_i and b_j are zero for all but a finite number of i and j, the sum $\sum_{g_i g_j = g_k} a_i b_j$ is a finite sum. Note that only finitely many g_k would actually appear on the right hand side of (13.1.1). Thus the multiplication is well-defined in $R(G)$. The multiplication is associative since

$$(a_i g_i)[(b_j g_j)(c_k g_k)] = (a_i(b_j c_k))(g_i(g_j g_k))$$
$$= [(a_i g_i)(b_j g_j)](c_k g_k) = ((a_i b_j)c_k)((g_i g_j)g_k).$$

We leave it to the reader to check that $R(G)$ satisfies all the requirements for a ring.

Exercises 13.1

1. Show that an invertible element is never a zero divisor.

2. Show that the cancellation law with respect to multiplication holds for elements which are not zero divisors. To be more precise, let a, b, c be elements of a ring R such that a is nonzero and not a zero divisor. Show that

$$ab = ac \implies b = c \quad \text{and} \quad ba = ca \implies b = c.$$

3. Let R' be a nonempty subset of a ring R. Show that R' is a subring of R if $a - b$ and $ab \in R'$ for all $a, b \in R'$.

4. Find all the zero divisors in $\mathbb{Z}_{24}$ and $\mathbb{Z}_{27}$.

5. Let $\mathbb{Z}[\sqrt{m}] = \{a + b\sqrt{m} \in \mathbb{C} : a, b \in \mathbb{Z}\}$ where m is a square-free integer other than 1. Show that $\mathbb{Z}[\sqrt{m}]$ is a commutative ring under the operations of addition and multiplication of complex numbers.

6. Let α be a zero of the polynomial $x^2 + x + 2$ in $\mathbb{C}$, and let

$$\mathbb{Z}[\alpha] = \{a + b\alpha \in \mathbb{C} : a, b \in \mathbb{Z}\}.$$

Show that $\mathbb{Z}[\alpha]$ is a commutative ring with identity under the addition and multiplication of complex numbers.

7. Let

$$K = \left\{ \begin{bmatrix} a & b \\ -b & a \end{bmatrix} \,\middle|\, a, b \in \mathbb{R} \right\}.$$

Show that K is a field.

8. In a ring with identity 1, show that if $1 - ab$ is invertible, so is $1 - ba$.

9. Show that any finite domain is a division ring.

10. Show that a division ring contains exactly two idempotent elements.

11. Show that 0 is the only nilpotent element in a domain.

12. Show that $a^2 - b^2 = (a + b)(a - b)$ for all a and b in a ring R if and only if R is a commutative ring.

13. Show that in a commutative ring R,

$$(a + b)^n = \sum_{i=0}^{n} \binom{n}{i} a^i b^{n-i}$$

for all $a, b \in R$ and $n \in \mathbb{Z}_+$.

14. Let $G = \{e, a\}$ be a cyclic group of order 2 and $\mathbb{Z}_2$ be the finite field with 2 elements. Give the additive and multiplicative tables for the group ring

$$\mathbb{Z}_2(G) = \{0, e, a, e + a\}.$$

13.2   Matrix Rings

Let R be a ring and n be a positive integer. The set of $n \times n$ matrices over R is denoted by $M_n(R)$ which consists of $n \times n$ matrices

$$A = \begin{bmatrix} a_{11} & a_{12} & \cdots & a_{1n} \\ a_{21} & a_{22} & \cdots & a_{2n} \\ \vdots & \vdots & \ddots & \vdots \\ a_{n1} & a_{n2} & \cdots & a_{nn} \end{bmatrix}$$

with $a_{ij} \in R$. Two matrices $A = \begin{bmatrix} a_{ij} \end{bmatrix}$ and $B = \begin{bmatrix} b_{ij} \end{bmatrix}$ are equal if and only if $a_{ij} = b_{ij}$ for every i and j.

The addition of matrices is defined entry-wise by the formula

$$\begin{bmatrix} a_{ij} \end{bmatrix} + \begin{bmatrix} b_{ij} \end{bmatrix} = \begin{bmatrix} a_{ij} + b_{ij} \end{bmatrix}.$$

It follows that $(M_n(R), +)$ is also an abelian group with the zero matrix as the additive identity and the inverse of A given by $-A = \begin{bmatrix} -a_{ij} \end{bmatrix}$.

The multiplication of matrices is defined by $AB = \begin{bmatrix} c_{ij} \end{bmatrix}$ with

$$c_{ij} = \sum_{k=1}^{n} a_{ik} b_{kj} = a_{i1} b_{1j} + a_{i2} b_{2j} + \cdots + a_{in} b_{nj}.$$

It is tedious but routine to check that multiplication is associative.

If R contains a unity 1, the identity matrix I_n given by

$$I_n = \begin{bmatrix} 1 & 0 & \cdots & 0 \\ 0 & 1 & \cdots & 0 \\ \vdots & \vdots & \ddots & \vdots \\ 0 & 0 & \cdots & 1 \end{bmatrix}$$

is the multiplicative identity or the unity of $M_n(R)$. Moreover, we can define E_{ij} to be the matrix having 1 for its (i,j)-entry and 0 for all other entries. These n^2 matrices E_{ij}, $1 \leq i, j \leq n$, are called **matrix units** of $M_n(R)$. It is easy to see that the matrix

$$A = \begin{bmatrix} a_{ij} \end{bmatrix} = \sum_{1 \leq i, j \leq n} a_{ij} E_{ij}$$

and the coefficients of the matrix units for A are uniquely determined.

One can verify the following multiplication rules

$$E_{ij}E_{k\ell} = \delta_{jk}E_{i\ell},$$

where δ_{jk} is the **Kronecker delta** defined by

$$\delta_{jk} = \begin{cases} 1, & \text{if } j = k; \\ 0, & \text{if } j \neq k. \end{cases}$$

In fact, one can see that the product of matrices are defined so that the distributive laws are satisfied:

$$AB = \left(\sum_{1 \leq i,j \leq n} a_{ij}E_{ij} \right) \left(\sum_{1 \leq i,j \leq n} b_{ij}E_{ij} \right)$$

$$= \sum_{1 \leq i,j,k,l \leq n} a_{ik}b_{lj}E_{ik}E_{lj} = \sum_{1 \leq i,j \leq n} \left(\sum_{1 \leq k \leq n} a_{ik}b_{kj} \right) E_{ij}.$$

Let $a \in R$. Even though a is not an element in $M_n(R)$, by an abuse of notation, we often write $aA = a\left[a_{ij}\right] = \left[aa_{ij}\right]$ for $a \in R$. This turns out to be not so much of an abuse since we can embed R into $M_n(R)$ and make the identification

$$a = \begin{bmatrix} a & 0 & \cdots & 0 \\ 0 & a' & \cdots & 0 \\ \vdots & \vdots & \ddots & \vdots \\ 0 & 0 & \cdots & a \end{bmatrix}_{n \times n}.$$

Recall from Linear Algebra we have the following results regarding units in a matrix ring.

Proposition 13.2.1. *Let $F = \mathbb{R}$ or $\mathbb{C}$. Let A and B be in $M_n(F)$. Then*

$$\det(AB) = \det A \det B.$$

Proposition 13.2.2. *Let $F = \mathbb{R}$ or $\mathbb{C}$ and let n be a positive integer. Then*

$$U(M_n(F)) = GL_n(F) = \{ A \in M_n(F) : \det A \neq 0 \}.$$

In §19.2 we will give a detailed treatment on matrices and determinant. From Theorem 19.2.10 we will see that Proposition 13.2.1 is true for any commutative ring with identity. We will also see in Proposition 19.2.11 that

$$U(M_n(R)) = \{\, A \in M_n(R) : \det A \text{ is invertible in } R \,\}$$

where R is a commutative ring with identity.

Next, we introduce the ring of **real quaternions**. Let

$$\mathbb{H} = \{\, a + bi + cj + dk \mid a, b, c, d \in \mathbb{R} \,\}.$$

The addition is given by

$$(a + bi + cj + dk) + (e + fi + gj + hk)$$
$$= (a + e) + (b + f)i + (c + g)j + (d + h)k.$$

Then $\mathbb{H}$ is a 4-dimensional vector space over $\mathbb{R}$ with $\{\, 1, \, i, \, j, \, k \,\}$ as a basis over $\mathbb{R}$. The multiplication in $\mathbb{H}$ is defined by the distributive laws satisfying the following rules:

(i) $i^2 = j^2 = k^2 = -1$;

(ii) $ij = k = -ji$, $jk = i = -kj$, $ki = j = -ik$.

Explicitly we have

$$(a + bi + cj + dk)(p + qi + rj + sk)$$
$$= (ap - bq - cr - ds) + (aq + bp + cs - dr)i$$
$$+ (ar - bs + cp + dq)j + (as + br - cq + dp)k.$$

One can check by brutal force that $\mathbb{H}$ is a non-commutative ring with 1 as the identity.

For $x = a + bi + cj + dk$, let $\overline{x} = a - bi - cj - dk$. Define the **norm** of x to be

$$N(x) = x\overline{x} = a^2 + b^2 + c^2 + d^2 \in \mathbb{R}.$$

With more brutal force one can verify that

$$N(xy) = N(x)N(y).$$

This is the well-known *Lagrange's Theorem* which asserts that the product of two four-squares is still a four-squares. Note that

$$x \neq 0 \quad \Longleftrightarrow \quad N(x) \neq 0.$$

Since $x\overline{x} = N(x)$, we have

$$x^{-1} = \frac{\overline{x}}{N(x)} = \frac{a - bi - cj - dk}{a^2 + b^2 + c^2 + d^2}.$$

when $x \neq 0$. Every nonzero element in $\mathbb{H}$ has a multiplicative inverse. Thus, $\mathbb{H}$ is a division ring.

To make things more transparent, we would like to point out that the real quaternions $a + bi + bj + ck \in \mathbb{H}$ can also be represented by the matrix

$$\begin{bmatrix} a + bi & c + di \\ -c + di & a - bi \end{bmatrix} \in M_2(\mathbb{C}).$$

Note that

$$\begin{bmatrix} a + bi & c + di \\ -c + di & a - bi \end{bmatrix} + \begin{bmatrix} e + fi & g + hi \\ -g + hi & e - fi \end{bmatrix}$$

$$= \begin{bmatrix} (a + e) + (b + f)i & (c + g) + (d + h)i \\ -(c + g) + (d + h)i & (a + e) - (b + f)i \end{bmatrix}$$

The addition is compatible with the addition in $\mathbb{H}$. To verify that the multiplication is also consistent, note that i, j, k in $\mathbb{H}$ are represented by

$$(13.2.1) \qquad I = \begin{bmatrix} i & 0 \\ 0 & -i \end{bmatrix}, \quad J = \begin{bmatrix} 0 & 1 \\ -1 & 0 \end{bmatrix} \quad \text{and} \quad K = \begin{bmatrix} 0 & i \\ i & 0 \end{bmatrix}$$

respectively. It is easy to verify that

(i) $I^2 = J^2 = K^2 = -I_2$;

(ii) $IJ = K = -JI, \ JK = I = -KJ, \ KI = J = -IK$

(see Exercise 4). Hence the multiplication is also compatible with the multiplication in $\mathbb{H}$. Thus $\mathbb{H}$ can be identified with the set

$$\left\{ \begin{bmatrix} z & w \\ -\overline{w} & \overline{z} \end{bmatrix} \in M_2(\mathbb{C}) : z, \ w \in \mathbb{C} \right\}$$

with inherited operations from $M_2(\mathbb{C})$. Observe that $\mathbb{H}$ is a subring of $M_2(\mathbb{C})$. Moreover, let $x \in \mathbb{H}$ and let A be its corresponding matrix. Then $N(x) = \det A$. Now it follows easily that $N(xy) = N(x)N(y)$ by Proposition 13.2.1.

In 1843, W. R. Hamilton (1805–1865) constructed the first example of non-commutative division rings. This was intended as an extension of the field of complex numbers whose elements are quadruples of real numbers expressed as

$$a_0 + a_1 i + a_2 j + a_3 k$$

with i, j and k behaving similarly as i in complex numbers. The quaternion group Q_8 is a finite subgroup of the multiplicative group $\mathbb{H}^*$.

To further extend $\mathbb{H}$, there are the **Cayley numbers** with 8 components of real numbers. These numbers are expressed as

$$x = \sum_{j=0}^{7} x_j e_j, \quad x_j \in \mathbb{R}.$$

The multiplication of Cayley numbers is based on the following rules:

(i) $e_j^2 = -e_0$, $j = 1, 2, \ldots, 7$;

(ii) $e_1 e_2 e_4 = e_2 e_3 e_5 = e_3 e_4 e_6 = e_4 e_5 e_7 = e_5 e_6 e_1 = e_6 e_7 e_2 = e_7 e_1 e_3 = -e_0$.

However, this operation is not associative since

$$(e_1 e_2) e_3 = e_4 e_3 = -e_6 \qquad \text{while} \qquad e_1 (e_2 e_3) = e_1 e_5 = e_6.$$

Exercises 13.2

1. Find the inverses of the following matrices over $\mathbb{R}$.

(a) $\begin{bmatrix} 0 & 1 & 0 \\ 0 & 0 & 1 \\ 2 & -3 & 1 \end{bmatrix}$

(b)
$$\begin{bmatrix} 1 & 2 & -1 & 1 \\ 2 & 1 & 1 & -1 \\ -1 & 1 & 1 & -2 \\ 1 & -1 & -2 & 1 \end{bmatrix}$$

2. Show that for any a in a ring R and $i \neq j$, the matrix $I_n + aE_{ij}$ is invertible in $M_n(R)$ with inverse $I_n - aE_{ij}$.

3. Find the following products of real quaternions.

   (a) $(j + k)(j + 2k)$

   (b) $(1 + i + j - k)(2 + i - j + k)$

   (c) $(i + j - k)(i - j + k)(-i + j + k)$

   (d) $(i + j + k)^4$

4. Let $\mathscr{H}$ be a subset of $M_2(\mathbb{C})$, consisting of 2×2 matrices in the form

$$\begin{bmatrix} z & w \\ -\overline{w} & \overline{z} \end{bmatrix} = \begin{bmatrix} a + bi & c + di \\ -c + di & a - bi \end{bmatrix}.$$

   In $\mathscr{H}$ let I, J and K be the three matrices as given in (13.2.1).

   (a) Show that $I^2 = J^2 = K^2 = -I_2$, where I_2 denotes the 2×2 identity matrix.

   (b) Show that $IJ = -JI = K$, $JK = -KJ = I$, $KI = -IK = J$.

5. For each real quaternion $x = a_0 + a_1 i + a_2 j + a_3 k$, let

$$N(x) = a_0^2 + a_1^2 + a_2^2 + a_3^2.$$

   Show that $N(xy) = N(x)N(y)$ using the previous exercise.

6. Show that the matrix $\begin{bmatrix} 3 & 6 \\ 2 & 4 \end{bmatrix}$ is a zero divisor in $M_2(\mathbb{Z})$.

Review Exercises for Chapter 13

1. Find all idempotents in $\mathbb{Z}_{30}$.

2. If $a^2 = 0$ for all a in a ring R, show that $ar + ra$ commutes with a for any r in R.

3. If R is a ring and a is an idempotent in R, show that $(ra - ara)^2 = (ar - ara)^2 = 0$ for every r in R.

4. Let R be a ring in which $r^3 = r$ for every $r \in R$. Show that R is a commutative ring.

5. Let R be a commutative ring with identity 1 and let u be a unit of R. Show that $u + a$ is invertible if a is nilpotent.

6. Show that if a and b are nilpotent elements in a commutative ring R, then $a + b$ is also nilpotent.

7. Let $G = \{e,\, a\}$ be the multiplicative group of order 2. Describe the elements and the operations in the group ring $\mathbb{Z}(G)$.

8. Let p be a prime integer ≥ 3 and

$$1 + \frac{1}{2} + \cdots + \frac{1}{p-1} = \frac{a}{b},$$

where a and b are relatively prime positive integers. Show that p is a divisor of a. (Hint: multiply on both sides by $(p-1)!$ and then consider the equation in $\mathbb{Z}_p$. Use Wilson's Theorem.)

9. Let $F = \mathbb{R}$ or $\mathbb{C}$. Show that $A \in M_n(F)$ is a zero divisor if and only if A is not invertible.

CHAPTER 14

Ideals and Quotient Rings

The most important substructures of a ring are actually its ideals. Ideals are additive subgroups of the ring which satisfy additional conditions so that the additive quotient groups would also possess a natural multiplication. With the addition of these quotient groups and the induced multiplication, we obtain natural ring structures called the quotient rings.

14.1 Ideals

To study a ring, it is more important to study its ideals rather than its subrings.

Definition 14.1.1. Let R be a ring. A *nonempty* subset I of R is called a **left** (or **right**, respectively) **ideal** of R if

(I1) $(I, +)$ is a subgroup of $(R, +)$, and

(I2) ra (or ar, respectively) $\in I$ for all $r \in R$ and $a \in I$.

A **two-sided** ideal is an ideal which is both a left ideal and a right ideal.

By an ideal we may mean a left, or right, or two-sided ideal. When R is a commutative ring, all ideals are two-sided.

You will notice that every time we mention some facts regarding ideals, we always have three cases to consider. It is tedious to define and prove everything three times. For the rest of this book all ideals will be assumed to be *left* ideals unless otherwise noted. All results regarding left ideals can be translated accordingly to those of right or two-sided ideals.

Lemma 14.1.2. *Let R be a ring with identity and let I be a nonempty subset of R. Then I is a left ideal if $a + b$ and $ra \in I$ for all $a, b \in I$ and $r \in R$.*

Proof. It remains to show that $(I, +)$ is a subgroup of $(R, +)$. And for this it remains to show that $-a \in I$ for all $a \in I$. Since $1 \in R$, R contains the element -1. Hence $-a = (-1)a \in I$, and we are done.   □

While the element 0 belongs to any ideal, the unity is seldom in one. The set $\{0\}$ is indeed an ideal in any ring, and it is called the **trivial ideal**. An ideal containing the unity is called the **unit** ideal. If a left ideal of the ring R contains the unity, then for any $r \in R$, we have $r = r1 \in I$. Thus this left ideal is in fact R. Similarly, if a right (or two-sided) ideal contains the unity, it is equal to R. In a division ring or in a field, the only ideals are the trivial ideal and the unit ideal. We will call any ideal which is a proper subset of a ring a **proper** ideal.

Example 14.1.3. In the ring $\mathbb{Z}$ the subset

$$n\mathbb{Z} = \{nk : k \in \mathbb{Z}\}$$

is an ideal for all $n \in \mathbb{Z}$.

Example 14.1.4. In the ring $M_2(\mathbb{Z})$,

$$I = \left\{ \begin{bmatrix} a & 0 \\ b & 0 \end{bmatrix} \middle| a, b \in \mathbb{Z} \right\}$$

is a left ideal, but not a right ideal. We check only (I2) in Definition 14.1.1 since the others are easy. For all p, q, r, $s \in \mathbb{Z}$, we have that

$$\begin{bmatrix} p & q \\ r & s \end{bmatrix} \begin{bmatrix} a & 0 \\ b & 0 \end{bmatrix} = \begin{bmatrix} pa + qb & 0 \\ ra + sb & 0 \end{bmatrix} \in I.$$

Hence I is a left ideal. On the other hand, we have

$$\begin{bmatrix} 1 & 0 \\ 0 & 0 \end{bmatrix} \in I \quad \text{but} \quad \begin{bmatrix} 1 & 0 \\ 0 & 0 \end{bmatrix} \begin{bmatrix} 0 & 1 \\ 0 & 0 \end{bmatrix} = \begin{bmatrix} 0 & 1 \\ 0 & 0 \end{bmatrix} \notin I.$$

Thus I is not a right ideal.

Proposition 14.1.5. *Let R be a ring and let $\{I_i : i \in \Lambda\}$ be a nonempty collection of ideals of R. Then $\bigcap_{i \in \Lambda} I_i$ is an ideal of R.*

Proof. Let $I = \bigcap_{i \in \Lambda} I_i$. First, note that $0 \in I_i$ for all $i \in \Lambda$. Hence 0 is in the intersection and I is nonempty. Let $a, b \in I$ and $r \in R$. Then a, $b \in I_i$ for all i. It follows that $a + b$ and ra are in I_i for all i. Thus, $a + b$ and ra are both in I. $\qquad\qquad\qquad\qquad\qquad\qquad\qquad\qquad\qquad\qquad\square$

Let S be a subset of a ring R. Then by Proposition 14.1.5, the intersection of all ideals of R containing S is the smallest ideal of R containing S. We define this ideal to be (S), **the ideal generated by S in R**. If you want to distinguish between left and right ideals, you can use $R(S)$ and $(S)R$. If an ideal $I = (S)$, we say that the set S **generates** I or S is a **generating set of I**. If S is finite, we say I is a **finitely generated** ideal. An ideal which can be generated by a single element is called a **principal** ideal.

We have a few remarks on ideals.

(1) If $S_1 \subseteq S_2$ then $(S_1) \subseteq (S_2)$.

(2) Even though $(\varnothing) = \{0\}$, we usually denote the trivial ideal by (0).

(3) When $S = \{a_1, a_2, \ldots, a_n\}$ is finite, we also use $(a_1, a_2, \ldots, a_n)$ to denote (S).

Theorem 14.1.6. *Let R be a ring with unity and S be a subset of R. Then*

$$(14.1.1) \quad R(S) = \left\{ \sum_{i=1}^{k} r_i a_i \ \middle| \ k \text{ is a nonnegative integer, } r_i \in R, \ a_i \in S \right\}.$$

By convention we will define the empty sum (the sum of no elements) to be 0.

Proof. Let I be the set on the right-hand side of (14.1.1). By definition of ideals we have that $I \subseteq R(S)$. Conversely, observe that I contains S. It remains to show that I is an ideal. It would then follow that $I \supseteq R(S)$.

Since 0 is an empty sum, it is in I and so I is nonempty. Let

$$a = \sum_{i=1}^{k} r_i a_i \quad \text{and} \quad b = \sum_{j=1}^{\ell} r'_j a'_j$$

be elements in I where $a_i, a'_j \in S$ and $r_i, r'_j \in R$. Then obviously $a + b \in I$ and

$$ra = \sum_{i=1}^{k} r r_i a_i \in I$$

for any $r \in R$. We conclude that I a left ideal of R. $\qquad\square$

Thanks to Theorem 14.1.6, we often write $R(a_1, a_2, \ldots, a_n)$ as

$$Ra_1 + Ra_2 + \cdots + Ra_n.$$

In the case of right ideals, $(a_1, a_2, \ldots, a_n)R$ is written as

$$a_1 R + a_2 R + \cdots + a_n R.$$

Of course, when R is a commutative ring the two notations would make no differences. It is also very common to write a left ideal I as

$$RI = \{ ri \in R : i \in I, \ r \in R \}.$$

This notation makes an emphasis on the base ring, which is sometimes very useful. For example, since the integer 2 can be considered as an integer or a rational number, the notation $\mathbb{Z}(2)$ is more specific than the notation (2). Obviously, $\mathbb{Z}(2) \subsetneq \mathbb{Z}$ and $\mathbb{Q}(2) = \mathbb{Q}$. Note that the ideal $\mathbb{Z}(2)$ can also be written as $\mathbb{Z}2$, $2\mathbb{Z}$ or $(2)\mathbb{Z}$.

Proposition 14.1.7. *Every ideal in $\mathbb{Z}$ is a principal ideal. In other words, every ideal in $\mathbb{Z}$ is of the form $k\mathbb{Z}$ for some k in $\mathbb{Z}$.*

Proof. If I is trivial then $I = (0)$ and we are done. We assume $I \neq (0)$. Find the least positive integer k in I. We claim that $I = (k)$.

First, note that $(k) \subseteq I$ since $k \in I$. Take any $a \in I$. Find $q, r \in \mathbb{Z}$ such that $a = kq + r$ where $0 \leq r < k$. Since $r = a - kq \in I$, we must have $r = 0$. It follows that $a \in (k)$ and thus $I \subseteq (k)$. $\square$

Exercises 14.1

1. Show that if an ideal contains an invertible element, it is the unit ideal.

2. Let I be the subset of $C[0, 1]$ (see Example 13.1.9) consisting of the f with $f(0) = 0$. Is I an ideal of $C[0, 1]$?

3. Let R be a ring with unity. Show that R is a division ring if and only if (0) and (1) are the only ideals of R.

4. Show that the set of all nilpotent elements in a commutative ring R is an ideal.

5. Let I and J be ideals of a ring R. Show that $I \cup J$ is an ideal of R if and only if either $I \subseteq J$ or $J \subseteq I$.

6. Let $I_1, I_2, \ldots, I_n$ be ideals of R. Show that

$$I_1 + I_2 + \cdots + I_n$$
$$= \{\, i_1 + i_2 + \cdots + i_n \in R : i_j \in I_j \text{ for } j = 1, 2, \ldots, n \,\}$$

is the smallest ideal of R containing $I_1 \cup \cdots \cup I_n$. Furthermore, show that

$$(S_1) + \cdots + (S_n) = (S_1 \cup \cdots \cup S_n).$$

The ideal $I_1 + I_2 + \cdots + I_n$ is called the **sum** of the ideals $I_1, \ldots, I_n$.

7. Let k and $\ell \in \mathbb{Z}$. Show that $k\mathbb{Z} + \ell\mathbb{Z} = d\mathbb{Z}$ and $k\mathbb{Z} \cap \ell\mathbb{Z} = m\mathbb{Z}$ where g is a g.c.d. of k and ℓ while m is a l.c.m. of k and ℓ.

8. Let $M_2(\mathbb{R})$ be the ring of 2×2 matrices over $\mathbb{R}$. Prove that $M_2(\mathbb{R})$ has no nontrivial proper two-sided ideals.

14.2   Quotient Rings

Let I be a two sided ideal of a ring R. Then $(I, +)$ is a subgroup of $(R, +)$. Thus, we have an additive group

$$R/I = \{ r + I : r \in R \} = \{ \overline{r} : r \in R \}$$

where

$$(r_1 + I) + (r_2 + I) = (r_1 + r_2) + I$$

for $r_1, r_2 \in R$. We now attempt to define

(14.2.1) $(r_1 + I)(r_2 + I) = r_1 r_2 + I.$

Proposition 14.2.1. *The multiplication in* (14.2.1) *is well-defined and* R/I *is a ring. The element* $\overline{0}$ *is the zero element of* R. *If* R *is a ring with unity* 1, *then* R/I *is a also a ring with unity* $\overline{1}$.

Proof. Suppose that $a + I = a' + I$ and $b + I = b' + I$. We need to show that $ab + I = a'b' + I$. Note that

$$a = a' + i \qquad \text{and} \qquad b = b' + j$$

for some i and j in I. It follows that

$$ab = a'b' + ib' + a'j + ij.$$

By definition $ab - a'b' = ib' + a'j + ij \in I$. Hence $ab + I = a'b' + I$. This proves that the multiplication in R/I is well-defined. Now it is routine to verify the rest of this proposition. $\square$

The ring R/I is called the **quotient ring** or the **factor ring** of R modulo I.

Example 14.2.2. For all integers $n \geq 2$, the ring $\mathbb{Z}_n$ is the quotient ring $\mathbb{Z}/n\mathbb{Z}$.

Example 14.2.3. Let $f(x)$ be a nonzero polynomial of degree n in $\mathbb{R}[x]$, the ring of polynomials in the variable x with real coefficients. Then any element in $\mathbb{R}[x]/(f)$ is represented by $\overline{g(x)}$ for some $g(x) \in \mathbb{R}[x]$. Using the division algorithm, we can find $q(x)$ and $r(x)$ in $\mathbb{R}[x]$ such that

$$g(x) = f(x)q(x) + r(x) \qquad \text{where } r(x) = 0 \text{ or } \deg r(x) < n.$$

Hence $\overline{g(x)} = \overline{r(x)}$ can be represented by the zero polynomial or a polynomial of degree $< n$. On the other hand, if $\overline{r(x)} = \overline{r'(x)}$ for another polynomial which is either zero or with degree $< n$, then $r(x) - r'(x) \in (f(x))$. This means that $r(x) - r'(x) = 0$ for otherwise $\deg r(x) - r'(x) < n$, which is a contradiction since $f(x)$ is a divisor of $r(x) - r'(x)$. We conclude that every element of $\mathbb{R}[x]/(f)$ is *uniquely* expressed as the coset of either the zero polynomial or of a polynomial of degree $< n$.

Let's study the ring of Gaussian integers $\mathbb{Z}[i]$. For any $z = a + bi \in \mathbb{C}$ where a and $b \in \mathbb{R}$, define the **conjugate** of z to be $\overline{z} = a - bi$ and the **norm** of z to be

$$N(z) = z\overline{z} = (a + bi)(a - bi) = a^2 + b^2.$$

Note that the norm of a Gaussian integer is always a nonnegative integer. Moreover, for z and $w \in \mathbb{C}$, we have the identity

$$N(zw) = (zw)\overline{zw} = z\overline{z}w\overline{w} = N(z)N(w).$$

Lemma 14.2.4. *For the ring of Gaussian integers, we have*

$$U(\mathbb{Z}[i]) = \{\pm 1, \pm i\}.$$

Proof. If $z = a + bi$, $a, b \in \mathbb{Z}$, is a unit, then there exists $w \in \mathbb{Z}[i]$ such that $zw = 1$. It follows that $N(z)N(w) = 1$ and thus $N(z) = 1$. Solving $a^2 + b^2 = 1$ for $a, b \in \mathbb{Z}$, we find that the only possible units in $\mathbb{Z}[i]$ are ± 1 and $\pm i$. Obviously, these four elements are indeed units of $\mathbb{Z}[i]$. $\square$

Example 14.2.5. How many elements are there in $\mathbb{Z}[i]/(2-i)$?

Solution. Let $R = \mathbb{Z}[i]/(2-i)$. Clearly, $\overline{2-i} = \overline{0}$ and thus $\overline{i} = \overline{2}$. Hence for any $a, b \in \mathbb{Z}$, $\overline{a+bi} = \overline{a+2b}$. Any element of R can be represented by an integer. Note that $5 = (2+i)(2-i) \in (2-i)$. Hence $\overline{5} = \overline{0}$ and $R = \{\overline{0}, \overline{1}, \overline{2}, \overline{3}, \overline{4}\}$. With a little work, one can check that these 5 elements are distinct from each other. Hence R contains exactly 5 elements.   ◇

For the rest of this section, we assume that R is a *ring with identity*. In the ring R, we have the following two special types of proper ideals.

(1) The proper ideal P is called a **prime** ideal of R if

$$ab \in P \implies a \in P \text{ or } b \in P.$$

(2) The proper ideal M is called a **maximal** ideal of R if for any ideal N of R such that $M \subseteq N \subseteq R$ we have $N = M$ or $N = R$.

We point out the relations between these special ideals and their quotient rings.

Proposition 14.2.6. *Let R be a ring with identity and P be a two-sided ideal of R. Then P is a prime ideal if and only if R/P is a domain.*

Proof. Assume P is a prime ideal. Let $\overline{a}$ and $\overline{b}$ be nonzero elements of R/P. Then neither a nor b is in P. It follows that $ab \notin P$. Hence $\overline{ab} \neq \overline{0}$. This proves that R/P is a domain.

Conversely, assume R/P is a domain. Let $ab \in P$. Then $\overline{ab} = \overline{0}$. Since R/P contains no zero divisors, $\overline{a}$ or $\overline{b} = \overline{0}$. Hence $a \in P$ or $b \in P$. This proves that P is a prime ideal.   □

Proposition 14.2.7. *Let R be a commutative ring with identity. Then M is a maximal ideal of R if and only if R/M is a field.*

Proof. Assume M is a maximal ideal and let $\overline{a}$ be a nonzero element in R/M. Then $a \notin M$. Consider the ideal $aR + M$. It is an ideal containing a and M and thus properly contains M. It implies that $aR + M = R$. In particular, $1 = ab + m$ for some $b \in R$ and $m \in M$. Therefore, $\overline{ab} = \overline{1}$. It follows that $\overline{a}$ is invertible and hence R/M is a field.

Conversely, assume R/M is a field. Let N be an ideal of R properly containing M. Find $a \in N \setminus M$. Since $\bar{a} \neq \bar{0}$ in the field R/M, there exists $b \in R \setminus M$ such that $\bar{a}\bar{b} = \bar{1}$. Hence

$$1 - ab \in M \subseteq N \quad \text{and} \quad 1 = ab + (1 - ab) \in N.$$

Thus, $N = R$. This proves that M is maximal. $\qquad\qquad\qquad\square$

Corollary 14.2.8. *In a commutative ring with identity, a maximal ideal is always a prime ideal.*

Proof. Let M be a maximal ideal of R. Then R/M is a field. It follows that R/M is an integral domain. Hence M is a prime ideal. $\qquad\square$

Proposition 14.2.9. *Let n be a nonzero integer. Then n is a prime integer if and only if $n\mathbb{Z}$ is a prime ideal if and only if $n\mathbb{Z}$ is a maximal ideal.*

Proof. By Theorem 14.1.7, an ideal of $\mathbb{Z}$ is of the form $n\mathbb{Z}$. Since $n\mathbb{Z} = (-n)\mathbb{Z}$, we may assume that an ideal of $\mathbb{Z}$ is of the form $n\mathbb{Z}$ for $n \geq 0$. Now assume $n > 0$. By Proposition 13.1.5, $\mathbb{Z}_n$ is a field if and only if it is an integral domain if and only if n is a prime integer. Hence, $n\mathbb{Z}$ is a maximal ideal of $\mathbb{Z}$ if and only if $n\mathbb{Z}$ is a prime ideal if and only if n is a prime integer by Propositions 14.2.6 and 14.2.7. $\qquad\qquad\qquad\square$

Example 14.2.10. In $\mathbb{Z}$, the trivial ideal (0) is a prime ideal but not a maximal ideal since $\mathbb{Z}/(0) = \mathbb{Z}$ is an integral domain but not a field. We conclude that the prime ideals of $\mathbb{Z}$ are (0) and $p\mathbb{Z}$ with p a prime integer. The maximal ideals of $\mathbb{Z}$ are $p\mathbb{Z}$ with p a prime integer.

Next, let's study the ring of Gaussian integers more closely.

Example 14.2.11. Let $M = 3\mathbb{Z}[i] = \{3a + 3bi \in \mathbb{C} : a, b \in \mathbb{Z}\}$. We first check that M is a maximal ideal of $\mathbb{Z}[i]$.

Suppose that $M \subseteq N \subseteq \mathbb{Z}[i]$ and $N \neq M$ is an ideal of $\mathbb{Z}[i]$. Then there exists $z = r + si \in N \setminus M$ where r and $s \in \mathbb{Z}$. Since $3 \nmid r$ or $3 \nmid s$, we have that modulo 3, r^2 and s^2 are congruent to 0 or 1 but not both to 0 at the same time. It follows that $r^2 + s^2 \not\equiv 0 \pmod 3$. Thus, $3 \nmid (r^2 + s^2)$. Let $t = r^2 + s^2 = z\bar{z} \in N$. Find two integers a and b such that $ta + 3b = 1$. This implies that $1 \in N$ and $N = \mathbb{Z}[i]$. Hence M is a maximal ideal.

On the other hand, $I = 5\mathbb{Z}[i] = \{a + bi : a, b \in \mathbb{Z},\ 5\,|\,a \text{ and } 5\,|\,b\}$ is not a maximal ideal of $\mathbb{Z}[i]$. Observe that

$$5\mathbb{Z}[i] \subseteq (1 + 2i)\mathbb{Z}[i] \subseteq \mathbb{Z}[i].$$

If $5\mathbb{Z}[i] = (1 + 2i)\mathbb{Z}[i]$, then $1 + 2i \in 5\mathbb{Z}[i]$. This would imply that

$$\frac{1 + 2i}{5} = \frac{1}{5} + \frac{2}{5}i \in \mathbb{Z}[i],$$

a contradiction. If $(1 + 2i)\mathbb{Z}[i] = \mathbb{Z}[i]$, then $1 \in (1 + 2i)\mathbb{Z}[i]$. This means that $1 = (1+2i)(a+bi)$ for some a and $b \in \mathbb{Z}$. It follows that $1+2i$ is a unit in $\mathbb{Z}[i]$, a contradiction by Lemma 14.2.4. Thus, $5\mathbb{Z}[i]$ is not a maximal ideal. Moreover, $5\mathbb{Z}[i]$ is neither a prime ideal. Note that $5 = (1+2i)(1-2i)$ but neither $1 + 2i$ nor $1 - 2i$ is in $5\mathbb{Z}[i]$.

Exercises 14.2

1. In a commutative ring with identity, show that $I = aR$ is the unit ideal if and only if a is a unit.

2. Show that the ideal I in Exercise 2, §14.1 is a maximal ideal.

3. Show that $\mathbb{Z}[i]/(1 - i)$ is a field. Hence $(1 - i)$ is a maximal (and prime) ideal in $\mathbb{Z}[i]$.

4. Let $M = 7\mathbb{Z}[i]$. Show that M is a maximal ideal of $\mathbb{Z}[i]$ and $\mathbb{Z}[i]/M$ is a field with 49 elements.

5. Let R be a commutative ring with identity and let $I \subseteq J$ be ideals of R.

   (a) Show that R/I is also commutative with identity.

   (b) Show that J/I is an ideal of R/I.

   (c) Show that J/I is a principal ideal of R/I if J is a principal ideal of R.

6. Let R be a commutative ring with identity. We say an element p is a **prime** element in R if (i) it is nonzero and non-unit, and (ii) $p \mid ab \Rightarrow p \mid a$ or $p \mid b$ for $a, b \in R$.

   Show that a nontrivial non-unit principal ideal (p) is a prime ideal if and only if p is a prime element.

Review Exercises for Chapter 14

1. Let I be an ideal of $\mathbb{Z}$. Show that

   $$I[x] = \left\{ a_n x^n + a_{n-1} x^{n-1} + \cdots + a_1 x + a_0 \in \mathbb{Z}[x] : a_i \in I \text{ for all } i \right\}$$

   is an ideal of $\mathbb{Z}[x]$.

2. Let I, J and K be ideals of a commutative ring R.

   (a) Show that $\{ij \in R : i \in I, \ J \in J\}$ is not an ideal in R by giving an counterexample. When we want to define the **product** of the ideals I and J, we define IJ to be the ideal *generated* by $\{ij \in R : i \in I, \ J \in J\}$. (This usage of notation is different from the product of groups.)

   (b) Suppose $I = (S)$ and $J = (T)$. Show that IJ is generated by $\{ st \in R : s \in S, \ t \in T \}$.

   (c) Does the distributive law $I(J + K) = IJ + IK$ hold?

   (d) Show that $IJ \subseteq I \cap J$.

   (e) Show that $IJ = I \cap J$ does not hold in general by giving a counterexample.

3. Let B be the set of 2×2 upper triangular matrices over the ring of integers, i.e.,

   $$B = \left\{ \begin{bmatrix} a & b \\ 0 & c \end{bmatrix} \ \middle|\ a, b, c \in \mathbb{Z} \right\}.$$

   Verify that B is a subring of $M_2(\mathbb{Z})$ and determine the ideals of B.

4. A ring which contains no nonzero nilpotent elements is called a **reduced** ring. On the other hand, we say an ideal I is an **radical** ideal if $a^n \in I$ implies that $a \in I$. Show that an ideal I of R is radical if and only if R/I is reduced.

5. Show that any domain is reduced. Show that any prime ideal is radical.

6. Let I be an ideal of a commutative ring R. Show that

$$\sqrt{I} = \{\, a \in R : a^n \in I \,\}$$

is a radical ideal of R containing I.

7. Let R be a commutative ring. Show that $\sqrt{(0)}$ is the ideal consisting of nilpotent elements in R.

8. In $\mathbb{Z}$, find $\sqrt{(360)}$.

9. In $\mathbb{Z}[x]$ find $\sqrt{(x^3 + x^2 - x - 1)}$.

10. Let $I_1 \subseteq I_2 \subseteq I_3 \subseteq \cdots$ be a possibly infinite sequence of ideals in a ring R. Show that $\bigcup_i I_i$ is an ideal of R.

CHAPTER 15

Ring Homomorphisms

Ring homomorphisms are maps which preserve ring structures. We will see that the fundamental theorem of homomorphisms and the three isomorphism Theorems for groups can be adapted to work for rings. As an application we will generalize the famous Chinese remainder theorem.

An integral domain can be naturally embedded into its field of fractions. This means that we can often solve a problem on integral domains by translating it into a problem on fields. If we are fortunate, this can turn an algebraic problem into a problem in Linear Algebra.

We will also discuss the concepts of prime rings and prime fields. These are the minimal rings with identity and minimal fields.

15.1   Ring Homomorphisms

Definition 15.1.1. Let R and R' be rings. A mapping $\varphi \colon R \to R'$ is a **ring homomorphism** if it preserves the operations of R, *i.e.*, for all a, $b \in R$ we have

(i) $\varphi(a + b) = \varphi(a) + \varphi(b)$, and

(ii) $\varphi(ab) = \varphi(a)\varphi(b)$.

Note that since φ is also a group homomorphism from $(R, +)$ to $(R', +)$, we have the additional properties

(a) $\varphi(0) = 0$, and

(b) $\varphi(-a) = -\varphi(a)$ for all $a \in R$.

When constructing ring homomorphisms between rings with identities, it often requires that the unity also be preserved. In this case we also have the following implication:

(c) $\varphi(1) = 1 \implies \varphi(u^{-1}) = \varphi(u)^{-1}$ for all units u of R.

Proposition 15.1.2. *If $\varphi \colon R \to R'$ is a ring homomorphism, then both*

$$\operatorname{Im} \varphi = \{\,\varphi(r) : r \in R\,\} \quad \text{and} \quad \ker \varphi = \{\,r \in R : \varphi(r) = 0\,\}$$

are subrings of R' and R, respectively. Furthermore, $\ker \varphi$ is a two-sided ideal of R.

Proof. We leave it to the reader to verify that $\operatorname{Im} \varphi$ is a subring of R' and $\ker \varphi$ is a subring of R. Suppose $k \in \ker \varphi$ and $r \in R$. Then

$$\varphi(rk) = \varphi(r)\varphi(k) = \varphi(r)0 = 0 \quad \text{and} \quad \varphi(kr) = \varphi(k)\varphi(r) = 0\varphi(r) = 0.$$

Thus both rk and $kr \in \ker \varphi$. The kernel is a two-sided ideal. $\square$

A surjective ring homomorphism is called a **ring epimorphism** and an injective ring homomorphism is called a **ring monomorphism**. Finally, a bijective ring homomorphism is called a **ring isomorphism**. If $\varphi \colon R \to S$ is a ring epimorphism, we say S is a **homomorphic image** of R.

Regarding ring homomorphisms we have the following properties which are straightforward to check.

(1) The ring homomorphism $\varphi \colon R \to S$ is onto if $\mathrm{Im}\,\varphi = S$, and is one-to-one if $\ker \varphi = \{\,0\,\}$.

(2) The identity function on a ring R is a ring isomorphism.

(3) The composite of ring monomorphisms (epimorphisms or isomorphisms, respectively) is still a ring monomorphism (epimorphism or isomorphism, respectively).

(4) The inverse of a ring isomorphism is a ring isomorphism.

(5) We say the ring R is **isomorphic to** the ring S, denoted $R \cong S$, if there exists a ring isomorphism from R onto S. The relation "$\cong$" is an equivalence relation.

Example 15.1.3. Let n be a positive integer. Define $\varphi \colon \mathbb{Z} \to \mathbb{Z}_n$ by letting $\varphi(m) = \overline{m}$ for each $m \in \mathbb{Z}$. Then φ is a ring epimorphism and $\ker \varphi = n\mathbb{Z}$.

Example 15.1.4. Let $\mathbb{Q}[x]$ be the polynomial ring in the variable x with rational coefficients. Define $\varphi_0 \colon \mathbb{Q}[x] \to \mathbb{C}$ to be the evaluation mapping $\varphi_0[f(x)] = f(0) = $ the constant term of $f(x)$. Then φ_0 is a ring homomorphism from $\mathbb{Q}[x]$ to $\mathbb{C}$, and the kernel of φ_0 is given by

$$\ker \varphi_0 = \{\, f(x) \in \mathbb{Q}[x] : \text{the constant term of } f(x) \text{ is } 0 \,\} = x\mathbb{Q}[x].$$

In general, the evaluation mapping $\varphi_a[f(x)] = f(a)$, $a \in \mathbb{Q}$, is a ring homomorphism and the kernel of φ_a contains $(x-a)\mathbb{Q}[x]$. Let $g(x) \in \mathbb{Q}[x]$. By the division algorithm, there exist $q(x) \in \mathbb{Q}[x]$ and $r \in \mathbb{Q}$ such that

$$g(x) = (x - a)q(x) + r.$$

If $g(a) = 0$, then $r = 0$ and $g(x)$ is a multiple of $x - a$. Hence the kernel of φ_a is $(x - a)\mathbb{Q}[x]$.

Example 15.1.5. Let $C[0,1]$ be the ring of all real-valued continuous functions. The map $\sigma \colon C[0,1] \to \mathbb{R}$ defined by

$$\sigma(f) = \int_0^1 f(x)\, dx$$

is a group homomorphism for the addition since

$$\sigma(f + g) = \int_0^1 (f(x) + g(x))\, dx = \int_0^1 f(x)\, dx + \int_0^1 g(x)\, dx = \sigma(f) + \sigma(g).$$

However, it is not a ring homomorphism because $\sigma(fg) \neq \sigma(f)\sigma(g)$ in general.

Example 15.1.6. Let $R = \mathbb{Q}[\sqrt{2}] = \{\, a + b\sqrt{2} \mid a, b \in \mathbb{Q} \,\}$. Define the function $\varphi \colon R \to R$ by letting $\varphi(a + b\sqrt{2}) = a - b\sqrt{2}$. Then

$$\varphi((a + b\sqrt{2}) + (c + d\sqrt{2})) = \varphi((a + c) + (b + d)\sqrt{2})$$
$$= (a + c) - (b + d)\sqrt{2} = (a - b\sqrt{2}) + (c - d\sqrt{2})$$
$$= \varphi(a + b\sqrt{2}) + \varphi(c + d\sqrt{2})$$

and

$$\varphi((a + b\sqrt{2})(c + d\sqrt{2})) = \varphi((ac + 2bd) + (bc + ad)\sqrt{2})$$
$$= (ac + 2bd) - (bc + ad)\sqrt{2} = (a - b\sqrt{2})(c - d\sqrt{2})$$
$$= \varphi(a + b\sqrt{2})\varphi(c + d\sqrt{2}).$$

Thus φ is a ring homomorphism from R into itself. We leave it to the reader to verify that φ is a ring isomorphism.

Example 15.1.7. The mapping $\det \colon M_n(\mathbb{R}) \to \mathbb{R}$ satisfies

$$\det(AB) = (\det A)(\det B)$$

for all A, B in $M_n(\mathbb{R})$. However, it is not a ring homomorphism of $M_n(\mathbb{R})$ to $\mathbb{R}$ since $\det(A + B) \neq \det A + \det B$ in general.

Let I be an ideal of R. We have the **canonical epimorphism** $\pi \colon R \twoheadrightarrow R/I$ sending r to $r + I$. This is indeed a group epimorphism if we treat R and R/I as additive groups. By (14.2.1), we can see that π is a ring homomorphism as well. Hence π is onto with kernel I.

Let $\varphi \colon R \to R'$ be a ring homomorphism and let I be an ideal contained in $\ker \varphi$. Since I is an additive subgroup of R, φ induces a group homomorphism $\overline{\varphi} \colon R/I \to R'$ as additive groups such that $\varphi = \overline{\varphi}\pi$. Remember that $\overline{\varphi}(\overline{a}) = \varphi(a)$ for $a \in R$. Since

$$\overline{\varphi}(\overline{a}\,\overline{b}) = \overline{\varphi}(\overline{ab}) = \varphi(ab) = \varphi(a)\varphi(b) = \overline{\varphi}(\overline{a})\overline{\varphi}(\overline{b})$$

for all $a, b \in R$, we see that $\overline{\varphi}$ is also a ring homomorphism. Hence we have the following extremely important result.

Theorem 15.1.8 (Fundamental Theorem of Ring Homomorphisms). *Let* $\varphi: R \to R'$ *be a ring homomorphism. Let I be a two-sided ideal of R contained in* $\ker \varphi$. *There is a unique ring homomorphism* $\overline{\varphi}: R/I \to R'$ *such that* $\varphi = \overline{\varphi}\pi$ *where π is the canonical epimorphism* $R \twoheadrightarrow R/I$.

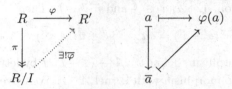

Moreover, if $I = \ker \varphi$, then $\overline{\varphi}$ is a monomorphism.

Proof. We have already established the existence of $\overline{\varphi}$. The rest follows from the fundamental theorem of group homomorphisms. □

Using similar reasoning as in the group case, we have the following theorems.

Theorem 15.1.9 (Correspondence Theorem). *Let I be a two-sided ideal of R and let $\pi: R \twoheadrightarrow R/I$ be the canonical epimorphism. Let $\mathscr{A}$ be the family of ideals of R containing I and let $\mathscr{B}$ be the family of ideals of R/I. There is an* order-preserving *one-to-one correspondence*

$$\Phi: \quad \mathscr{A} \quad \longrightarrow \quad \mathscr{B}$$
$$J \quad \longmapsto \quad J/I.$$

The inverse of Φ is given by

$$\Psi: \quad \mathscr{B} \quad \longrightarrow \quad \mathscr{A}$$
$$\mathfrak{J} \quad \longmapsto \quad \pi^{-1}(\mathfrak{J}).$$

The map Φ also induces an order-preserving one-to-one correspondence between the prime (or maximal) ideals of R containing I and the prime (or maximal) ideals of R/I.

Theorem 15.1.10 (First Isomorphism Theorem). *Let $\varphi: R \to R'$ be a ring epimorphism. Then*

$$\frac{R}{\ker \varphi} \cong R'.$$

Theorem 15.1.11 (Second Isomorphism Theorem). *Let A be a subring of a ring R and I be a two-sided ideal of R. Then $A+I = \{\, a+i \mid a \in A, i \in I \,\}$ is a subring of R, I is a two-sided ideal of $A+I$, $A \cap I$ is a two-sided ideal of A, and*

$$\frac{A+I}{I} \cong \frac{A}{A \cap I}.$$

Proof. To show that $A + I$ is a subring of R, notice that $A + I$ is closed under multiplication. Let $a,\, a' \in A$ and $i,\, i' \in I$. Then

$$(a+i)(a'+i') = aa' + (ia' + ai' + ii') \in A+I.$$

For the final isomorphism, define $\varphi \colon A \to (A+I)/I$ by letting $\varphi(a) = a+I$. Then φ is a ring epimorphism with kernel $A \cap I$. We leave it as an exercise for the reader to complete the proof. $\qquad\square$

Theorem 15.1.12 (Third Isomorphism Theorem). *Suppose that R is a ring, and that $J \subseteq I$ are two-sided ideals of R. Then*

$$\frac{R}{I} \cong \frac{R/J}{I/J}.$$

Now let's look at a few applications.

Example 15.1.13. Let $R = C[0,1]$, the ring of continuous functions from $[0,1]$ to $\mathbb{R}$. We may use $\alpha \in \mathbb{R}$ to denote the constant function which sends every element of $[0,1]$ to α. Define $\varphi \colon R \to \mathbb{R}$ by letting $f(x) \mapsto f(1/2)$. Then φ is a ring homomorphism and

$$\ker \varphi = \left\{\, f \in R \;\middle|\; f\left(\frac{1}{2}\right) = 0 \,\right\} \quad \text{and} \quad \operatorname{Im} \varphi = \mathbb{R}.$$

Note that $R/\ker\varphi \cong \mathbb{R}$. Since $f(x) - f(1/2) \in \ker\varphi$, every element $f(x) + \ker\varphi$ in $R/\ker\varphi$ can be written as $f(1/2) + \ker\varphi$.

Example 15.1.14. Let

$$S = \left\{\, \begin{bmatrix} a & b \\ 0 & a \end{bmatrix} \;\middle|\; a,b \in \mathbb{R} \,\right\}.$$

Then S is a subring of $M_2(\mathbb{R})$. Define a map φ from S to $\mathbb{R}$ sending $\begin{bmatrix} a & b \\ 0 & a \end{bmatrix}$ to a. Since

$$\varphi\left(\begin{bmatrix} a & b \\ 0 & a \end{bmatrix} + \begin{bmatrix} a' & b' \\ 0 & a' \end{bmatrix} \right) = \varphi\left(\begin{bmatrix} a+a' & b+b' \\ 0 & a+a' \end{bmatrix} \right)$$

$$= a + a' = \varphi \left(\begin{bmatrix} a & b \\ 0 & a \end{bmatrix} \right) + \varphi \left(\begin{bmatrix} a' & b' \\ 0 & a' \end{bmatrix} \right)$$

and

$$\varphi \left(\begin{bmatrix} a & b \\ 0 & a \end{bmatrix} \begin{bmatrix} a' & b' \\ 0 & a' \end{bmatrix} \right) = \varphi \left(\begin{bmatrix} aa' & ab' + ba' \\ 0 & aa' \end{bmatrix} \right) = aa'$$

$$= \varphi \left(\begin{bmatrix} a & b \\ 0 & a \end{bmatrix} \right) \varphi \left(\begin{bmatrix} a' & b' \\ 0 & a' \end{bmatrix} \right),$$

we have that φ is a ring homomorphism which is clearly onto. The kernel of φ is

$$I = \left\{ \begin{bmatrix} 0 & b \\ 0 & 0 \end{bmatrix} \;\middle|\; b \in \mathbb{R} \right\}$$

and $S/I \cong \mathbb{R}$.

Exercises 15.1

1. Complete the proofs of Theorems 15.1.9–15.1.12.

2. Find a polynomial $f(x)$ in $\mathbb{Q}[\,x\,]$ such that $\mathbb{Q}[\sqrt{2}\,] \cong \mathbb{Q}[\,x\,]/(f(x))$.

3. Find all the ideals of $\mathbb{Z}/36\mathbb{Z}$. Find also the prime and maximal ideals in $\mathbb{Z}/36\mathbb{Z}$.

4. Show that every nontrivial ring homomorphism from a division ring into a ring R is a monomorphism.

5. Show that $I = (2+i)\mathbb{Z}[\,i\,]$ is a maximal ideal of $\mathbb{Z}[\,i\,]$ and $\mathbb{Z}[i]/I \cong \mathbb{Z}_5$.

15.2   Direct Products of Rings

Let $R_1, \ldots, R_n$ be n rings. Remember that the direct product $R = R_1 \times \cdots \times R_n$ is an additive group. Define

$$(a_1, \ldots, a_n)(b_1, \ldots, b_n) = (a_1 b_1, \ldots, a_n b_n)$$

where a_i, $b_i \in R$ for all i. It is easy to show that $(R, +, \cdot)$ is a ring and we call it the **direct product** of $R_1, \ldots, R_n$.

Below we make a few remarks on the product rings $R = R_1 \times \cdots \times R_n$.

(1) If all the R_i's are rings with identity, then R is also a ring with identity $(1_{R_1}, 1_{R_2}, \ldots, 1_{R_n})$.

(2) If all the R_i's are commutative then so is R.

(3) If all the R_i's are rings with identity, then

$$U(R) = U(R_1) \times \cdots \times U(R_n).$$

An element $(a_1, \ldots, a_n)$ is a unit in R if and only if there is some $(b_1, \ldots, b_n) \in R$ such that

$$(a_1, \ldots, a_n)(b_1, \ldots, b_n) = (b_1, \ldots, b_n)(a_1, \ldots, a_n) = (1, \ldots, 1).$$

This means that $a_i b_i = b_i a_i = 1$ and a_i is a unit of R_i for each i.

(4) If all the R_i's are rings with identity then R is never a domain for $n \geq 2$ even when all the R_i's are domains. For example, the relation

$$(1, 0, \ldots, 0)(0, 1, 0, \ldots, 0) = (0, \ldots, 0)$$

gives us two zero divisors of R.

(5) Let S be another ring. For each i, let φ_i be a ring homomorphism from S to R_i. The map from S to R sending $s \in S$ to $(\varphi_1(s), \ldots, \varphi_n(s))$ is a ring homomorphism.

In a commutative ring with identity, we say two ideals I and J are **comaximal** in R if $I + J = R$. The following result is a generalized version of the Chinese remainder theorem.

Theorem 15.2.1 (Chinese Remainder Theorem). *Let I and J be comaximal ideals in a commutative ring with identity R. Then $I \cap J = IJ$ and the map*

$$\varphi: \quad \frac{R}{IJ} \quad \longrightarrow \quad \frac{R}{I} \times \frac{R}{J}$$
$$r + IJ \quad \longmapsto \quad (r + I, \ r + J)$$

is a ring isomorphism.

Proof. Recall that $IJ \subseteq I \cap J$ (see Exercise 2(d), P. 219). Now let $a \in I \cap J$. Find $i \in I$ and $j \in J$ such that $1 = i + j$. Then $a = a(i + j) = ai + aj$. Since $ai = ia$ and aj are both in IJ, we have that $a \in IJ$. Thus we also have $I \cap J \subseteq IJ$.

To prove the second part, note that we have a ring homomorphism $\widetilde{\varphi}$ from R to $R/I \times R/J$ sending r to $(r + I, \ r + J)$. Clearly $r \in \ker \widetilde{\varphi}$ if and only if $r + I = I$ and $r + J = J$ if and only if $r \in I \cap J$. Thus $\ker \widetilde{\varphi} = I \cap J = IJ$. This induces a ring monomorphism φ from $R/(IJ)$ to $R/I \times R/J$ as stated in the statement. It remains to show that φ is onto. Let $(r + I, \ r' + J) \in R/I \times R/J$. Let $1 = i + j$ where $i \in I$ and $j \in J$. Then

$$\begin{cases} rj + r'i + I = rj + I = rj + ri + I = r + I; \\ rj + r'i + J = r'i + J = r'i + r'j + J = r' + J. \end{cases}$$

Hence $\varphi(rj + r'i + I \cap J) = (r + I, \ r' + J)$ and φ is onto. $\qquad\square$

The classical Chinese Remainder Theorem is a special case of this theorem.

Lemma 15.2.2. *Two integers m and n are relatively prime if and only if $m\mathbb{Z}$ and $n\mathbb{Z}$ are comaximal.*

Proof. Two integers are relatively prime if and only $1 = ma + nb$ for some integers a and b if and only if $m\mathbb{Z} + n\mathbb{Z} = (1) = \mathbb{Z}$ if and only if $m\mathbb{Z}$ and $n\mathbb{Z}$ are comaximal in $\mathbb{Z}$. $\qquad\square$

Corollary 15.2.3 (Chinese Remainder Theorem). *Let m and n be relatively prime integers. Then the map*

$$\varphi: \quad \begin{array}{ccc} \dfrac{\mathbb{Z}}{mn\mathbb{Z}} & \longrightarrow & \dfrac{\mathbb{Z}}{m\mathbb{Z}} \times \dfrac{\mathbb{Z}}{n\mathbb{Z}} \\ k + mn\mathbb{Z} & \longmapsto & (k + m\mathbb{Z}, \ k + n\mathbb{Z}) \end{array}$$

is a ring isomorphism.

Proof. Since $mn\mathbb{Z} = (m\mathbb{Z})(n\mathbb{Z})$, the result follows from Lemma 15.2.2, and Theorem 15.2.1. $\qquad\square$

Corollary 15.2.4. *Let* $n = p_1^{\alpha_1} p_2^{\alpha_2} \cdots p_s^{\alpha_s}$ *be a product of powers of distinct prime integers. Then*

$$\varphi: \quad \begin{array}{ccc} \dfrac{\mathbb{Z}}{n\mathbb{Z}} & \longrightarrow & \dfrac{\mathbb{Z}}{p_1^{\alpha_1}\mathbb{Z}} \times \cdots \times \dfrac{\mathbb{Z}}{p_s^{\alpha_s}\mathbb{Z}} \\[2mm] k + n\mathbb{Z} & \longmapsto & (k + p_1^{\alpha_1}\mathbb{Z}, \ldots, k + p_1^{\alpha_s}\mathbb{Z}) \end{array}$$

is a ring isomorphism.

Proof. Since $p_1^{\alpha_1}$ is relatively prime with $p_2^{\alpha_2} \cdots p_s^{\alpha_s}$, we have that

$$\varphi: \quad \begin{array}{ccc} \dfrac{\mathbb{Z}}{n\mathbb{Z}} & \longrightarrow & \dfrac{\mathbb{Z}}{p_1^{\alpha_1}\mathbb{Z}} \times \dfrac{\mathbb{Z}}{p_2^{\alpha_2} \cdots p_s^{\alpha_s}\mathbb{Z}} \\[2mm] k + n\mathbb{Z} & \longmapsto & (k + p_1^{\alpha_1}\mathbb{Z}, \ k + p_2^{\alpha_2} \cdots p_s^{\alpha_s}\mathbb{Z}) \end{array}$$

is an isomorphism. The rest follows from induction on s. $\qquad\square$

As an application of Chinese remainder theorem, we have the formula for the well-known Euler φ-function (see the discussions on PP. 24–25).

Corollary 15.2.5. *Let* $n = p_1^{\alpha_1} p_2^{\alpha_2} \cdots p_s^{\alpha_s}$ *be a product of powers of distinct positive prime integers. Then*

$$\varphi(n) = n \prod_{i=1}^{s} \left(1 - \frac{1}{p_i}\right).$$

Proof. It is easy to see that $\varphi(p_i^{\alpha_i}) = p_i^{\alpha_i} - p_i^{\alpha_i - 1}$ since among the $p_i^{\alpha_i}$ integers in between and including 1 and $p_i^{\alpha_i}$ there are exactly $p_i^{\alpha_i - 1}$ multiples of p. We have that

$$\varphi(n) = |U(\mathbb{Z}_n)| = |U(\mathbb{Z}_{p_1^{\alpha_1}})| \times \cdots \times |U(\mathbb{Z}_{p_s^{\alpha_s}})| = \varphi(p_1^{\alpha_1})\varphi(p_2^{\alpha_2}) \cdots \varphi(p_s^{\alpha_s})$$

by Corollary 15.2.4 and the third remark on P. 228. $\qquad\square$

Example 15.2.6. Solve the quadratic congruence equation

(15.2.1) $x^2 + x + 7 \equiv 0 \pmod{189}$.

Solution. First make the computation $189 = 27 \cdot 7$. Find the solutions for

$$x^2 + x + 7 \equiv 0 \pmod{27} \quad \text{and} \quad x^2 + x + 7 \equiv 0 \pmod{7}$$

by trial and error. We will find the solutions are

$$\begin{cases} x \equiv 4, \ 13, \ -5, & (\text{mod } 27), \\ x \equiv 0, \ -1, & (\text{mod } 7). \end{cases}$$

Next find

$$4 \cdot 7 + 27 \cdot (-1) = 28 - 27 = 1.$$

By the proof in Theorem 15.2.1 we see that the solutions to (15.2.1) are the integers which are congruent to

$$28 \cdot 4 - 27 \cdot 0 = 112,$$
$$28 \cdot 13 - 27 \cdot 0 = 364 \equiv 175,$$
$$28 \cdot (-5) - 27 \cdot 0 = -140 \equiv 49,$$
$$28 \cdot 4 - 27 \cdot (-1) = 139,$$
$$28 \cdot 13 - 27 \cdot (-1) = 391 \equiv 13,$$
$$28 \cdot (-5) - 27 \cdot (-1) = -113 \equiv 76$$

modulo 189. ◇

Exercises 15.2

1. Let I_i be an ideal of the ring R_i for $i = 1, \ldots, n$. Show that $I_1 \times \cdots \times I_n$ is an ideal of $R_1 \times \cdots \times R_n$.

2. Let R_i be a ring with identity for each i. Let I be an ideal of $R_1 \times \cdots \times R_n$. Show that $I = I_1 \times \cdots \times I_n$ where I_i is an ideal of R_i for each i.

3. Let R_i be commutative rings with identity for each i. Let $I = I_1 \times \cdots \times I_n$ be an ideal of $R_1 \times \cdots \times R_n$.

   (a) Show that $\dfrac{R_1 \times \cdots \times R_n}{I_1 \times \cdots \times I_n} \cong \dfrac{R_1}{I_1} \times \cdots \times \dfrac{R_n}{I_n}$.

   (b) Show that I is a prime ideal if and only if there exists i with $1 \leq i \leq n$ such that I_i is prime and $I_k = R_k$ for all $k \neq i$.

(c) Show that I is a maximal ideal if and only if there exists i with $1 \leq i \leq n$ such that I_i is maximal and $I_k = R_k$ for all $k \neq i$.

4. Find all prime ideals and maximal ideals of $\mathbb{Z}_4 \times \mathbb{Z}_6$.

15.3   The Quotient Field of an Integral Domain

Recall that an integral domain is a commutative ring with identity which is without any zero divisors. The construction of rational numbers from integers is not an isolated phenomenon. This can be done to any integral domain.

Let D be an integral domain and D^* be the set of its nonzero elements. Define a relation $\sim$ on $D \times D^*$ by

$$(a, b) \sim (c, d) \quad \text{if and only if} \quad ad = bc.$$

We check that $\sim$ is an equivalence relation on $D \times D^*$.

(i) Reflexivity: $(a, b) \sim (a, b)$ since $ab = ba$.

(ii) Symmetry: if $(a, b) \sim (c, d)$, then $ad = bc$. This implies $(c, d) \sim (a, b)$ since $cb = da$.

(iii) Transitivity: if $(a, b) \sim (c, d)$ and $(c, d) \sim (e, f)$, then $ad = bc$ and $cf = de$. It follows

$$adf = bcf = bde.$$

Since $d \neq 0$, we get $af = be$. Hence $(a, b) \sim (e, f)$.

Let Q be the set of all equivalence classes

$$[a, b] = \{ (u, v) \in D \times D^* : (u, v) \sim (a, b) \}$$

where $(a, b) \in D \times D^*$. Define the addition and multiplication in Q as

$$[a, b] + [c, d] = [ad + bc, bd] \quad \text{and} \quad [a, b] \cdot [c, d] = [ac, bd].$$

Note that since D is an integral domain, we do have that $bd \neq 0$.

Lemma 15.3.1. *The structure $(Q, +)$ described as above is an abelian group in which the additive identity is $[0,1]$ and the additive inverse of $[a, b]$ is $[-a, b]$.*

Proof. Indeed, it is routine to prove this lemma provided that the operation $+$ is well-defined. Suppose that $[a, b] = [a', b']$ and $[c, d] = [c', d']$, we want to prove that

$$[ad + bc, bd] = [a'd' + b'c', b'd'].$$

From $ab' = ba'$ and $cd' = dc'$, we have

$$(ad + bc)b'd' = ab'dd' + bb'cd' = ba'dd' + bb'dc' = bd(a'd' + b'c').$$

Our assertion hence follows. The rest is trivial. $\qquad\square$

Lemma 15.3.2. *The element $[a, b] \neq 0$ in Q if and only if $a \neq 0$.*

Proof. We have that $[a, b] = [0, 1]$ if and only if $a = a \cdot 1 = b \cdot 0 = 0$. $\qquad\square$

We leave the following result as an exercise (see Exercise 1).

Lemma 15.3.3. *Let Q^* be the set of nonzero elements in Q. Then $(Q^*, \cdot)$ is an abelian group in which the identity is $[1, 1]$ and the multiplicative inverse of $[a, b]$ is $[b, a]$.*

To finish proving that Q is a field, we need to verify that the distributive law is satisfied. Note that

$$([a, b] + [c, d])[e, f] = [ade + bce, bdf].$$

Since $bdf \neq 0$ we also have

$$[a, b][e, f] + [c, d][e, f] = [ae, bf] + [ce, df] = [ade, bdf] + [bce, bdf]$$
$$= [(ade + bce)bdf, (bdf)^2] = [ade + bce, bdf].$$

We have thus proved

Theorem 15.3.4. *The triple $(Q, +, \cdot)$ is a field.*

The field Q is called the **quotient field** or the **field of fractions** of the integral domain D. It is customary to denote the equivalence class $[a, b]$ by a/b.

Define a map i from D to Q be letting $\varphi(a) = a/1$. It is easy to check that this is a ring homomorphism. Moreover, it is one-to-one since $i(a) = a/1 = 0/1$ if and only if $a = 0$.

A ring monomorphism is also called an **embedding**. We have thus embedded the integral domain into its quotient field. The integral domain D is isomorphic to a subring of its quotient field Q. We often identify D with its image and simply consider D as a subring of Q.

There is an essential property regarding quotient fields.

Theorem 15.3.5. *Let Q be the quotient field of D defined above. Suppose there is a ring monomorphism φ from D to another field F. Then there exists a unique ring monomorphism ψ from Q to F such that $\varphi = \psi i$.*

Before we prove this theorem we would like to remark that φ sends the identity of D to the identity of F.

Lemma 15.3.6. *Let φ be a ring homomorphism from a ring with identity R to a domain D. Then the following properties hold:*

(a) $\varphi(1) = 1$;

(b) *if u is a unit in R, then $\varphi(u)$ is a unit of D and $\varphi(u^{-1}) = \varphi(u)^{-1}$.*

Proof. (a) Since $\varphi(1) = \varphi(1 \cdot 1) = \varphi(1)^2$, we have that $\varphi(1) = 1$ by cancellation law in D.

(b) Note that $\varphi(u)\varphi(u^{-1}) = \varphi(uu^{-1}) = \varphi(1) = 1$ and $\varphi(u^{-1})\varphi(u) = 1$. Hence $\varphi(u^{-1}) = \varphi(u)^{-1}$. $\qquad\qquad\square$

Proof of Theorem 15.3.5. First we attempt to define a map ψ from Q to F by letting $\psi(a/b) = \varphi(a)\varphi(b)^{-1}$. Let's verify that ψ is indeed well-defined.

First note that since φ is one-to-one and $b \neq 0$, $\varphi(b)^{-1}$ exists in F. Next, suppose $a/b = c/d$ in Q. Then $ad = bc$. This implies that $\varphi(a)\varphi(d) = \varphi(b)\varphi(c)$ and hence $\varphi(a)\varphi(b)^{-1} = \varphi(c)\varphi(d)^{-1}$. The map ψ is indeed well-defined. Now it is easy to verify that ψ is a ring homomorphism and we skip the proof. Of course, ψ is a one-to-one since the domain of ψ is a field (*Cf.* Exercise 4, §15.1). It is also easy to see that by Lemma 15.3.6 we have $\psi i(a) = \psi(a/1) = \varphi(a)1^{-1} = \varphi(a)$ for all $a \in D$ and so $\psi i = \varphi$.

At last we check that ψ is unique. Let ψ' be any ring homomorphism from Q to F. Then by Lemma 15.3.6.

$$\psi'\left(\frac{a}{b}\right) = \psi'\left(\frac{a}{1}\right)\psi'\left(\frac{1}{b}\right) = \varphi(a)\psi'\left(\frac{b}{1}\right)^{-1} = \varphi(a)\varphi(b)^{-1} = \psi\left(\frac{a}{b}\right)$$

for all $a/b \in Q$. $\qquad\square$

The previous theorem tells us that Q is the *smallest* field containing the integral domain D. Next we will see that any field with this property is "the same" as the quotient field Q.

Corollary 15.3.7. *Let Q' be another field with an embedding $i' \colon D \hookrightarrow Q'$ such that for all embedding $\varphi \colon D \hookrightarrow F$ there exists a unique monomorphism $\psi' \colon Q' \hookrightarrow F$ such that $\varphi = \psi'i'$. Then $Q \overset{f}{\cong} Q'$, where f is unique if one requires that $i' = fi$.*

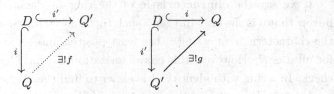

Proof. Using the property of Q, there exists a unique ring monomorphism $f \colon Q \to Q'$ such that $i' = fi$. On the other hand there exists a unique ring monomorphism $g \colon Q' \to Q$ such that $i = gi'$. Hence we have that $i = gi' = gfi$. However, we also have $i = 1_Q i$. This implies that $1_Q = gf$. Similarly we have $fg = 1_{Q'}$. Thus f is a ring isomorphism. $\qquad\square$

The property in Theorem 15.3.5 is called the **universal property** of the quotient fields and it can be used to characterize quotient fields. In other words, any field satisfying the property in Theorem 15.3.5 (that the quotient field is the smallest field containing D) is called a quotient field of D. Quotient fields are unique up to canonical isomorphisms.

The construction in §2.3 clearly shows that $\mathbb{Q}$ is the quotient fields of $\mathbb{Z}$. If F is a field, then F itself is the quotient field of F since among fields containing F there are no smaller fields than F.

Example 15.3.8. Consider $\mathbb{Z}[i]$, the ring of Gaussian integers. The field $\mathbb{C}$ must contain a copy of the quotient field Q of $\mathbb{Z}[i]$. Let $a + bi$ and $c + di \in \mathbb{Z}[i]$ where $a, b, c, d \in \mathbb{Z}$. Then

$$\frac{a + bi}{c + di} = \frac{(ac + bd) + (bc - ad)i}{c^2 + d^2} = r + si$$

where $r = \dfrac{ac + bd}{c^2 + d^2}$ and $s = \dfrac{bc - ad}{c^2 + d^2} \in \mathbb{Q}$. Hence every element of Q is an element in

$$\mathbb{Q}[i] = \{\, r + si \in \mathbb{C} : r, s \in \mathbb{Q} \,\}.$$

Every element in $\mathbb{Q}[i]$ is also of the form

$$\frac{a}{b} + \frac{c}{d}i = \frac{ad + bci}{bd}, \qquad a, b, c, d \in \mathbb{Z},$$

which is an element of Q. Thus $\mathbb{Q}[i]$ is the quotient field of $\mathbb{Z}[i]$.

If we say the **characteristic** of the ring R, denoted Char R, is 0, it means that 0 is the only integer k such that $ka = 0$ for all $a \in R$. Otherwise, the characteristic of a ring is the least positive integer k such that $ka = 0$ for all $a \in R$. However, this characterization is somewhat cumbersome to check. In a ring with identity, it is easier to find the characteristic of a ring.

Let 1 be the unity of the ring R. Consider the set

$$S = \{i \in \mathbb{Z}_+ : i = \underbrace{1 + 1 + \cdots + 1}_{i \text{ times}} = 0 \text{ in } R\}.$$

Let $k = \inf S$ if $S \neq \varnothing$. Then we have $k = 0$ in R (and hence $ka = 0$ for all $a \in R$) and $i \neq 0$ in R for all $i = 1, \ldots, k - 1$. In this case Char $R = k$. If the set S is empty then obviously Char $R = 0$.

Let R be a ring with unity. Consider the ring homomorphism $\varphi \colon \mathbb{Z} \to R$ sending k in $\mathbb{Z}$ to k in R (see the discussions on P. 198). Actually, the kernel of φ is a principal ideal generated by Char R. Thus, if Char $R = 0$, we see that $\mathbb{Z}$ is embedded into R and we may consider $\mathbb{Z}$ as a subring of R. On the other hand, if Char $R = n$, φ induces a ring monomorphism from $\mathbb{Z}_n$

into R. In this case we may consider $\mathbb{Z}_n$ as a subring of R. Thus, we can see that $\mathbb{Z}$ and $\mathbb{Z}_n$ are the minimal rings among all rings with unity. For this reason, we call $\mathbb{Z}$ and $\mathbb{Z}_n$, $n \in \mathbb{Z}_+$, the **prime rings**. Every ring with unity contains one and only one prime ring which is determined by its characteristic.

Let F be a field. Then the characteristic of a field is either 0 or a positive prime integer p (see Exercise 5). When Char $F = p$, F contains $\mathbb{Z}_p$ not only as a subring, but also as a subfield. When Char $F = 0$, F contains $\mathbb{Z}$ as a subring. By Theorem 15.3.5, F contains $\mathbb{Q}$ as a subfield. Thus, we call $\mathbb{Q}$ and $\mathbb{Z}_p$, p any positive prime integer, the **prime fields**. Prime fields are the minimal fields among fields. Every field contains one and only one prime field which is determined by its characteristic.

Exercises 15.3

1. Prove Lemma 15.3.3.

2. Find the quotient field of $\mathbb{Z}[\sqrt{2}] = \left\{\, a + b\sqrt{2} \in \mathbb{Q} : a, b \in \mathbb{Z} \,\right\}$.

3. Describe the quotient field of the polynomial ring $\mathbb{Q}[x]$.

4. Let D be an integral domain, which is also a subring of a field F. Let $Q = \left\{\, ab^{-1} \in F : a, b \in D,\ b \neq 0 \,\right\}$.

   (a) Show that Q is a subfield of F.

   (b) Show that Q is the smallest subfield in F containing D.

   (c) Show that Q is the quotient field of D.

5. Show that the characteristic of a domain is either 0 or a positive prime integer p.

6. If R is a commutative ring of characteristic p where p is a prime integer.

   (a) Show that $(a + b)^p = a^p + b^p$ for all $a,\ b \in R$.

   (b) Show that $(a + b)^{p^n} = a^{p^n} + b^{p^n}$ for all $n \in \mathbb{Z}_+$.

(c) Show that $\varphi \colon R \to R$ defined by $\varphi(a) = a^p$ for all $a \in R$ is a ring homomorphism. The map φ is called the **Frobenius map** on R.

7. Let $\operatorname{Char} R = m$ and let $\operatorname{Char} S = n$. Find $\operatorname{Char} R \times S$.

Review Exercises for Chapter 15

1. Let R be a commutative ring with identity and let $a \in R$.

   (a) Show that
   $$\operatorname{ann}_R a = \{\, r \in R : ar = 0 \,\}$$
   is an ideal of R, called the **annihilator** of a.

   (b) Since Rx is an ideal of R, it is also a subring of R. Show that
   $$\frac{R}{\operatorname{ann}_R a} \cong Rx.$$

2. Show that $\mathbb{Z}[x]/(2, x) \cong \mathbb{Z}_2$. Hence $(2, x)$ is a maximal ideal of $\mathbb{Z}[x]$.

3. Show that the only ring homomorphism of $\mathbb{R}$ into $\mathbb{R}$ is the identity function.

4. If $p(x)$ and $q(x)$ are relatively prime polynomials in $\mathbb{Q}[x]$, show that
   $$\frac{\mathbb{Q}[x]}{(p(x)q(x))} \cong \frac{\mathbb{Q}[x]}{(p(x))} \times \frac{\mathbb{Q}[x]}{(q(x))}.$$

5. Solve the following system of equations in $\mathbb{Q}[x]$:
   $$\begin{cases} f(x) \equiv x - 1 & (\bmod \ (x+1)^2), \\ f(x) \equiv 2x + 3 & (\bmod \ x^2 + 1). \end{cases}$$

6. Suppose two integral domains are isomorphic. Show that their quotient fields are also isomorphic to each other.

7. Let F be a field of characteristic $p > 0$. Show that
   $$F^p = \{\, a^p \in F : a \in F \,\} \quad \text{and} \quad K = \{\, a \in F : a^p = a \,\}$$
   are both subfields of F.

CHAPTER 16

Polynomial Rings

In this chapter we formally construct polynomial rings with coefficients in any given ring. Especially, we will review many of the properties that we are familiar with regarding polynomial rings such as $\mathbb{R}[x]$ or $\mathbb{Z}[x]$. Some of them are still true in the more general setting, while some of them will surprise us with counterexamples.

Polynomial rings in general are of great interest to algebraists. Basically speaking, a polynomial ring is the smallest ring containing its generators and its base ring. One is interested in studying the relationship between a polynomial ring and its base ring.

16.1 Polynomial Rings in the Indeterminates

Let R be a ring. We now want to formally construct a ring $R[x]$, called the **polynomial ring in the indeterminate** (or **variable**) x **over** R.

The elements of $R[x]$ are the expressions

$$f = a_0 + a_1 x + \cdots + a_n x^n,$$

where n is a nonnegative integer and $a_i \in R$. We call f a **polynomial** of x over R. The term $a_i x^i$ is called the **degree** i **term** of the polynomial f. The element a_i is called the **coefficient** of x^i (or of the degree i term) in the polynomial f. The degree 0 term a_0 is also called the **constant term** of f. If the polynomial $f = \sum a_i x^i$ is such that $a_i = 0$ for all $i > 0$ then f is called a **constant polynomial**. A polynomial can be expressed as $\sum_{i=0}^{\infty} a_i x^i$, or simply $\sum a_i x^i$. It is implicitly understood that in this case $a_i = 0$ for all $i < 0$ and for all sufficiently large i. In other words, there is an N such that $a_i = 0$ for all $i \geq N$ and $i < 0$.

Define

$$\sum a_i x^i = \sum b_i x^i \qquad \Longleftrightarrow \qquad a_i = b_i \quad \text{for all } i.$$

Define the addition and multiplication in $R[x]$ as follows:

$$(16.1.1) \qquad \begin{cases} \sum a_i x^i + \sum b_i x^i = \sum (a_i + b_i) x^i; \\ \left(\sum a_i x^i \right) \left(\sum b_i x^i \right) = \sum \left(\sum_k a_k b_{i-k} \right) x^i. \end{cases}$$

It is tedious but routine to check that $(R[x], +, \cdot)$ is a ring, and so we skip the details. Its zero element is the **zero polynomial** $f = \sum a_i x^i$ where $a_i = 0$ for all i. It is a commutative ring if R is commutative. If R is with identity 1, then $R[x]$ is also with identity which is the constant polynomial 1.

Our main interest lies in the case when R is *commutative*, and this what we will assume for the rest of this book unless otherwise noted.

We may adjoin a new variable y to the polynomial ring of one variable $R[x]$ and form a new polynomial ring $R[x][y]$. An element in $R[x][y]$ is an expression

$$(16.1.2) \qquad \sum_j f_j(x) y^j,$$

where $f_j(x) = \sum_i a_{ij} x^i \in R[x]$. This expression can be reexpressed as

$$\sum a_{ij} x^i y^j, \qquad a_{ij} \in R$$

where $a_{ij} = 0$ for all $i < 0$ or $j < 0$ and for all sufficiently large i, j. This can be done since the $f_j(x)$ in (16.1.2) can be recovered from the coefficients a_{ij}. Now suppose in $R[x][y]$

$$\sum a_{ij} x^i y^j = \sum b_{ij} x^i y^j, \qquad a_{ij}, \ b_{ij} \in R.$$

Then

$$\sum_j f_j(x) y^j = \sum_j g_j(x) y^j$$

where $f_j = \sum_i a_{ij} x^i$ and $g_j(x) = \sum_i b_{ij} x^i$. This implies that $f_j(x) = g_j(x)$ for all j, and hence $a_{ij} = b_{ij}$ for all i, j. We conclude that in $R[x][y]$,

$$\sum a_{ij} x^i y^j = \sum b_{ij} x^i y^j \qquad \Longleftrightarrow \qquad a_{ij} = b_{ij} \quad \text{for all } i, j.$$

If we go through the same process with $R[y][x]$, we will see that $R[y][x]$ consists of the same expressions as those of $R[x][y]$. In both rings the addition and multiplication are defined the same way. Inductively, we have the following result.

Proposition 16.1.1. *Let R be a commutative ring and let $x_1, x_2, \ldots, x_n$ be indeterminates. Then*

$$R[x_1][x_2] \cdots [x_n] = R[x_{\sigma(1)}][x_{\sigma(2)}] \cdots [x_{\sigma(n)}]$$

for all $\sigma \in S_n$.

We will use $R[x_1, \ldots, x_n]$ to denote the ring $R[x_1][x_2] \cdots [x_n]$. The ring $R[x_1, \ldots, x_n]$ is called the **polynomial ring of n variables** (or **in n indeterminates**) **over** R.

Next, let's discuss the general polynomial rings. Let R be a subring of a commutative ring S and let $U \subseteq S$. We will use $R[U]$ to denote the *smallest* subring of S containing R and U (see Exercise 1), and it is called the **polynomial ring** generated by U over R. If U is finite, we say $R[U]$ is **finitely generated** over R as a ring.

It is easy to see that the following properties hold for commutative rings $R \subseteq S$.

(1) For U and $V \subseteq S$, we have $R[U][V] = R[U \cup V]$.

(2) If $U = \{u_1, u_2, \ldots, u_n\} \subseteq S$ we write $R[u_1, u_2, \ldots, u_n]$ for $R[U]$. We have that
$$R[u_1, u_2, \ldots, u_n] = R[u_1][u_2] \cdots [u_n].$$

In light of this it is natural to study the ring $R[u]$ since all finitely generated polynomial rings over R can be obtained by adjoining one generator at a time.

It is clear that $R[u]$ contains all elements of the form
$$a_0 + a_1 u + a_2 u^2 + \cdots + a_n u^n,$$
which are **polynomials in u with coefficients in R**. Let $f = \sum_0^n a_i u^i$ and $g = \sum_0^m b_i u^i$ be two such polynomials. We might as well assume that $n \geq m$. It is also clear that

(16.1.3)
$$
\begin{cases}
f + g = (a_0 + b_0) + (a_1 + b_1)u + (a_2 + b_2)u^2 \\
\qquad\quad + \cdots + (a_m + b_m)u^m + a_{m+1}u^{m+1} + \cdots + a_n u^n, \\
-f = (-a_0) + (-a_1)u + \cdots + (-a_n)u^n, \text{ and} \\
fg = p_0 + p_1 u + p_2 u^2 + \cdots + p_{m+n}u^{m+n}, \\
\qquad\qquad\qquad \text{where } p_i = \sum_{j=0}^{i} a_j b_{i-j} = \sum_{j+k=i} a_j b_k
\end{cases}
$$

are all polynomials in u over R. We see that the set of all polynomials in u with coefficients in R form a subring of S. Hence we have the following proposition.

Proposition 16.1.2. *The polynomial ring $R[u]$ consists of all polynomials in u with coefficients in R.*

Example 16.1.3. The polynomial ring $\mathbb{Z}[i]$ supposedly is the ring consisting of all polynomials in i with coefficients in $\mathbb{Z}$. However, since $i^2 = -1$ is an integer, we have that
$$\mathbb{Z}[i] = \{a + bi \in \mathbb{C} : a, b \in \mathbb{Z}\}.$$

This is the ring of **Gaussian integers** which we have previously encountered.

The most special characterization about the polynomial rings generated by indeterminates can be described by the following theorem.

Theorem 16.1.4. *Let R and S be two rings and let u be an element in S. Let η be a ring homomorphism from R to S. Let $R[x]$ be the polynomial ring of one variable over R. Then η has one and only one extension to a ring homomorphism $\widetilde{\eta}$ from $R[x]$ to S sending x to u.*

Proof. Define $\widetilde{\eta}\colon R[x] \to S$ by sending $\sum_{i=0}^{n} a_i x^i$ to $\sum_{i=0}^{n} a_i u^i$. This is a map extending η and sending x to u. Furthermore, the relations in (16.1.1) and (16.1.3) guarantee that $\widetilde{\eta}$ is a ring homomorphism. $\square$

Using induction on the number of the indeterminates, we have the following generalization.

Corollary 16.1.5. *Let R and S be two rings and let $u_1, u_2, \ldots, u_n$ be (possibly repeated) elements in S and let η be a ring homomorphism from R to S. Let $R[x_1, x_2, \ldots, x_n]$ be the polynomial ring of n variables over R. Then η has one and only one extension to a ring homomorphism $\widetilde{\eta}$ from $R[x_1, \ldots, x_n]$ to S sending x_i to u_i.*

We have two immediate applications.

Corollary 16.1.6. *Let R be a ring and let $x_1, \ldots, x_n$ and $y_1, \ldots, y_n$ be indeterminates over R. Then $R[x_1, \ldots, x_n] \cong R[y_1, \ldots, y_n]$.*

Proof. By Corollary 16.1.5, there is a ring homomorphism from $R[x_1, \ldots, x_n]$ to $R[y_1, \ldots, y_n]$ fixing R and sending x_i to y_i. There is also a ring homomorphism from $R[y_1, \ldots, y_n]$ to $R[x_1, \ldots, x_n]$ fixing R and sending y_i to x_i. These two homomorphisms are inverse to each other. $\square$

Corollary 16.1.7. *Let S be a commutative ring containing $u_1, \ldots, u_n$ and containing R as a subring. Then*

$$R[u_1, \ldots, u_n] \cong \frac{R[x_1, x_2, \ldots, x_n]}{I}$$

for some ideal I of $R[x_1, x_2, \ldots, x_n]$ such that $I \cap R = (0)$.

Proof. Let $\varphi \colon R[x_1, \ldots, x_n] \to R[u_1, \ldots, u_n]$ be the ring homomorphism fixing R and sending x_i to u_i for each i. Then φ is clearly onto and $R[u_1, \ldots, u_n] \cong R[x_1, \ldots, x_n]/\ker \varphi$. Since φ restricted to R is one-to-one, we have $\ker \varphi \cap R = \{0\}$. $\square$

Exercises 16.1

In all the following exercises, x denotes an indeterminate. All the rings here are commutative.

1. Let R be a ring and let U be a subset of R. Show that there is a smallest subring containing U in R.

2. Describe the elements in $xR[x]$.

3. In the ring $\mathbb{Z}_8[x]$, show that $1 + 2x$ is a unit.

4. The ring R can be embedded naturally into the ring $R[x]$ by viewing elements of R as constant polynomials. Show that Char $R[x] =$ Char R.

5. Show that $R[x]/(x) \cong R$.

6. Let R be a ring and I an ideal of R. Suppose that $R[x]$ and $I[x]$ are the polynomial rings over R and I, respectively.

   (a) Show that $I[x]$ is an ideal of $R[x]$.

   (b) Show that $R[x]/I[x] \cong (R/I)[x]$.

7. Let F be a field. We use $F(x)$ to denote the quotient field of $F[x]$. It consists of the **rational functions** $f(x)/g(x)$ where $f(x), g(x) \in F[x]$ and $g(x) \neq 0$. The field $F(x)$ is call the **rational field** of x over F.

   Let D be an integral domain and let Q be its quotient field. Show that the $Q(x)$ is the quotient field of $D[x]$.

8. For each square free integer m, show that $\mathbb{Q}[\sqrt{m}]$ is the quotient field of $\mathbb{Z}[\sqrt{m}]$.

16.2   Properties of the Polynomial Rings of One Variable

Throughout this section, we assume that R is a commutative ring with identity and $x, x_1, \ldots, x_n$ are indeterminates over R.

Let $f = \sum_i a_i x^i \in R[x]$. We define the **degree** of the zero polynomial to be $-\infty$. Otherwise, we define the degree of f, denoted $\deg f$, to be the highest n such that $a_n \neq 0$. And the coefficient a_n is called the **leading coefficient** of f. The term $a_n x^n$ is called the **leading term** of $f(x)$. If $a_n = 1$, the polynomial $f(x)$ is called a **monic** polynomial.

Example 16.2.1. Let $f(x) = 1 + 2x$ and $g(x) = 3 + x + 3x^2$ be polynomials in $\mathbb{Z}_6[x]$. Find the degree of $f(x)g(x)$ and its leading coefficient.

Solution. Since $f(x)g(x) = 3 + x + 5x^2$, the degree of fg is 2 and the leading coefficient is 5.   ◇

Let $f(x)$ and $g(x)$ be two polynomials in $R[x]$. We sum up the two properties on degrees below:

(i) $\deg(f(x) + g(x)) \leq \max\{\deg f(x), \ \deg g(x)\}$, where equality holds if and only if either $\deg f(x) \neq \deg g(x)$ or when $\deg f(x) = \deg g(x)$ the leading coefficients do not cancel each other;

(ii) $\deg(f(x)g(x)) \leq \deg f(x) + \deg g(x)$, where equality holds if and only if the product of the leading coefficients is nonzero.

In particular, $\deg(f(x)g(x)) = \deg f(x) + \deg g(x)$ if R is an *integral domain*.

Proposition 16.2.2. *If D is an integral domain then $D[x]$ is an integral domain and so is $D[x_1, \ldots, x_n]$ for every n.*

Proof. Let f and g be two nonzero polynomials in $D[x]$. Then $\deg fg = \deg f + \deg g \geq 0$ and so fg is a nonzero polynomial. The rest follows from induction on n.   □

Proposition 16.2.3. *Let D be an integral domain. Then the units of $D[x_1, \ldots, x_n]$ are the units of D. In other words, $U(D[x_1, \ldots, x_n]) = U(D)$.*

In particular, if F be a field, the units in $F[x_1, \ldots, x_n]$ are exactly the nonzero constant polynomials. In other words, $U(F[x_1, \ldots, x_n]) = F^*$.

Proof. It suffices to check the ring case. We first consider the case $n = 1$. Remember that elements of D are considered constant polynomials in $D[x]$. Clearly, $U(D) \subseteq U(D[x])$. Let $f \in U(D[x])$. Then there is a g in $D[x]$ such that $fg = 1$. Since $\deg f + \deg g = \deg fg = 0$, we have $\deg f = \deg g = 0$. Thus f and g are both in D and f is a unit in D.

In general, since $D[x_1, \ldots, x_n] = D[x_1, \ldots, x_{n-1}][x_n]$, we have that

$$U(D[x_1, \ldots, x_n]) = U(D[x_1, \ldots, x_{n-1}]) = U(D)$$

by induction on n.   □

Next we will give the property that make the polynomial rings of one variable behave so well.

Theorem 16.2.4 (Division Algorithm). *Let R be a commutative ring with identity. Let $g(x)$ be a polynomial in $R[x]$ whose leading coefficient is a unit. Then there is a $q(x)$ and an $r(x)$ in $R[x]$ such that*

$$f(x) = g(x)q(x) + r(x)$$

where $\deg r(x) < \deg g(x)$.

Proof. We prove the assertion by induction on the degree of $f(x)$. If $\deg f(x) < \deg g(x)$, we simply choose $q(x) = 0$ and $r(x) = f(x)$. Thus, we may assume $\deg f(x) \geq \deg g(x)$. Suppose that the assertion is true for all polynomials with degrees less than $n = \deg f(x)$. Let

$$\begin{cases} f(x) = a_0 + a_1 x + \cdots + a_n x^n, & a_n \neq 0, \\ g(x) = b_0 + b_1 x + \cdots + b_m x^m, & b_m \text{ is a unit of } R, \end{cases}$$

with $n \geq m$. Then

$$f_1(x) = f(x) - \frac{a_n}{b_m} x^{n-m} g(x)$$

is a polynomial with $\deg f_1(x) < n$. By the induction hypothesis, there exist $q_1(x)$ and $r(x)$ in $R[x]$ such that

$$f_1(x) = g(x)q_1(x) + r(x), \qquad \deg r(x) < \deg g(x).$$

It follows that

$$f(x) = g(x) \left[\frac{a_n}{b_m} x^{n-m} + q_1(x) \right] + r(x), \qquad \deg r(x) < \deg g(x). \quad \square$$

Example 16.2.5. Let $\mathbb{R}[x]$ be the polynomial ring of one variable over $\mathbb{R}$. The inclusion map $\mathbb{R} \hookrightarrow \mathbb{C}$ can be extended to a ring homomorphism $\varphi \colon \mathbb{R}[x] \to \mathbb{C}$ sending x to i. The map φ is onto since for all $a + bi \in \mathbb{C}$, $a, b \in \mathbb{R}$, we have $\varphi(a + bx) = a + bi$. Clearly, $\varphi(x^2 + 1) = i^2 + 1 = 0$. We have $(x^2 + 1) \subseteq \ker \varphi$. On the other hand, let $f(x) \in \ker \varphi$. Using the division algorithm, there exist $q(x) \in \mathbb{R}[x]$ and $a, b \in \mathbb{R}$ such that

$$f(x) = (x^2 + 1)q(x) + ax + b.$$

Thus, $\varphi(f(x)) = ai + b = 0$. It follows that $a = b = 0$ and $f(x) \in (x^2 + 1)$. We have shown that $\ker \varphi = (x^2 + 1)$. Hence by the first isomorphism theorem, $\mathbb{R}[x]/(x^2 + 1) \cong \mathbb{C}$.

Let R and S be commutative rings and let $\eta \colon R \to S$ be a ring homomorphism. Let $s \in S$. There is a (unique) ring homomorphism $\eta_s \colon R[x] \to S$ sending $r \in R$ to $\eta(r)$ and x to s. The map η_s is called a **substitution** or **evaluation homomorphism**. As a convention, we write $f_\eta(s)$ for the image of $f(x)$ under η_s. Naturally,

$$f_\eta(s) = \eta(a_0) + \eta(a_1)s + \cdots + \eta(a_n)s^n$$

if $f(x) = a_0 + a_1 x + \cdots + a_n x^n$. In particular, if $\eta \colon R \hookrightarrow S$ is the inclusion map, that is, $\eta(r) = r$ for all $r \in R$, we will simply write $f(s)$ for $f_\eta(s)$. In other words,

$$f(s) = a_0 + a_1 s + \cdots + a_n s^n.$$

We say s is **substituted** into the polynomial $f(x)$. Obviously, using the properties of homomorphisms we have

$$(f + g)_\eta(s) = \eta_s(f + g) = \eta_s(f) + \eta_s(g) = f_\eta(s) + g_\eta(s)$$

and

(16.2.1) $$(fg)_\eta(s) = \eta_s(fg) = \eta_s(f)\eta_s(g) = f_\eta(s)g_\eta(s)$$

for all $f, g \in R[x]$.

When R is a subring of S, we say that $a \in S$ is a **zero** of $f(x) \in R[x]$ or that a is a **root** of $f(x) = 0$ if $f(a) = 0$. We also say that $g(x)$ is a **factor** of $f(x)$ in $R[x]$ (or over R) if $f(x) = g(x)q(x)$ for some polynomial $q(x)$ in $R[x]$.

We have the following results well-known to high school students, but in a much more general context.

Corollary 16.2.6 (Remainder Theorem). *Let R be a commutative ring with identity. For $f(x) \in R[x]$ and $a \in R$, there exists $q(x) \in R[x]$ such that*

$$f(x) = (x - a)q(x) + f(a).$$

Proof. Now that the divisor $x - a$ is monic, we can apply Theorem 16.2.4 to obtain that

$$f(x) = (x - a)q(x) + r, \qquad q(x) \in R[x] \text{ and } r \in R.$$

After substituting in a, we have that $f(a) = (a - a)q(a) + r = r$. □

Corollary 16.2.7 (Factor Theorem). *Let R be a commutative ring with identity. For $f(x) \in R[x]$ and $a \in R$, $x - a$ is a factor of $f(x)$ over R if and only if $f(a) = 0$.*

Proof. If $x - a$ is a factor of $f(x)$, then $f(x) = (x - a)q(x)$ for some $q(x) \in R[x]$. Thus $f(a) = 0$. Conversely, if $f(a) = 0$, $f(x) = (x - a)q(x)$ for some $q(x) \in R[x]$ by Theorem 16.2.6. □

Corollary 16.2.8. *Let D be an integral domain. Let $f(x)$ be a nonzero polynomial of degree n over D. Then $f(x) = 0$ has at most n distinct roots in D.*

Proof. Let $a_1, \ldots, a_r$ be r distinct roots of $f(x)$ in D. By Theorem 16.2.7, we have $f(x) = (x - a_1)q_1(x)$. Substituting in a_2 we have that $(a_2 - a_1)q_1(a_2) = 0$. Since $a_2 - a_1 \neq 0$ and D is a integral domain, $q_1(a_2) = 0$. Hence $q_1(x) = (x - a_2)q_2(x)$ for some $q_2(x) \in D[x]$. Continue in this fashion and we have that $f(x) = (x - a_1)(x - a_2) \cdots (x - a_r)q_r(x)$ where $q_r(x) \in D[x]$. By comparing the degrees on both sides we see that $n = r + \deg q_r(x) \geq r$. □

Note that the corollary above is NOT true if the ring D is not an integral domain. For example, the polynomial $f(x) = (x - 1)^3$ in $\mathbb{Z}_8[x]$ has four solutions: 1, 3, 5 and 7.

Exercises 16.2

In the following exercises, x denotes an indeterminate. All rings are commutative with identity.

1. In $\mathbb{Z}_3[x]$, show that $x^4 + x$ and $x^2 + x$ determine the same function from $\mathbb{Z}_3$ to $\mathbb{Z}_3$.

2. In $\mathbb{Z}_7[x]$, find $q(x)$ and $r(x)$ such that

$$5x^4 + 3x^3 + 1 = (3x^2 + 2x - 1)q(x) + r(x)$$

   and $\deg r(x) < 2$.

3. Let $f(x) = x^n + a_1 x^{n-1} + \cdots + a_n$, where $n > 0$ and a_i are elements in a field F. Let $u = x + (f(x)) \in F[x]/(f(x))$. Show that every element of $F[u]$ can be written in one and only one way in the form

$$b_0 + b_1 u + \cdots + b_{n-1} u^{n-1}, \qquad b_i \in F.$$

4. Find the number of elements in $\mathbb{Z}_3[x]/(x^3 + x + 1)$.

5. Show that the $q(x)$ and $r(x)$ in Theorem 16.2.4 are unique.

6. Let p be a positive prime integer and let $f(x) = x^{p-1} - \overline{1} \in \mathbb{Z}_p[x]$.

   (a) Show that $f(\overline{n}) = 0$ for all $\overline{n} \in \mathbb{Z}_p \setminus \{0\}$.

   (b) Show that

$$x^{p-1} - \overline{1} = (x - \overline{1})(x - \overline{2}) \ldots (x - \overline{p-1}).$$

   (c) Use (b) to show that $(p - 1)! \equiv -1 \pmod{p}$. (This gives a different proof of Wilson's theorem from the one given in Theorem 2.2.6.)

7. Let F be a finite field such that $F^* = \{a_1, a_2, \ldots, a_{q-1}\}$. Prove that $a_1 a_2 \cdots a_{q-1} = -1$. This is a generalization of Wilson's theorem.

16.3   Principal Ideal Domains and Euclidean Domains

For now we will focus our attention on the polynomial ring of one variable over a field. In this section F denotes a field and x denotes an indeterminate.

We call a commutative ring with identity in which every ideal is a principal ideal a **principal ideal domain** or simply a **PID**. Remember that $\mathbb{Z}$ is a PID by Proposition 14.1.7. Next we will use the division algorithm to prove that this is also true for $F[x]$.

Proposition 16.3.1. *Let $F[x]$ be a polynomial ring of one variable over the field F. Then $F[x]$ is a PID.*

Proof. Let I be a nonzero ideal in $F[x]$. Choose a nonzero polynomial $f(x)$ in I so that it has the least possible degree. We claim that $I = f(x)F[x]$. Of course, $f(x)F[x] \subseteq I$. Now we prove the opposite inclusion.

By the division algorithm, for any $g(x) \in I$ there exist $q(x)$ and $r(x)$ in $F[x]$ such that

$$g(x) = f(x)q(x) + r(x), \qquad \deg r(x) < \deg f(x).$$

However, $r(x) = g(x) - f(x)q(x) \in I$. This implies that $r(x) = 0$. Hence $g(x) = f(x)q(x) \in f(x)F[x]$. This proves $I \subseteq f(x)F[x]$.   □

From the proof above, we can see that the most important reason that makes $F[x]$ a PID is the division algorithm. Hence, we make the following definition.

Definition 16.3.2. An integral domain D is a **Euclidean domain** if there exists a function v from the set of nonzero elements in D to the set of nonnegative integers such that the following properties are satisfied:

  (i) for all $a, b \in D$ such that $b \neq 0$, there exist q and r in D such that

$$a = bq + r, \qquad \text{where } r = 0 \text{ or } v(r) < v(b);$$

  (ii) for all nonzero $a, b \in D$, we have $v(a) \leq v(ab)$.

The function v is called a **Euclidean valuation**.

The division algorithm guarantees that $F[x]$, the polynomial ring of one variable over a field, is a Euclidean domain since the degree of a polynomial is a Euclidean valuation. The ring $\mathbb{Z}$ is a Euclidean domain with the absolute value as its Euclidean valuation.

Next we show that the ring of Gaussian integers is also a Euclidean domain. Remember that the norm of $a + bi \in \mathbb{C} = N(a + bi) = a^2 + b^2$. Remember also that $N(zz') = N(z)N(z')$ for $z,\ z' \in \mathbb{C}$. Next we will show that N is a Euclidean valuation for $\mathbb{Z}[i]$.

Proposition 16.3.3. *For any $u,\ v \in \mathbb{Z}[i]$ with $v \neq 0$, there exist q and $r \in \mathbb{Z}[i]$ such that*

$$u = qv + r, \qquad \text{where } N(r) \leq \frac{N(v)}{2} < N(v).$$

Hence, $\mathbb{Z}[i]$ is a Euclidean domain.

Note that in the proof we will see that q and r are not unique in general.

Proof. Let $u/v = c + di$ where $c,\ d \in \mathbb{Q}$. Choose integers c_1 and d_1 so that

$$|c - c_1| \leq \frac{1}{2} \quad \text{and} \quad |d - d_1| \leq \frac{1}{2}.$$

Set $q = c_1 + d_1 i$ and $r = u - qv$. Then

$$N(r) = N(v)N\left(\frac{u}{v} - q\right) = N(v)N\big((c - c_1) + (d - d_1)i\big)$$

$$\leq N(v)\left[\left(\frac{1}{2}\right)^2 + \left(\frac{1}{2}\right)^2\right] = \frac{N(v)}{2} < N(v).$$

since $v \neq 0$ implies that $N(v) \neq 0$. The rest of the proposition is easy. $\quad\square$

The proof of Proposition 16.3.1 can be modified to give the following more general result.

Theorem 16.3.4. *Every Euclidean domain is a PID.*

Proof. Let D be a Euclidean domain with v as its Euclidean valuation and let I be a nonzero ideal in D. Choose a nonzero element a in I so that $v(a)$ is minimal. We claim that $I = aD$. Of course, $aD \subseteq I$. Now we prove the opposite inclusion.

By the division algorithm, for any $b \in I$ there exist q and r in D such that

$$b = aq + r, \qquad v(r) < v(a).$$

However, $r = b - aq \in I$. This implies that $r = 0$. Hence $b = aq \in aD$. This proves that $I \subseteq aD$. $\qquad\qquad\qquad\qquad\qquad\qquad\qquad\qquad\qquad\qquad\square$

Corollary 16.3.5. *Let D be the Euclidean domain with v as the Euclidean valuation. Then a nonzero ideal of D is generated by any nonzero element a in I such that $v(a)$ is minimal.*

In particular, let $F[x]$ be the polynomial ring of one variable over a field F. Then any nonzero ideal I in $F[x]$ is generated by a nonzero polynomial of the lowest degree in I. Any nonzero ideal I in $\mathbb{Z}$ is generated by the least positive integer in I. Any nonzero ideal I in $\mathbb{Z}[i]$ is generated by a nonzero element with the least possible norm in I.

Although Euclidean domains are PID's, there are many PID's which are not Euclidean domains. For example, the ring

$$D = \left\{ \frac{a + b\sqrt{19}i}{2} : a \text{ and } b \text{ are either both even or both odd integers} \right\}$$

$$= \left\{ a + b\sqrt{19}i : a,\ b \in \mathbb{Z} \right\} \cup \left\{ a + b\sqrt{19}i + \frac{1 + \sqrt{19}i}{2} : a,\ b \in \mathbb{Z} \right\}.$$

is a PID but not a Euclidean domain. The proof of this result is beyond the scope of this course.[1]

Example 16.3.6. Let $\mathbb{Z}[x]$ be the polynomial ring of one variable over $\mathbb{Z}$. The ring homomorphism $\mathbb{Z} \twoheadrightarrow \mathbb{Z}_2$ can be extended to a ring homomorphism from $\mathbb{Z}[x]$ to $\mathbb{Z}_2$ sending x to 0. The homomorphism φ is clearly onto. It is also obvious that

$$\ker \varphi = \{ f = a_0 + a_1 x + \cdots + a_n x^n \in \mathbb{Z}[x] : a_0 \text{ is even} \}.$$

Note that clearly $(2, x) \subseteq \ker \varphi$. If $f \in \ker \varphi$ then $f = 2(a_0/2) + x(a_1 + a_2 x + \cdots + a_n x^{n-1}) \in (2, x)$. Hence $\ker \varphi \subseteq (2, x)$. It follows that $\ker \varphi = (2, x)$

---

[1] The interested reader can refer to J. C. Wilson, *A principal ideal domain that is not a Euclidean ring*, Mathematics Magazine 46 (1973), 34–38.

and $\mathbb{Z}[x]/(2,x) \cong \mathbb{Z}_2$. Since $\mathbb{Z}_2$ is a field, we have that $(2,x)$ is a maximal ideal of $\mathbb{Z}[x]$.

However, the ideal $(2,x)$ is not principal. Suppose $(2,x) = (f)$. Since $f \mid 2$, we know that f is a constant polynomial by comparing the degree. If $f = \pm 2$, then $x \notin (2)$, a contradiction. If $f = \pm 1$, then $(2,x) = \mathbb{Z}[x]$, another contradiction. Hence, $\mathbb{Z}[x]$ is neither a PID nor a Euclidean domain.

Exercises 16.3

1. Find α and $\gamma \in \mathbb{Z}[i]$ such that
$$3 + 8i = (-2 + 3i)\alpha + \gamma, \qquad \text{where } N(\gamma) < N(-2 + 3i).$$

2. Show that $\mathbb{Z}[\sqrt{2}i]$ is a Euclidean domain.

3. Show that $\mathbb{Z}[\sqrt{2}]$ is a Euclidean domain. (Hint: Use the norm $N(a + b\sqrt{2}) = |(a + b\sqrt{2})(a - b\sqrt{2})| = |a^2 - 2b^2|$.)

4. We say that a commutative ring R is a **principal ideal ring** if every ideal in R is principal.

   Show that $\mathbb{Z}_n$ is a principal ideal ring for all positive integers n.

5. Show that the mapping defined by $v(a) = a^2$ is a Euclidean valuation on $\mathbb{Z}$. Hence Euclidean valuations on a Euclidean domain are not unique.

6. Show that $F[x,y]$, the polynomial ring of two variables over the field F, is not a principal ideal domain.

Review Exercises for Chapter 16

1. Let I be an ideal of the commutative ring R and let $x_1, \ldots, x_n$ be indeterminates. Let $I[x_1, \ldots, x_n]$ denote the subset of $R[x_1, \ldots, x_n]$ of polynomials with coefficients in I.

   (a) Show that $I[x_1, \ldots, x_n]$ is an ideal of $R[x_1, \ldots, x_n]$.

   (b) Show that $R[x_1, \ldots, x_n]/I[x_1, \ldots, x_n] \cong (R/I)[x_1, \ldots, x_n]$.

2. Let F be an infinite field. If a polynomial $f(x) \in F[x]$ has infinitely many zeros in F then $f(x)$ is the zero polynomial.

3. Find a polynomial with integer coefficients which has $1/2$ and $-4/5$ as zeros.

4. Suppose R is a principal ideal ring.

   (a) Let $R \twoheadrightarrow S$ be an epimorphism. Is it true that S is always a principal ideal ring?

   (b) Let $S \hookrightarrow R$ be a monomorphism. Is it true that S is always a principal ideal ring?

5. Let D be a principal ideal domain. Show that D contains a maximal ideal in the following steps.

   (a) Suppose D does not contain a maximal ideal. Show that one can find a infinite sequence of proper ideals

$$I_1 \subsetneq I_2 \subsetneq I_3 \subsetneq \cdots.$$

   (b) Use Exercise 10 on P. 220 to derive a contradiction.

6. Let D be a PID which is not a field. Show that $D[x]$ is never a PID.

CHAPTER 17

Factorization

In 1637, French mathematician Pierre de Fermat (1601–1665) claimed he could prove that there were no integer solutions to $x^n + y^n = z^n$ for $n \geq 3$. However he lamented "I have discovered a truly marvelous demonstration which this margin is too narrow to contain!" Since then many mathematicians tried to conquer this problem which for a long time was referred to "Fermat's last problem". In 1847, Gabriel Lamé (1795–1870) stirred excitement by claiming that he had solved the problem by factoring the polynomial $x^p + y^p$, p any odd prime, into

$$(x + y)(x + \zeta y) \cdots (x + \zeta^{p-1}y)$$

in $\mathbb{Z}[\zeta]$ where $\zeta = \cos(2\pi/p) + i\sin(2\pi/p)$. However, Ernst Eduard Kummer (1810–1893) soon pointed out that Lamé had made the mistake of assuming that factorization in $\mathbb{Z}[\zeta]$ is unique and thus rendered Lamé's proof useless.

In this chapter we will look closely at what factorization is and find out what requirements an integral domain must satisfy to have unique factorization. By the way, Fermat's last problem was finally solved by Andrew Wiles (1953–) and Richard Taylor (1962–) in September, 1994. It is now referred to as "Fermat's Last Theorem".

255

17.1   Irreducible and Prime Elements

In $\mathbb{Z}$ when we factor

$$6 = 2 \cdot 3 = (-2) \cdot (-3),$$

few people would consider these two factorizations fundamentally different. The reason is that -1 is a unit in $\mathbb{Z}$, and therefore it is a factor of any element in $\mathbb{Z}$. Exactly for this reason when we factor an integer, we do not need to distinguish between integers such as 2 and -2 or between 3 and -3. We generalize this idea into the following concept.

In R we say an element b is an **associate** of a if $b = au$ for some unit u in R. In this case, we will write $b \sim a$. It is easy to check that $\sim$ is an equivalence relation.

Lemma 17.1.1. *Let D be an integral domain. Then $a \mid b$ and $b \mid a$ if and only if $a \sim b$ in D.*

Note that $aD = bD$ if and only if $a \mid b$ and $b \mid a$ by definition.

Proof. The "if" part is trivial. To prove the "only if" part, observe that if $b \mid a$ then $b = au$ for some u of D, and if $a \mid b$ then $a = bv$ for some v of D. It follows that $a = bu = avu$. This implies that $vu = 1$ since D is an integral domain. Hence, u and v are both units of D.   □

Let $a \in R$ be *nonzero* and *non-unit*. We say a is **reducible** in R if $a = bc$ for some non-units b and c in R. Otherwise we say that a is **irreducible** in R.

When $f(x)$ is an irreducible polynomial in $R[x]$, we also say that $f(x)$ is irreducible *over R*.

How do we determine when a polynomial is irreducible over a field F? First recall that in the polynomial ring $F[x]$ of one variable over a field F, the units are exactly the nonzero constant polynomials by Proposition 16.2.3.

Proposition 17.1.2. *In $F[x]$ where F is a field, any linear polynomial is irreducible. On the other hand, $ax - b$ is a linear factor of $f(x)$ in $F[x]$ if and only if b/a is a zero of $f(x)$.*

Proof. Let $f(x) \in F[x]$. If $f = gh$ in $F[x]$, then $\deg g + \deg h = 1$. It follows that $\deg g = 0$ or $\deg h = 0$. This means that g or h is a unit. Hence f is irreducible over F.

If $ax - b$ is a linear factor of $f(x)$, then $f(x) = (ax - b)g(x)$ for some $g \in F[x]$. It follows that $f(b/a) = (a(b/a) - b)g(b/a) = 0$. Conversely, suppose $f(b/a) = 0$. By Theorem 16.2.7, $f(x) = (x - b/a)g(x) = (ax - b)(1/a)g(x)$. Thus $ax - b$ is a factor of $f(x)$. $\qquad\square$

The next result makes it easy to determine whether a polynomial of degree ≤ 3 is irreducible over a field.

Corollary 17.1.3. *Let $f(x) \in F[x]$ be a nonzero polynomial.*

(a) *If $\deg f(x) = 0$ then $f(x)$ is a unit.*

(b) *If $\deg f(x) = 1$ then $f(x)$ is irreducible over F.*

(c) *If $\deg f(x) = 2$ or 3 then $f(x)$ is irreducible over F if and only if $f(x)$ has no zeros in F.*

Proof. Part (c) holds because if a polynomial of degree 2 or 3 is reducible, it must have a linear factor. The result follows from Proposition 17.1.2. $\qquad\square$

Example 17.1.4. Find all units of $\mathbb{Z}_2[x]$. Determine which of $f = x^2 + x + 1$, $g = x^3 + x^2 + x + 1$ and $h = x^4 + x^3 + x^2 + x + 1$ are irreducible.

Solution. First note that $U(\mathbb{Z}_2[x]) = \{1\}$. Since $f(0) = f(1) = 1$, f is irreducible over $\mathbb{Z}_2$. Since $g(1) = 0$, $g(x)$ is reducible over $\mathbb{Z}_2$. Now check that $h(0) = h(1) = 1$. This does not mean that h is irreducible over $\mathbb{Z}_2$. It simply means that h does not have a linear factor. If h is still reducible over $\mathbb{Z}_2$, it is a product of two irreducible polynomials of degree 2. The polynomials of degree 2 over $\mathbb{Z}_2$ are $x^2 + 1 = (x+1)^2$, $x^2 + x = x(x+1)$ and $f = x^2 + x + 1$, and among them the only irreducible polynomial is f. Hence if h is reducible, we have $h = f^2$. Now $f^2 = (x^2 + x + 1)^2 = x^4 + x^2 + 1 \neq h$. Thus, h is also irreducible over $\mathbb{Z}_2$. $\qquad\diamond$

By the fundamental theorem of algebra, every polynomial in $\mathbb{C}[x]$ can be factored into a product of linear factors. Thus the only irreducible polynomials in $\mathbb{C}[x]$ are polynomials of degree 1. On the other hand, irreducible polynomials in $\mathbb{R}[x]$ are of degree 1 or degree 2. However, it is difficult to determine which polynomials in $\mathbb{Q}[x]$ are irreducible in general.

For the following type of integral domains, norm is very useful.

Proposition 17.1.5. *Let* $D = \mathbb{Z}[i\sqrt{m}]$ *where* m *is a square-free positive integer. The following statements are true.*

(a) *The element* $z \in D$ *is a unit in* D *if and only if* $N(z) = 1$. *Hence;*

$$U(D) = \begin{cases} \{\pm 1, \pm i\}, & \text{if } m = 1; \\ \{\pm 1\}, & \text{if } m \geq 2. \end{cases}$$

(b) *If* $N(z)$ *is a prime integer then* z *is irreducible in* D.

Proof. Remember that $N(z)$ is a positive integer for $z \in D \setminus \{0\}$.

(a) Let $z \in D$. If $N(z) = 1$ then $z\overline{z} = 1$ and $\overline{z}$ is the inverse of z. Conversely, let z be a unit. Find $w \in D$ such that $zw = 1$. Then $N(z)N(w) = N(1) = 1$. Since $N(z)$ is a positive integer, $N(z) = 1$. To find the elements in $U(D)$, we need to solve $N(a + bi\sqrt{m}) = a^2 + b^2m = 1$ for a and $b \in \mathbb{Z}$. The rest is easy.

(b) Suppose $z = xy$ for x and $y \in D$. Then $N(x)N(y) = N(z) = p$ is a prime integer. It follows that $N(x)$ or $N(y) = 1$. Hence x or y is a unit. $\square$

Note that when z is irreducible in D, $N(z)$ is not necessarily a prime integer.

Example 17.1.6. Is 2 irreducible in $\mathbb{Z}[i]$?

Solution. First compute that $N(2) = 4$. If $z = xy$ is reducible in $\mathbb{Z}[i]$, then $N(x) = 2$. Let $x = ai + b$ where $a, b \in \mathbb{Z}$, we need to solve $a^2 + b^2 = 2$ over $\mathbb{Z}$. It is easy to see that $a = \pm 1$ and $b = \pm 1$. We can see that $2 = (1 + i)(1 - i)$ is reducible in $\mathbb{Z}[i]$. ◇

Example 17.1.7. Which of $6 + i\sqrt{5}$ and 3 are irreducible in $\mathbb{Z}[i\sqrt{5}]$?

Solution. Since $N(6 + i\sqrt{5}) = 41$ is a prime integer, it is irreducible in $\mathbb{Z}[i\sqrt{5}]$. On the other hand, $N(3) = 9$. If $z = xy$ is reducible in $\mathbb{Z}[i\sqrt{5}]$, then $N(x) = 3$. However, there are no integer solutions for $a^2 + 5b^2 = 3$. Hence 3 is also irreducible in $\mathbb{Z}[i\sqrt{5}]$. ◇

Let R be a commutative ring with identity. Let $a \in R$ be nonzero and non-unit. We say a is a **prime** element in R if whenever $a \mid bc$ in R, we have that $a \mid b$ or $a \mid c$.

Proposition 17.1.8. *Any prime element in an integral domain is irreducible.*

Proof. Let p be a prime element in the integral domain D. Suppose $p = ab$. Since $p \mid ab$, we have $p \mid a$ and $p \mid b$. Without loss of generality, we may assume $p \mid a$. But we also have $a \mid p$. By Lemma 17.1.1, $a \sim p$ and b is a unit. $\square$

However, even in integral domains irreducible elements are not necessarily prime.

Example 17.1.9. In Example 17.1.7 we have seen that 3 is irreducible in $\mathbb{Z}[i\sqrt{5}]$. Note that
$$3 \cdot 3 = (2 + i\sqrt{5})(2 - i\sqrt{5}).$$
Since $3 \nmid 2 + i\sqrt{5}$ and $3 \nmid 2 - i\sqrt{5}$, 3 is not prime in $\mathbb{Z}[i\sqrt{5}]$.

Nevertheless, irreducible elements behave extremely well in PID's.

Theorem 17.1.10. *Let D be a PID and let $I = (a)$ be a nonzero ideal of D. Then the following conditions are equivalent:*

(i) *I is a maximal ideal;*

(ii) *I is a prime ideal;*

(iii) *a is a prime element in D;*

(iv) *a is an irreducible element in D.*

Proof. We already know that (ii) $\Leftrightarrow$ (iii) (see Exercise 6, §14.2).
(i) $\Rightarrow$ (ii): This is Corollary 14.2.8.
(iii) $\Rightarrow$ (iv): This follows from Proposition 17.1.8.
(iv) $\Rightarrow$ (i): Suppose that a is irreducible in D. Let N be any proper ideal of D such that $I = (a) \subseteq N \subsetneq D$. We can find $b \in D$ such that $N = (b)$. Since $a \in N$, we can find $c \in D$ such that $a = bc$. Since N is proper, b is not a unit. It follows that c is a unit. Hence $I = (a) = (bc) = (b) = N$. $\square$

Corollary 17.1.11. *Let F be a field. Any irreducible polynomial over F is a prime in $F[x]$.*

Recall that the only irreducible polynomials in $\mathbb{C}[x]$ are polynomials of degree 1, that is, the polynomials $u(x - a)$ where u is a nonzero complex number. By Theorem 17.1.10 there is a one-to-one correspondence between the point a in $\mathbb{C}$ and the maximal ideals $(x - a)$ in $\mathbb{C}[x]$. Such a correspondence can be extended to the polynomial ring of several variables $\mathbb{C}[x_1, x_2, \ldots, x_n]$. Hilbert's Nullstellensatz (Hilbert's Zero Point Theorem) asserts that there is a one-to-one correspondence between the point $(a_1, a_2, \ldots, a_n)$ in $\mathbb{C}^n$ and the maximal ideals $(x_1 - a_1, \ldots, x_n - a_n)$ of the polynomial ring $\mathbb{C}[x_1, x_2, \ldots, x_n]$.

For any $(a_1, a_2, \ldots, a_n) \in \mathbb{C}^n$, define the evaluation mapping

$$\varphi \colon \mathbb{C}[x_1, x_2, \ldots, x_n] \to \mathbb{C}$$

by sending x_i to a_i and thus $f(x_1, x_2, \ldots, x_n)$ to $f(a_1, a_2, \ldots, a_n)$. Since φ is onto, $\mathbb{C}[x_1, x_2, \ldots, x_n]/\ker \varphi \cong \mathbb{C}$ where $\ker \varphi = (x_1 - a_1, \ldots, x_n - a_n)$ is a maximal ideal of $\mathbb{C}[x_1, x_2, \ldots, x_n]$ (see Exercise 8). The hard part is to show that any maximal ideal of $\mathbb{C}[x_1, x_2, \ldots, x_n]$ is an ideal of this type, which is beyond the scope of this course.

Exercises 17.1

In all of the following problems $x, x_1, \ldots, x_n$ denote indeterminates.

1. Show that an associate of an irreducible element is still irreducible.

2. Find all $a \in \mathbb{Z}_3$ so that $\mathbb{Z}_3[x]/(x^2 + x + a)$ is a field.

3. Find all $a \in \mathbb{Z}_5$ so that $\mathbb{Z}_5[x]/(x^2 + ax + 2)$ is a field.

4. Is $\mathbb{Q}[x]/(x^2 - 4x + 3)$ a field? Is $\mathbb{R}[x]/(x^2 + x + 1)$ a field? Justify your answers.

5. Show that the following polynomials are irreducible over the field indicated.

    (a) $x^2 + x + 7$ over $\mathbb{R}$.

    (b) $x^3 - 2$ over $\mathbb{Q}$.

    (c) $x^3 + x + 1$ over $\mathbb{Z}_2$.

6. Factor the polynomial $x^3 - 2x^2 + 2x - 1$ into a product of irreducible factors in $\mathbb{Z}_7[x]$.

7. Factor the polynomial $2x^3 + 3x^2 - 7x - 5$ into a product of irreducible factors in $\mathbb{Z}_{11}[x]$.

8. Let $\varphi \colon \mathbb{C}[x_1, \ldots, x_n] \to \mathbb{C}$ be the ring homomorphism fixing $\mathbb{C}$ and sending x_i to a_i and thus sending $f(x_1, \ldots, x_n)$ to $f(a_1, \ldots, a_n)$. Show that

$$\ker \varphi = (x_1 - a_1, \ldots, x_n - a_n).$$

17.2   Unique Factorization Domains

We say a nonzero non-unit element a in an integral domain D has a **factorization into irreducible elements** if

$$a = u p_1 p_2 \cdots p_s$$

where u is a unit and all the p_i's are irreducible in D. Let $u_1, u_2, \ldots, u_s$ be units of D. Then

$$a = u(u_1 u_2 \cdots u_s)^{-1}(u_1 p_1)(u_2 p_2) \cdots (u_s p_s)$$

is considered to be *essentially the same* factorization into irreducible elements of a since $p_i \sim u_i p_i$ for all i. Recall that the associate of an irreducible element remains irreducible (see Exercise 1, §17.1). For example, in $\mathbb{Z}$,

$$6 = 2 \cdot 3 = (-2)(-3) = (-1)2(-3) = (-1)(-2)3$$

are all essentially the same factorization into irreducible elements.

Definition 17.2.1. Let D be an integral domain. We say $a = u p_1 p_2 \cdots p_s$, u unit, is an **essentially unique factorization** into irreducible elements in D if for any other factorization into irreducibles

$$a = u p_1 p_2 \cdots p_s = v p_1' p_2' \cdots p_t',$$

where v is a unit, we have $s = t$ and $p_i \sim p_i'$ after a suitable reordering of the p_i''s.

Definition 17.2.2. We say an integral domain D is a **factorial** domain or a **unique factorization domain**, or simply a **UFD**, if every nonzero non-unit element in D has an *essentially unique* factorization into irreducible elements.

In our experience $\mathbb{Z}$, $\mathbb{C}[x]$ and $\mathbb{R}[x]$ all seem to be UFD's. We intend to find out if this is true.

We have the following familiar result on UFD's.

Proposition 17.2.3. *In a factorial domain D, let*

$$(17.2.1) \qquad \begin{cases} a = up_1^{r_1} p_2^{r_2} \cdots p_t^{r_t}, & u \in U(D) \text{ and } r_i \geq 0 \text{ for all } i, \\ b = vp_1^{s_1} p_2^{s_2} \cdots p_t^{s_t}, & v \in U(D) \text{ and } s_i \geq 0 \text{ for all } i \end{cases}$$

where the p_i's are distinct irreducible elements. Then

$$a \mid b \quad \Longleftrightarrow \quad r_i \leq s_i \text{ for all } i.$$

When we say distinct irreducible elements we mean that these irreducible elements are not associate with each other. For example, 3 and -3 are not considered distinct irreducible elements in $\mathbb{Z}$. For convenience' sake, we allow the exponents of the p_i's to be 0 so as to include unnecessary but helpful irreducible factors.

Proof. The "$\Leftarrow$" part is clear. We prove the "$\Rightarrow$" part. Let $b = ac$. Factor c as

$$c = wp_1^{e_1} p_2^{e_2} \cdots p_t^{e_t} p_{t+1}^{e_{t+1}} \cdots p_f^{e_f}$$

where w is a unit, the p_i's are distinct irreducible elements and $e_i \geq 0$ for all i. Thus we have

$$vp_1^{s_1} p_2^{s_2} \cdots p_t^{s_t} = uwp_1^{r_1+e_1} p_2^{r_2+e_2} \cdots p_t^{r_t+e_t} p_{t+1}^{e_{t+1}} \cdots p_f^{e_f}.$$

By the essential uniqueness of factorization, we have that $r_i + e_i = s_i$ for $1 \leq i \leq t$ and $e_{t+1} = \cdots = e_f = 0$. Hence $r_i \leq s_i$ for all i. $\qquad \square$

The following criterion is the main reason for the existence of factorization in an integral domain.

- **Ascending chain condition (ACC) for principal ideas.** Every ascending chain of principal ideals eventually stabilizes. To be more precise, for each ascending chain of principal ideals

$$(a_1) \subseteq (a_2) \subseteq (a_3) \subseteq \cdots\cdots,$$

there is a positive integer N such that

$$(a_N) = (a_{N+1}) = (a_{N+2}) = \cdots.$$

Remember that for any a and b in an integral domain D, we have the following properties:

(i) $(a) \subseteq (b) \iff b \mid a$, and

(ii) $(a) = (b) \iff a \sim b$.

Hence the ACC for principal ideals is equivalent to the following condition.

- **Divisor chain condition.** If we are given a divisor chain, that is, a sequence of elements $a_1, a_2, a_3, \ldots$ in D such that

$$a_2 \mid a_1, \ a_3 \mid a_2, \ a_4 \mid a_3, \ a_5 \mid a_4, \ a_6 \mid a_5, \ \ldots\ldots$$

then there is a positive integer N (large enough) such that

$$a_N \sim a_{N+1} \sim a_{N+2} \sim \cdots.$$

Next is a criterion for the essential uniqueness of factorization under ACC for principal ideals.

- **Primeness condition.** Every irreducible element is prime.

Corollary 17.1.10 says that a PID satisfies the primeness condition.

Now we have the main theorem of this section.

Theorem 17.2.4. *An integral domain is a UFD if and only if it satisfies the divisor chain condition and the primeness condition.*

Proof. The "only if" part: Let D be a UFD. Let $a_1, a_2, a_3, \ldots$ be a divisor chain in D such that

$$a_2 \mid a_1, \ a_3 \mid a_2, \ a_4 \mid a_3, \ a_5 \mid a_4, \ a_6 \mid a_5, \ \ldots\ldots$$

Let the *length* of an element be the number of irreducible factors (counting multiplicity) in an essentially unique factorization. Proposition 17.2.3 tells us that the length of the a_i's is decreasing. Since the length of the a_i's cannot decrease indefinitely, there must be an N such that the length of a_i stabilizes for $i \geq N$. It follows that $a_N \sim a_{N+1} \sim \cdots$.

Next we verify the primeness condition. Let $p \mid ab$. Factor a and b as in (17.2.1) where $p_1 = p$. Then by Proposition 17.2.3, $1 \leq r_1 + s_1$. It follows that $r_1 > 0$ or $s_1 > 0$ and so $p \mid a$ or $p \mid b$.

The "if" part: Let the integral domain D satisfies the divisor chain condition and the primeness condition. Suppose there is a nonzero nonunit a in D that is not a finite product of irreducible elements in D. Then a is reducible over D. Let $a = a_1 b_1$ where a_1 and b_1 are both non-units. Then a_1 and b_1 cannot both be expressed as finite products of irreducible elements. Without loss of generality we may assume a_1 is not a finite product of irreducible elements. Repeat this process. For all i we have that $a_i = a_{i+1} b_{i+1}$ such that a_{i+1} and b_{i+1} are both non-units and a_{i+1} is not a finite product of irreducible elements. Thus, we have a divisor chain of elements $a_1, a_2, a_3, \ldots$ in D such that $a_i \nsim a_{i+1}$ for all i. This contradicts the divisor chain condition. Hence every element in D has a factorization into irreducible elements. For the uniqueness part, let

$$a = u p_1 p_2 \cdots p_s = v p_1' p_2' \cdots p_t',$$

be two factorizations into irreducible elements. Without loss of generality, we may assume $s \leq t$. We prove by induction on s. If $s = 1$, then $a \sim p_1$ is irreducible. This implies that $t = 1$, for otherwise $v p_1' p_2' \cdots p_t'$ is reducible. We now assume $s > 1$. Since p_s is a prime by the primeness condition and p_s is not a divisor of v, we must have that $p_s \mid p_i'$ for some i. We might as well assume that $p_s \mid p_t'$. Let $p_t' = p_s w$ for some w. Since p_t' is irreducible and p_s is not a unit, we have that w is a unit. Thus $p_s \sim p_t'$. Canceling p_s from both factorizations of a we have that

$$u p_1 p_2 \cdots p_{s-1} = v w p_1' p_2' \cdots p_{t-1}'$$

where vw is a unit. Now by the induction hypothesis we have that $s - 1 = t - 1$ and $p_i \sim p_i'$ for $i = 1, \ldots, s - 1$ after a reordering of the p_i's. □

For any nonzero elements a and b in an integral domain D, we say an element c is a **common divisor** if $c \,|\, a$ and $c \,|\, b$ in D. An element d is a **greatest common divisor**, or simply a **g. c. d.**, of a and b if

(i) d is a common divisor of a and b, and

(ii) if c is a common divisor of a and b, then $c \,|\, d$.

If d is a greatest common divisor of a and b, then so is any associate of d. The greatest common divisor is not unique if it exists. However, it is unique up to associates. We use (a, b) to denote the class of all the associates of d and we write $(a, b) \sim d$. In particular, if $(a, b) \sim 1$ we say a and b are **relatively prime**.

In an integral domain D, We say n is a **common multiple** of a and b if $a \,|\, n$ and $b \,|\, n$. We call m a **least common multiple**, or simply a **l. c. m.**, of a and b if

(i) m is a common multiple of a and b, and

(ii) if n is any common multiple of a and b then $m | n$.

The least common multiple is unique up to associates. When m is a l. c. m., we denote it by $[a, b] \sim m$.

Proposition 17.2.5. *Suppose D is a UFD and let a, $b \in D$. Suppose there are distinct irreducible elements $p_1, p_2, \ldots, p_t$ in D such that*

$$a = u p_1^{e_1} p_2^{e_2} \cdots p_t^{e_t} \qquad and \qquad b = v p_1^{f_1} p_2^{f_2} \cdots p_t^{f_t}$$

where u, v are units and $e_i, f_i \geq 0$. Then

$$(a, b) \sim p_1^{g_1} p_2^{g_2} \cdots p_t^{g_t} \qquad and \qquad [a, b] \sim p_1^{h_1} p_2^{h_2} \cdots p_t^{h_t}$$

where $g_i = \min\{e_i, f_i\}$ and $h_i = \max\{e_i, f_i\}$.

This is an immediate result of Proposition 17.2.3, and we leave it as an exercise (see Exercise 2).

We now isolate the property on g. c. d. in Proposition 17.2.5 into the following condition.

- **The g. c. d. condition.** Any two elements have a g. c. d..

Lemma 17.2.6. *Let D be an integral domain with the g. c. d. condition. Any finite number of nonzero elements $a_1, \ldots, a_r$ in D have a g. c. d., that is, there exists d in D such that $d \mid a_i$, $i = 1, \ldots, r$, and if $e \in D$ is such that $e \mid a_i$, $i = 1, \ldots, r$, then $e \mid d$.*

Proof. Let $d_1 = (a_1, a_2)$, $d_2 = (d_1, a_2), \ldots, d_r = (d_{r-1}, a_r)$. Then it follows from definition that d_r is the g. c. d. of $a_1, \ldots, a_r$. $\qquad\square$

We use $(a_1, a_2, \ldots, a_r)$ to denote the g. c. d. of $a_1, \ldots, a_r$. Observe that

$$(17.2.2) \qquad\qquad ((a, b), c) \sim (a, b, c) \sim (a, (b, c)).$$

Lemma 17.2.7. *Suppose a, b and c are elements in an integral domain with the g. c. d. condition. Then the following statements are true:*

(a) $c(a, b) \sim (ca, cb)$;

(b) *if $(a, b) \sim 1$ and $(a, c) \sim 1$ then $(a, bc) \sim 1$.*

Proof. (a) Let $(a, b) \sim d$ and $(ca, cb) \sim e$. Since $d \mid a$ and $d \mid b$ we have that $cd \mid ca$ and cb. Thus, $cd \mid e$. Conversely, since c is a common divisor of ca and cb, $c \mid e$. Let $e = cf$. It follows that $f \mid a$ and b. Thus, $f \mid d$ and $e = cf \mid cd$. We have proved that $e \sim cd$.

(b) Using (a) we have that

$$1 \sim (a, c) \sim (a, (a, b)c) \sim (a, (ac, bc)) \sim ((a, ac), bc) \sim (a, bc). \qquad\square$$

Lemma 17.2.8. *In an integral domain, the g. c. d. condition implies the primeness condition.*

Proof. Let D be an integral domain with the g. c. d. condition. Let p be an irreducible element in D and let a, $b \in D$. If $p \nmid a$ and $p \nmid b$ then we should have $(p, a) \sim 1$ and $(p, b) \sim 1$. By Lemma 17.2.7(b), it follows that $(p, ab) \sim 1$. Hence $p \nmid ab$. $\qquad\square$

Theorem 17.2.9. *Let D be an integral domain. Then D is a UFD if and only if the divisor chain condition and the g. c. d. condition hold in D.*

Proof. The "only if" part follows from Theorem 17.2.4 and Proposition 17.2.5. The "if" part follows from Theorem 17.2.4 and Lemma 17.2.8. □

Lemma 17.2.10. *Let a and b be elements in an integral domain D. If (a, b) is the principal ideal (d), then d is a g. c. d. of a and b.*

In particular, the g. c. d. condition is satisfied in a PID.

Proof. Since $a, b \in (d)$ we have that $a = dr$ and $b = dr'$ for some $r, r' \in D$. Thus, d is a common divisor of a and b. If e is a common divisor of a and b, then $a, b \in (e)$. This implies that $(d) = (a, b) \subseteq (e)$. Hence $d \in (e)$ and e is a divisor of d. The rest is easy. □

Proposition 17.2.11. *A PID is a UFD.*

Proof. Let D be a PID. Thanks to Lemma 17.2.10 (or Corollary 17.1.11) and Theorem 17.2.9 it remains to show that D satisfies the ACC for principal ideals. Let

$$(a_1) \subseteq (a_2) \subseteq (a_3) \subseteq \cdots\cdots,$$

be an ascending chain of principal ideals in D. Then $I = \bigcup_{i=1}^{\infty}(a_i)$ is also an ideal of D (see Exercise 10 on P. 220). Since D is a PID, $I = (a)$ for some $a \in I$. Since a is in the union of the (a_i)'s, there exists N such that $a \in (a_N)$. It then follows that $I = (a) \subseteq (a_N)$. Hence we have that $(a_i) = (a)$ for all $i \geq N$. □

Now we have given the full reasons why rings such as $\mathbb{Z}$, $\mathbb{R}[x]$ and $\mathbb{C}[x]$ are UFD's. We have also found other UFD's such as $\mathbb{Z}[i]$.

Corollary 17.2.12. *In a PID, such as $F[x]$ where F is a field, two elements a and b are relatively prime if and only if there are r and $s \in D$ such that $1 = ar + bs$.*

Proof. Let D be the given PID. Assume $(a, b) = (d)$. By Lemma 17.2.10 we have that d is a g. c. d. of a and b. Hence a and b are relatively prime if and only if $(a, b) = (1)$ if and only if $1 = ar + bs$ for some r and s in D. □

To conclude this section we provide a criterion for finding zeros of polynomials over a UFD.

Proposition 17.2.13. *Let D be a UFD and Q be its quotient field. Let*

$$f(x) = a_0 + a_1 x + \cdots + a_n x^n \in D[x]$$

where a_0, $a_n \neq 0$ and $n \geq 1$. Let $v/u \in Q$, where $(u, v) \sim 1$, be a zero of $f(x)$. Then $u \mid a_n$ and $v \mid a_0$.

Proof. Since $f(v/u) = 0$, we have the relation

$$a_0 u^n + a_1 v u^{n-1} + a_2 v^2 u^{n-2} + \cdots + a_{n-1} v^{n-1} u + a_n v^n = 0.$$

This implies that u is a factor of $a_0 u^n + a_1 v u^{n-1} + a_2 v^2 u^{n-2} + \cdots + a_{n-1} v^{n-1} u = -a_n v^n$. It follows that $u \mid a_n$ (see Exercise 1(b)). Similarly we have $v \mid a_0$. $\square$

Exercises 17.2

1. In a UFD D, let $(a, b) \sim 1$ and let $c \in D$.

   (a) Show that if $a \mid c$ and $b \mid c$ then $ab \mid c$.

   (b) Show that if $a \mid b^n c$ for some positive integer n then $a \mid c$.

2. Prove Proposition 17.2.5.

3. Show that in a UFD, two elements are relatively prime if and only if either a or b is a unit or no irreducible factor of either one is a factor of both.

4. Show that $p(x) = x^3 - 2x + 2$ and $q(x) = x^3 + 3x - 2$ are irreducible in $\mathbb{Q}[x]$.

5. Show that in a UFD, every nonzero non-unit element has only finitely many non-associated irreducible factors.

6. Let F be a field and let x be an indeterminate over F. Show that $F[x]/(f(x))$ contains a nonzero nilpotent element if and only if $f(x) = g(x)^2 q(x)$ for some $g(x)$ with $\deg g(x) > 0$.

7. Let D be a UFD and let Q be its quotient field. Suppose that $f(x) \in D[x]$ is monic. Show that any zero of $f(x)$ in Q is in D.

8. Let $R = \{a_0 + a_1 x + \cdots a_n x^n \in \mathbb{Q}[x] : a_0 \in \mathbb{Z}\}$.

   (a) Show that R is a subring of $\mathbb{Q}[x]$.

   (b) Show that R is not a UFD. (Hint: Show that R does not satisfy the divisor chain condition.)

17.3 Polynomial Extensions of Factorial Domains

Throughout this section x denotes an indeterminate.

The following theorem is the main result in this section.

Theorem 17.3.1. *If D is a UFD then so is $D[x]$.*

In proving this theorem, we will learn also how to handle polynomials over a UFD such as $\mathbb{Z}$. Rings such as $\mathbb{Z}[x]$ is more difficult since they are not PID's. First, we introduce a simple yet useful notion.

Definition 17.3.2. Let $f(x) = a_0 + a_1 x + \cdots + a_n x^n$ be a nonzero polynomial over a UFD D. We define the **content** $c(f)$ of f over D to be the g.c.d. of $a_0, a_1, \ldots, a_n$ in D.

Let $c(f) \sim d$ and let $a_i = da_i'$ for $i = 1, 2, \ldots, n$. Evidently $f = df_1$ where

$$f_1(x) = a_0' + a_1' x + \cdots + a_n' x^n.$$

From (17.2.2) and Lemma 17.2.7 we have that

$$d \sim (a_0, a_1, \ldots, a_n) \sim (da_0', da_1', \ldots, da_n') \sim d(a_0', a_1', \ldots, a_n').$$

Hence $(a_0', a_1', \ldots, a_n') \sim 1$.

We say a nonzero polynomial f over a UFD D is **primitive** if $c(f) \sim 1$.

Lemma 17.3.3. *Let D be a UFD and let Q be its quotient field. Let $f(x)$ be a nonzero polynomial in $Q[x]$. There exists a primitive polynomial $f_1(x)$ in $D[x]$ and $c \in Q$ such that $f(x) = cf_1(x)$. Moreover, the choice of c and $f_1(x)$ is unique up to unit multipliers in D.*

The c (unique up to a unit multiplier in D) here is called the **content** of $f(x)$ over D, which is also denoted by $c(f)$.

Proof. Let
$$f(x) = q_0 + q_1 x + \cdots + q_n x^n$$
where $q_i = b_i/a_i$ with $b_i \in D$ and $a_i \in D^*$ for $i = 0, 1, \ldots, n$. Then
$$g(x) = (a_0 a_1 \cdots a_n) f(x) \in D[x].$$

Let $g(x) = c(g) f_1(x)$ where $f_1(x)$ is primitive over D. Then $f(x) = c f_1(x)$ where $c = c(g)(a_0 a_1 \cdots a_n)^{-1} \in Q$ as required.

Now suppose $f(x) = c' f_2(x)$ for another $c' \in Q$ and another primitive polynomial $f_2(x)$ in $D[x]$. Let $c = b/a$ and $c' = b'/a'$ where $b, b' \in D$ and $a, a' \in D^*$. Then $a'b f_1(x) = ab' f_2(x)$ in $D[x]$. Since
$$a'b \sim c(a'b f_1(x)) \sim c(ab' f_2(x)) \sim ab',$$
we can find a unit u in D such that $a'b = uab'$. It follows that $c = uc'$ and $f_1(x) = u^{-1} f_2(x)$. $\hspace{2cm}$ □

For all the following three lemmas, we assume that D is a UFD and Q is its quotient field.

Lemma 17.3.4 (Gauss's Lemma). *Let $f(x)$ and $g(x) \in D[x]$ be primitive polynomials. Then $f(x)g(x)$ is primitive over D.*

Proof. Suppose that $f(x)g(x)$ is not primitive over D. Then there is a prime element p in D which is a factor of $c(fg)$. Consider the ring epimorphism $\pi: D \to D/(p)$. It can be extended to a ring homomorphism from $D[x]$ to $(D/(p))[x]$ sending x to x. This homomorphism sends $f(x) = \sum_i a_i x^i$ to $\overline{f(x)} = \sum_i \overline{a_i} x^i$. Then $\overline{f(x)g(x)} = \overline{f(x)}\ \overline{g(x)} = \overline{0}$. Since p is a prime, (p) is a prime ideal. Thus $D/(p)$ is an integral domain. It follows that $(D/(p))[x]$ is also a domain. Hence $\overline{f(x)} = 0$ or $\overline{g(x)} = 0$. Assume $\overline{f(x)} = 0$. Then p is a factor of $c(f)$ and $f(x)$ is not primitive over D. $\hspace{1cm}$ □

Lemma 17.3.5. *Let $f(x)$, $g(x) \in D[x]$ and $g(x)$ be primitive over D. If $g(x) \mid f(x)$ in $Q[x]$ then $g(x) \mid f(x)$ in $D[x]$.*

Proof. Let $f(x) = g(x)q(x)$ for some $q(x) \in Q[x]$. Let $q(x) = c(q)q_1(x)$ where $q_1(x)$ is primitive over D. Since $f(x) = c(q)g(x)q_1(x)$ and $g(x)q_1(x)$ is primitive by Gauss Lemma, we have that $c(q) \sim c(f) \in D$ by Lemma 17.3.3. Hence $q(x) \in D[x]$. $\qquad\square$

Lemma 17.3.6. *If $f(x) \in D[x]$ is of positive degree and irreducible in $D[x]$, then $f(x)$ is irreducible in $Q[x]$.*

Proof. If $f(x)$ is not irreducible in $Q[x]$, then $f(x) = g(x)h(x)$ for some $g(x), h(x) \in Q[x]$ with $\deg g(x)$, $\deg h(x) < \deg f(x)$. Let $g(x) = c(g)g_1(x)$ where $g_1(x)$ is primitive over D. Then $g_1(x) \mid f(x)$ in $Q[x]$. It follows that $g_1(x) \mid f(x)$ in $D[x]$ by Lemma 17.3.5. Hence $f(x)$ is reducible over D. $\quad\square$

Remember that any polynomial of degree 0 is a unit in $Q[x]$ while it is not necessarily a unit in $D[x]$. For example, $2x + 4 = 2(x + 2)$ is a meaningful factorization in $\mathbb{Z}[x]$ since 2 is not a unit in $\mathbb{Z}[x]$. Thus $2x + 4$ is reducible over $\mathbb{Z}$. However, $2x + 4$ is irreducible in $\mathbb{Q}[x]$ since it is of degree 1. This says that the converse of Lemma 17.3.6 is not true.

Proposition 17.3.7. *Let D be a UFD and Q be its quotient field. Let $f(x) \in D[x]$. Then $f(x)$ is irreducible over D if and only if one of the following conditions holds:*

(i) $f(x)$ *is a prime in D;*

(ii) $f(x)$ *is primitive over D and irreducible over Q.*

Proof. When $f(x) \in D$ the case is clear. We assume that $\deg f(x) > 0$. In this case, an irreducible polynomial over D is obviously primitive since its content must be a unit in D. The "only if" part now follows from Lemma 17.3.6. To prove the "if" part, let $f(x) = g(x)h(x)$ in $D[x]$. Since $f(x)$ is irreducible over Q, one of $g(x)$ or $h(x)$, say $g(x)$, must be of degree 0. Now $g(x) = a \in D$. It follows that $a \mid c(f) \sim 1$. Hence a is a unit in D. We have proved that $f(x)$ is irreducible over D. $\qquad\square$

Thus, to factor $f(x)$ in $D[x]$ we first find

$$f(x) = c(f)f_1(x), \qquad c(f) \in D.$$

Then we factor $c(f)$ in D and $f_1(x)$ in $Q[x]$.

Example 17.3.8. Is $f(x) = 3x^2 + 3$ irreducible over $\mathbb{Q}$? Over $\mathbb{Z}$? Over $\mathbb{Z}_2$?

Solution. Over $\mathbb{Q}$, $f(a) > 0$ for all $a \in \mathbb{Q}$. Thus $f(x)$ has no linear factors over $\mathbb{Q}$ and $f(x)$ is irreducible over $\mathbb{Q}$. Over $\mathbb{Z}$, $f(x) = 3(x^2 + 1)$ is reducible. Over $\mathbb{Z}_2$, $f(1) = 0$. Hence $f(x) = x^2 + 1 = (x + 1)^2$ is reducible. ◇

We are now ready to prove the main theorem of this section.

Proof of Theorem 17.3.1. Let Q be the quotient field of D. Let f_1, f_2, f_3, ... be a divisor chain in $D[x]$. Consider this chain in the factorial domain $Q[x]$. There exists N such that $f_N \sim f_{N+1} \sim \cdots$ in $Q[x]$. In other words, there is a primitive polynomial f such that $f_i = c(f_i)f$ for all $i \geq N$. It follows that $c(f_N)$, $c(f_{N+1})$, $c(f_{N+2})$, ... is a divisor chain in D. Since D is a UFD, there exists M such that $c(f_{N+M}) \sim c(f_{N+M+1}) \sim \cdots$ in D. We conclude that $f_{N+M} \sim f_{N+M+1} \sim \cdots$ in D. Thus D satisfies the divisor chain condition.

Next let $p(x)$ be an irreducible polynomial in $D[x]$ and let $f(x)$, $g(x)$ be polynomials over D. Suppose $p(x) \mid f(x)g(x)$ over D. If $p(x) = a \in D$, then $a \mid c(fg) = c(f)c(g)$ (see Exercise 1) in D. Since a is irreducible over the factorial domain D, we have that $a \mid c(f)$ or $a \mid c(g)$ in D by Proposition 17.3.7. Hence $p(x) = a \mid f(x)$ or $g(x)$ over D. We now assume that $p(x)$ is of positive degree. Then $p(x)$ is also a divisor of $f(x)g(x)$ over Q. By Lemma 17.3.6 $p(x)$ is irreducible over Q. Since $p(x)$ is a prime in $Q[x]$, we may assume without loss of generality that $p(x) \mid f(x)$ over Q. Since $p(x)$ is primitive by Proposition 17.3.7, we now have $p(x) \mid f(x)$ over D by Lemma 17.3.5. This proves the primeness condition. □

Corollary 17.3.9. *Let x_1, x_2, ..., x_n be indeterminates over D. If D is a UFD then so is $D[x_1, x_2, \ldots, x_n]$.*

Corollary 17.3.10. *Let $x_1, \ldots, x_n$ be indeterminates.*

(a) *Let F be any field. The polynomial ring $F[x_1, x_2, \ldots, x_n]$ is a UFD.*

(b) *The polynomial ring $\mathbb{Z}[x_1, x_2, \ldots, x_n]$ is a UFD.*

(c) *The polynomial ring $\mathbb{Z}[i][x_1, x_2, \ldots, x_n]$ is a UFD.*

We are now in a position to give one more criterion for testing irreducibility of polynomials.

Theorem 17.3.11 (Eisenstein Criterion). *Let D be a UFD and let Q be its quotient field. Let*

$$f(x) = a_n x^n + a_{n-1} x^{n-1} + \cdots + a_1 x + a_0$$

be a polynomial of degree $n \geq 1$ in $D[x]$. Suppose that there is a prime element p in D such that

(i) $p | a_i$ *for* $i = 0, 1, \ldots, n-1$,

(ii) $p \nmid a_n$, *and*

(iii) $p^2 \nmid a_0$.

Then $f(x)$ is irreducible in $Q[x]$.

Proof. Assume that $f(x)$ is reducible in $Q[x]$. By Lemma 17.3.5, we have that $f(x)$ is reducible over D, and we can find $u(x)$ and $v(x)$ in $D[x]$ such that

$$f(x) = u(x)v(x), \qquad \text{where } \deg u(x) \text{ and } \deg v(x) > 0.$$

Map this relation into $(D/pD)[x]$, we have

$$\overline{a_n} x^n = \overline{u(x)} \ \overline{v(x)}$$

in $(D/pD)[x]$. Note that $\overline{a_n} \neq 0$ since $p \nmid a_n$. Suppose the lowest terms of $\overline{u(x)}$ and $\overline{v(x)}$ are of degrees k and ℓ respectively. Since D/pD is an integral domain, the lowest term of $\overline{u(x)} \ \overline{v(x)}$ is of degree $k + \ell$. Thus, $k + \ell = n$ and $\overline{u(x)} = \overline{b} x^k$ and $\overline{v(x)} = \overline{c} x^\ell$ for some $b, c \in D$. Since $k, \ell < n$, this also implies that $k = \deg u(x) > 0$ and $\ell = \deg v(x) > 0$. It follows that the constant terms of $u(x)$ and $v(x)$ are both multiples of p. Thus, p^2 divides the product of the constant terms of $u(x)$ and $v(x)$, which is also the constant term of $f(x)$. We have reached a contradiction. $\square$

Example 17.3.12. Let m be any square-free integer other than ± 1. Find any p which is a prime factor of m. Then $x^n - m$ and p satisfies Eisenstein criterion. Hence, $x^n - m$ is an irreducible polynomial in $\mathbb{Q}[x]$. Let α be any zero of $x^n - m$ in $\mathbb{C}$. Let $\varphi: \mathbb{Q}[x] \to \mathbb{Q}[\alpha]$ be the ring homomorphism sending $\mathbb{Q}$ onto $\mathbb{Q}$ and x to α. Then φ is clearly onto and $(x^n - m) \subseteq \ker \varphi$. However, since φ is non-trivial, $\ker \varphi$ must be a proper ideal. By Theorem 17.1.10, $(x^n - m)$ is a maximal ideal in $\mathbb{Q}[x]$. We have that $\ker \varphi = (x^n - m)$. Hence

$$\mathbb{Q}[x]/(x^n - m) \cong \mathbb{Q}[\alpha].$$

We have also shown that $\mathbb{Q}[\alpha]$ is a field.

Observe that $x^n - m$ is primitive over $\mathbb{Z}$. Thus by Proposition 17.3.7, $x^n - m$ is irreducible over $\mathbb{Z}$ as well.

For any prime integer $p \geq 2$, the polynomial

$$\Phi_p(x) = \frac{x^p - 1}{x - 1} = 1 + x + x^2 + \cdots + x^{p-1} \in \mathbb{Z}[x]$$

is called the p-th **cyclotomic polynomial**.[1]

Corollary 17.3.13. *For a positive prime integer p, the p-th cyclotomic polynomial $\Phi_p(x)$ is irreducible over $\mathbb{Q}$ and over $\mathbb{Z}$.*

Proof. If Φ_p is reducible over $\mathbb{Q}$, then so is

$$f(y) = \Phi_p(y + 1) = \frac{(y+1)^p - 1}{y} = \sum_{j=1}^{p} \binom{p}{j} y^j$$

(see Exercise 4). The leading coefficient of $f(y)$ is 1 while $p \mid \binom{p}{j}$ for $1 \leq j \leq p - 1$. The constant term is $\binom{p}{1} = p$. Hence $f(y)$ is irreducible in $\mathbb{Q}[y]$ by Eisenstein criterion, a contradiction. Thus Φ_p is irreducible over $\mathbb{Q}$. Since Φ_p is primitive over $\mathbb{Z}$, it is also irreducible over $\mathbb{Z}$. $\square$

Exercises 17.3

In the following exercises, D always denotes a UFD and Q is its quotient field.

---

[1]The general n-th cyclotomic polynomial $\Phi_n(x)$ is defined to be the monic polynomial of degree $\varphi(n)$ with $e^{2k\pi i/b}$, $1 \leq k \leq n - 1$ and $(n, k) \sim 1$, as its zeros.

1. Let $f(x)$ and $g(x)$ be polynomials over Q. Show that $c(fg) = c(f)c(g)$.

2. Show that $p(x) = x^4 - 2x^2 + 8x + 1$ is irreducible in $\mathbb{Q}[x]$.

3. Show that $p(x) = x^3 - 3x^2 + 8$ and $q(x) = x^4 - x^2 + 1$ are irreducible in $\mathbb{Q}[x]$.

4. Let D be a UFD. Let $f(x)$ be a polynomial over D and let $c \in D$. Show that $f(x)$ is irreducible over D if and only if $f(x-c)$ is irreducible over D.

5. Show that $\mathbb{R}[x]/(x^2 + x + 1) \cong \mathbb{C}$.

6. Let p be a positive prime integer and $n \geq 2$ be an integer. Show that the polynomial
$$f(x) = \frac{x^{p^n} - 1}{x^{p^{n-1}} - 1}$$
is irreducible in $\mathbb{Q}[x]$.

Review Exercises for Chapter 17

1. Find all the irreducible polynomials of degree 3 and of degree 4 in $\mathbb{Z}_2[x]$.

2. Show that $\mathbb{Z}[\omega] = \{m + n\omega \mid m, n \in \mathbb{Z}\}$, where $\omega = (1 + \sqrt{-3})/2$. Show that $\mathbb{Z}[\omega]$ is a Euclidean domain.

3. Show that $2\sqrt{2} + i$ is a zero of the polynomial $f(x) = x^4 - 14x^2 + 81$. Show that $f(x)$ is irreducible in $\mathbb{Q}[x]$.

4. Let $f(x)$ be an irreducible polynomial of degree n over the finite field $\mathbb{Z}_p$. Show that $\mathbb{Z}_p[x]/(f(x))$ is a field with p^n elements.

5. Construct a field with 343 elements.

The following is a mini-project for deciding which prime integers are irreducible in $\mathbb{Z}[i]$. In the following problems, p denotes a positive prime integer.

We say an integer a is a **quadratic residue** modulo p if

$$x^2 \equiv a \pmod p$$

has a solution. Otherwise we say a is a **quadratic nonresidue** modulo p. Define the **Legendre symbol** by

$$\left(\frac{a}{p}\right) = \begin{cases} 0, & \text{if } p \mid a; \\ 1, & \text{if } a \text{ is a quadratic residue modulo } p \text{ and } p \nmid a; \\ 1, & \text{otherwise.} \end{cases}$$

6. Show that p is reducible in $\mathbb{Z}[i]$ if and only if $a^2 + b^2 = p$ has integer solutions.

7. Show that $U(\mathbb{Z}_p) = \mathbb{Z}_p^*$ is a cyclic group.

8. Let p be a prime integer ≥ 3. Show that there are exactly $(p-1)/2$ quadratic residues in $\mathbb{Z}_p^*$.

9. Show that $a^2 + b^2 = p$ has an integer solution if and only if -1 is a quadratic residue modulo p.

10. Show that

$$\left(\frac{-1}{p}\right) = \begin{cases} 1, & \text{if } p = 2 \text{ or } p \equiv 1 \pmod 4; \\ -1, & \text{if } p \equiv 3 \pmod 4. \end{cases}$$

11. Conclude that p is a prime in $\mathbb{Z}[i]$ if and only if $p \equiv 3 \pmod 4$.

CHAPTER 18

Introduction to Modules

In this book there is not enough space to cover Module Theory. However, we think it is worthwhile to give a brief introduction. Vector spaces are really just modules over a field.

18.1   Modules and Submodules

In this section, we *do not* assume that the rings involved to be commutative. However, we *do* assume that R contains the unity.

Definition 18.1.1. Let R be a ring with unity. A **left R-module** (or a **left module over** R) is an *abelian* group $(M, +)$ together with a map

$$
\begin{aligned}
R \times M &\longrightarrow M \\
(a, m) &\longmapsto a \cdot m
\end{aligned}
$$

satisfying the following properties:

(i) $a \cdot (m + n) = a \cdot m + a \cdot n$;

(ii) $(a + b) \cdot m = a \cdot m + b \cdot m$;

(iii) $(ab) \cdot m = a \cdot (b \cdot m)$;

(iv) $1 \cdot m = m$

for $m, n \in M$ and $a, b \in R$. This map is referred to as the **scalar multiplication**. The scalar multiplication $a \cdot m$ is often simply written as am when no confusion arises.

If in particular the ring $R = F$ is a field, we usually call an F-module an **F-vector space** or a **vector space over** F.

Condition (iv) in Definition 18.1.1 is optional for the more general module theory. However, we elect to add it in for this class of modules is at the center of module theory. Modules satisfying Definition 18.1.1 are called **unitary** modules. Throughout the whole book we will assume our modules to be unitary.

Example 18.1.2. Let R be a ring with unity. If I is a left ideal of R, then I is also a left module over R.

Example 18.1.3. Remember that in the abelian group

$$
R^n = \underbrace{R \times \cdots \times R}_{n \text{ copies}}
$$

we have the addition

$$(a_1, \ldots, a_n) + (b_1, \ldots, b_n) = (a_1 + b_1, \ldots, a_n + b_n)$$

for a_i, $b_j \in R$. It becomes a left R-module if we let

$$a(a_1, \ldots, a_n) = (aa_1, \ldots, aa_n),$$

where $a, a_1, \ldots, a_n \in R$.

In particular, R itself may be considered as a left R-module where the scalar multiplication is given by the left multiplication in R.

Example 18.1.4. Let A be a commutative ring with unity and let $R = M_n(A)$ be the matrix ring. Let M be the abelian group of column n-vector in A^n. Then M can be made into a left R-module by using the matrix multiplication as the scalar multiplication.

Example 18.1.5. Let F be a field. In a course on Linear Algebra one has that $M_{m \times n}(F)$ is an F-vector space. Let R be a commutative ring with unity. In a similar fashion, $M_{m \times n}(R)$ is a left R-module.

Proposition 18.1.6. *Let M be a left R-module. Then the following statements are true:*

(i) $0_R \cdot m = 0_M$;

(ii) $a \cdot 0_M = 0_M$;

(iii) $(-a) \cdot m = -(a \cdot m) = a \cdot (-m)$ *and in particular* $(-1) \cdot m = -m$

for all $m \in M$ and $a \in R$.

Proof. (i) Since

$$0_R \cdot m = (0_R + 0_R) \cdot m = 0_R \cdot m + 0_R \cdot m,$$

we deduce that $0_R \cdot m = 0_M$ by the cancellation law for addition in M.

(ii) Since

$$a \cdot 0_M = a \cdot (0_M + 0_M) = a \cdot 0_M + a \cdot 0_M,$$

we deduce that $a \cdot 0_M = 0_M$ by the cancellation law for addition in M.

(iii) We need to check that $(-a) \cdot m$ is the additive inverse of $a \cdot m$. For this we verify that

$$a \cdot m + (-a) \cdot m = \big(a + (-a)\big) \cdot m = 0_R \cdot m = 0_M.$$

Hence $(-a) \cdot m = -(a \cdot m)$. Similarly, one may check that $a \cdot (-m) = -(a \cdot m)$.
For the last statement, note that $(-1) \cdot m = -(1 \cdot m) = -m$. $\qquad\square$

Proposition 18.1.7. *Let M be a left R-module. Define*

$$\operatorname{ann}_R M = \{a \in R : a \cdot m = 0 \text{ for all } m \in M\}.$$

*Then $\operatorname{ann}_R M$, or simply $\operatorname{ann} M$ if R is understood, is a left ideal in R. This is called the **annihilator** of M.*

Furthermore, if I is any left ideal contained in $\operatorname{ann} M$ then M becomes a left R/I-module by defining $(a + I) \cdot m = a \cdot m$.

Proof. Clearly, $0 \in \operatorname{ann} M$. Let a and $b \in \operatorname{ann} M$. Then

$$(a + b) \cdot m = a \cdot m + b \cdot m = 0 + 0 = 0,$$

and we have $a + b \in \operatorname{ann} M$. Let $r \in R$. Then

$$(ra) \cdot m = r \cdot (am) = r0 = 0,$$

and we have $ra \in \operatorname{ann} M$. We have shown that $\operatorname{ann} M$ is a left ideal of R.

Now let I be an ideal contained in $\operatorname{ann} M$. For $a \in R$ and $i \in I$, we have $(a + i) \cdot m = a \cdot m + i \cdot m = a \cdot m + 0 = a \cdot m$. Thus the scalar multiplication $(a + I) \cdot m = a \cdot m$ is well-defined. We leave it as an exercise that this makes M into a left R/I-module (see Exercise 1). $\qquad\square$

The notion of *right* R-modules is dual to that of the *left* R-modules.

Definition 18.1.8. Let R be a ring. A **right R-module** (or a **right module over R**) is an *abelian* group M together with a map

$$\begin{aligned} M \times R &\longrightarrow M \\ (m, a) &\longmapsto m \cdot a \end{aligned}$$

satisfying the following properties:

(i) $(m + n) \cdot a = ma + na$

(ii) $m \cdot (a + b) = ma + mb$

(iii) $m \cdot (ab) = (mb)a$

(iv) $m \cdot 1 = m$

for m, $n \in M$ and a, $b \in R$. The scalar multiplication $m \cdot a$ is often simplified as ma when no confusion arises.

Example 18.1.9. Contrary to Example 18.1.3, right multiplication does not make R a right R-module if R is not commutative. To make R into a right R-module, we define the scaler multiplication as $m \cdot a = ma^{-1}$ where a is the scalar. We leave the details as an exercise. See Exercise 2.

For every property that left modules possess, right modules also have its corresponding property. We will not repeat every theorem for the right modules. However, if you look closely you will see all the arguments presented work for both left and right modules as long as you adapt the statement accordingly.

When R is commutative, there is no distinction between left and right modules. In this case, we will simply call the left (right) modules *modules*! In fact, this is often the case that we are really interested in.

Example 18.1.10. Every abelian group is a $\mathbb{Z}$-module. To see this, we need to define the scalar multiplication.

To simplify matter, we will assume the abelian group G is an additive group. Let $g \in G$. We define $0_{\mathbb{Z}}g = 0_G$. Let k be a positive integer. Define

$$kg = \underbrace{g + \cdots + g}_{k \text{ times}};$$

$$(-k)g = \underbrace{(-g) + \cdots + (-g)}_{k \text{ times}}.$$

We leave it as an exercise to verify that this makes G into a $\mathbb{Z}$-module. See Exercise 3.

As we shall see in Exercise 8 of § 18.2, problems on abelian groups can be treated as problems on $\mathbb{Z}$-modules. Sometimes this is a better approach.

Definition 18.1.11. Let M be a module over R and let N be a subset. We say that N is a **submodule** of M over R, or an R-submodule of M, if N is a module itself under the inherited operations. In particular, a submodule of a vector space is often called a **subspace** instead.

Proposition 18.1.12. *A subset N of M is an R-submodule of M if and only if the following conditions are satisfied:*

(i) $N \neq \varnothing$,

(ii) $n + n' \in N$, *and*

(iii) $an \in N$

for all $a \in R$ and n, $n' \in N$.

Proof. The "only if" part: A module is by definition nonempty. Thus $N \neq \varnothing$. The operations of N is inherited from M, this forces conditions (ii) and (iii) must hold in N.

The "if" part: Conditions (ii) and (iii) guarantee that the nonempty set N may inherit a module addition and a scalar multiplication from M. Under these operations we have that $-n = (-1)n \in N$ for all $n \in N$. This says that N is an additive group. The rest of the requirements for a module hold in N since it is already so in M. $\qquad\square$

Example 18.1.13. Any ring may be viewed as a module over itself. The submodules of R are the ideals of R.

Example 18.1.14. Remember that any additive group G is a $\mathbb{Z}$-module. The submodules of G over $\mathbb{Z}$ are the subgroups H of G.

Example 18.1.15. Let V be a vector space over a field K. The submodules of V are the subspaces of V.

Lemma 18.1.16. *Let $\{N_i\}_{i \in \Lambda}$ be a nonempty family of submodules of the R-module M. Then $\bigcap_{i \in \Lambda} N_i$ is a submodule of M.*

Proof. Let $N = \bigcap_{i \in \Lambda} N_i$. Clearly, $0 \in N$ since 0 is contained in any submodule of M. Let n, $n' \in N$ and $a \in R$. Then n, $n' \in N_i$ for all $i \in \Lambda$. It follows that $n + n'$ and an are both in N_i for all $i \in \Lambda$. Thus $n + n'$ and an are both in N. The result follows from Proposition 18.1.12. $\qquad\square$

Lemma 18.1.17. *Let S be a subset of an R-module M. There is a smallest submodule N in M which contains S. More precisely, if P is any submodule containing S, we have $N \subseteq P$.*

Proof. Let $\mathscr{C} = \{P \subseteq M : P \text{ is a submodule of } M \text{ and } S \subseteq P\}$. Since $M \in \mathscr{C}$, $\mathscr{C}$ is nonempty. Let $N = \bigcap \mathscr{C}$. The N is a submodule of M by Lemma 18.1.16. It is clear that N contains S and N is the smallest submodule in $\mathscr{C}$. $\square$

Let S be a subset of an R-module M. By virtue of Lemma 18.1.17, we will use $\langle S \rangle$ to denote *the* smallest submodule of M containing S. If $S = \{m_1, \ldots, m_r\}$ is a finite set, we also use $\langle m_1, \ldots, m_r \rangle$ to denote $\langle S \rangle$.

If $M = \langle S \rangle$, we say that S forms a **generating set** for M or S **generates** M over R. If M can be generated by a finite set, we say that M is **finitely generated** over R. A module generated by a single element is called a **cyclic** module.

In the case of F-vector spaces, $\langle S \rangle$ is often called the **linear span** of S over F. If $V = \langle S \rangle$, we also say V is **spanned** by S over F. If S is a finite set, we say V is a **finite dimensional vector space** over F.

Example 18.1.18. Note that $\langle \varnothing \rangle$ is the trivial module since it is the smallest module of all. Certainly $\langle 0 \rangle$ is also the trivial module.

In the following discussion, we will encounter the awkward situation that we need to add "nothing" together. We call this the empty sum. For the sake of convenience, we define the empty sum to be the module element 0!

Proposition 18.1.19. *Let S be a subset of an R-module M. Then setwise*

$$(18.1.1) \quad \langle S \rangle = \{a_1 m_1 + \cdots + a_r m_r \in M : r = 0, 1, 2, 3, \ldots,$$
$$a_1, \ldots, a_r \in R, \ m_1, \ldots, m_r \in S\}.$$

Proof. Let N be the set in (18.1.1). It is clear that $N \subseteq \langle S \rangle$ by Proposition 18.1.12. It is also clear that $S \subseteq N$. If we can show that N is a submodule of M, then we also have $\langle S \rangle \subseteq N$ by Lemma 18.1.17.

To show that N is a submodule, first note that N contains 0 since it contains the empty sum. Let n and $n' \in N$ and $a \in R$. We may write

$$\begin{cases} n = a_1 m_1 + a_2 m_2 + \cdots + a_r m_r, \\ n' = b_1 m_1' + b_2 m_2' + \cdots + b_s m_s', \end{cases}$$

where r and s are nonnegative integers, a_i, $b_j \in R$ and m_i, $m_j' \in S$. Hence, we have

$$n + n' = a_1 m_1 + a_2 m_2 + \cdots + a_r m_r + b_1 m_1' + b_2 m_2' + \cdots + b_s m_s' \in N;$$
$$an = a a_1 m_1 + a a_2 m_2 + \cdots + a a_r m_r \in N.$$

We conclude that N is a submodule of M by Proposition 18.1.12. $\square$

Let $a_i \in R$ and $m_i \in M$ for all i. An element of the form

$$a_1 m_1 + a_2 m_2 + \cdots + a_r m_r$$

is called an R-**linear combination** or a **linear combination** of $m_1, \ldots, m_r$ over R. Proposition 18.1.19 says that $\langle S \rangle$ consists of R-linear combinations of elements in S. We often use

(18.1.2) $$R m_1 + R m_2 + \cdots + R m_r$$

to denote $\langle m_1, m_2, \ldots, m_r \rangle$. This is in fact the left module notation. If M is a right module, we leave it to reader to verify that it makes sense to write

$$\langle m_1, \ldots, m_r \rangle = m_1 R + m_2 R + \cdots + m_r R.$$

The fact that submodules are closed under scalar multiplication can be expressed by the relation $RN = N$. We have the following result.

Corollary 18.1.20. *Let* $\{N_i\}_{i \in \Lambda}$ *be a family of submodules of the R-module* M. *Then*

$$\sum_{i \in \Lambda} N_i = \{ m_{i_1} + \cdots + m_{i_r} : r \in \mathbb{N}, \ m_{i_j} \in N_{i_j} \}$$

is the smallest submodule of M containing N_i for all i.

Corollary 18.1.20 says that $\langle \bigcup_{i\in\Lambda} N_i \rangle = \sum_{i\in\Lambda} N_i$. If $\Lambda = \{1, 2, \ldots, n\}$ is a finite set, $\sum_{i\in\Lambda} N_i$ is often written as $N_1 + \cdots + N_r$ (Cf. (18.1.2)).

Exercises 18.1

1. Complete the proof of Proposition 18.1.7.

2. Let R be a non-commutative ring with unity. Define the scalar multiplication as the function

$$
\begin{array}{ccc}
M \times R & \longrightarrow & R \\
(m, a) & \longmapsto & ma^{-1}
\end{array}
$$

   where $M = R$. Show that R is a right R-module.

3. Show that any abelian group is a $\mathbb{Z}$-module.

4. Let M be a $\mathbb{Q}$-module. Show that the given scalar multiplication is the only way that can be used to make M a $\mathbb{Q}$-module.

5. Let M be a non-trivial finite abelian group. Can M be made into a $\mathbb{Q}$-module?

6. Let R be a ring with unity and let I be a subset of R. Using the multiplication in R as the scalar multiplication, show that I is a left R-module if and only if I is a left ideal of R.

7. Let R be a commutative ring with unity. Let x be an indeterminate over R.

   (a) Show that $R[x]$ is an R-module.

   (b) Let I be an ideal of $R[x]$. Show that I is an R-module.

   (c) Let $\mathscr{P}_n$ be the set of polynomials of degree at most n in $R[x]$. Show that $\mathscr{P}_n$ is an R-module.

18.2   Linear Maps and Quotient Modules

In this section we discuss the homomorphisms for modules.

Definition 18.2.1. Let M and N be R-modules. A map $\eta : M \to N$ is called an R-**linear map** if

(i) $\eta(m + m') = \eta(m) + \eta(m')$, and

(ii) $\eta(am) = a\eta(m)$,

for all $a \in R$ and for all $m, m' \in M$. If in particular $R = F$ is a field, we also call an F-linear map an F-**linear transformation** or a linear transformation over F.

From definition we can see that an R-linear map from M to N is a group homomorphism from $(M, +)$ to $(N, +)$ which respect the scalar multiplication on M and in N.

The following results are some basic properties of R-linear maps.

Lemma 18.2.2. *Let $\eta : M \to N$ be an R-linear map. Then*

(a) $\eta(0_M) = 0_N$;

(b) $\eta(-m) = -\eta(m)$ *for all $m \in M$.*

Proof. (a) Since $\eta(0_M) + 0_N = \eta(0_M) = \eta(0_M + 0_M) = \eta(0_M) + \eta(0_M)$, we have that $\eta(0_M) = 0_N$ by the cancellation law for additive groups.

   (b) This follows from the fact that

$$\eta(-m) = \eta((-1)m) = (-1)\eta(m) = -\eta(m)$$

by Proposition 18.1.6.   □

Let $\eta : M \to N$ be an R-linear map. The **image** of η is given by

$$\operatorname{Im} \eta = \eta(M) = \{\eta(m) \in N : m \in M\}.$$

The **kernel** of η is given by

$$\ker \eta = \{m \in M : \eta(m) = 0\}.$$

Proposition 18.2.3. *Let* $\eta : M \to N$ *be an* R-*linear map. Then* $\ker \eta$ *is a submodule of* M *and* $\operatorname{Im} \eta$ *is a submodule of* N.

Proof. First note that $\eta(0_M) = 0_N$ by Lemma 18.2.2. We have $0_M \in \ker \eta$. Let $m, m' \in \ker \eta$ and $a \in R$. Then

$$\eta(m + m') = \eta(m) + \eta(m') = 0 + 0 = 0.$$

This says that $m + m' \in \ker \eta$. We also have

$$\eta(am) = a\eta(m) = a0_N = 0_N.$$

This says that $am \in \ker \eta$. We conclude that $\ker \eta$ is a submodule of M over R.

For the second part of this proposition, first note that $0_N = \eta(0_M)$ is in $\eta(M)$. Let $n, n' \in \eta(M)$ and $a \in R$. Find $m, m' \in M$ such that $\eta(m) = n$ and $\eta(m') = n'$. Then

$$n + n' = \eta(m) + \eta(m') = \eta(m + m'),$$
$$an = a\eta(m) = \eta(am).$$

This says that $n + n'$ and an are both in $\eta(M)$. Hence $\eta(M)$ is a submodule of N over R. $\qquad\square$

An injective R-linear homomorphism is called a **monomorphism**. A surjective R-linear homomorphism is called an **epimorphism**. A bijective R-linear homomorphism is called an **isomorphism**. Two modules M and N are called isomorphic, denoted $M \cong N$, if there is an isomorphism from M to N.

Proposition 18.2.4. *Let* M, N *and* P *be* R-*modules. Let* $\eta \colon M \to N$ *and* $\zeta \colon N \to P$ *be* R-*linear maps. Then* $\zeta\eta \colon M \to P$ *is also an* R-*linear map. Furthermore, if* η *and* ζ *are isomorphisms, so are* $\zeta\eta$ *and* η^{-1}.

Proof. Let m and $m' \in M$ and let $a \in R$. We have that

$$\zeta\eta(m + m') = \zeta(\eta(m + m')) = \zeta(\eta(m) + \eta(m'))$$
$$= \zeta(\eta(m)) + \zeta(\eta(m')) = \zeta\eta(m) + \zeta\eta(m'),$$
$$\zeta\eta(am) = \zeta(\eta(am)) = \zeta(a\eta(m)) = a\zeta(\eta(m)) = a\zeta\eta(m).$$

Hence $\zeta\eta$ is also an R-linear map.

Suppose η and ζ are both isomorphisms. We have that $\zeta\eta$ is an isomorphism by Corollary 1.2.4.

By Proposition 7.2.3, we have seen that η^{-1} is an isomorphism of M and N as additive groups. It remains to show that η^{-1} also respects the scalar multiplication in M and in N. Let $a \in R$ and $n \in N$. Find $m \in M$ such that $\eta(m) = n$. Then $\eta(am) = a\eta(m) = an$. This implies that

$$\eta^{-1}(an) = am = a\eta^{-1}(n).$$

Hence η^{-1} is an R-linear map. $\square$

The following corollary is an immediate result of the proposition above. We leave it as an exercise (see Exercise 3).

Corollary 18.2.5. *The relation "isomorphic" is an equivalence relation.*

Important! For the rest of this book, all rings are assumed to be *commutative with unity* unless otherwise noted. All modules are assumed to be *unitary.*

Let N be a submodule of an R-module M. We can consider the quotient group $\overline{M} = M/N$ since N is an additive subgroup (hence normal) of M. There is a natural R-module structure on $\overline{M}$. The addition came from its group structure:

(18.2.1) $(m + N) + (m' + N) = m + m' + N$

for $m, m' \in M$. The scalar multiplication can be given by

(18.2.2) $a(m + N) = am + N$

for $a \in R$ and $m \in M$. The element $m + N$ is also often written as $\overline{m}$. We leave it as an exercise to check that this does satisfy the conditions for modules (see Exercise 4). The R-module M/N is called the **quotient module** of M respect to N.

Example 18.2.6. The quotient module $\mathbb{Z}/n\mathbb{Z}$ is denoted as $\mathbb{Z}_n$. It may be viewed as a $\mathbb{Z}$-module. In this case $\operatorname{ann} \mathbb{Z}_n = n\mathbb{Z}$. From Proposition 18.1.7, it may also be viewed as a $\mathbb{Z}_n$-module.

Let M be an R-module and N be a submodule of M. There is a **canonical epimorphism** $\pi\colon M \to M/N$ sending $m \in M$ to $\overline{m}$. This is indeed R-linear since from (18.2.1) and (18.2.2) we have

$$\pi(m + m') = \overline{m + m'} = \overline{m} + \overline{m'} = \pi(m) + \pi(m'),$$
$$\pi(am) = \overline{am} = a\overline{m} = a\pi(m)$$

for $a \in R$ and $m, m' \in M$.

Just as the cases for groups and rings, we have the fundamental theorem for R-linear maps.

Theorem 18.2.7 (Fundamental theorem of R-linear maps). *Let $\eta\colon M \to N$ be an R-linear map and let $K \subseteq \ker \eta$ be a submodule of M. Let $\pi\colon M \twoheadrightarrow M/K$ be the canonical epimorphism. Then there is a unique R-linear map $\overline{\eta}\colon M/K \to N$ such that $\eta = \overline{\eta}\pi$.*

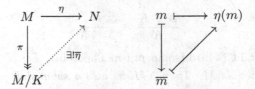

Moreover, $\overline{\eta}$ is a monomorphism if and only if $K = \ker \eta$.

Proof. This theorem is true at least at the group level (see Theorem 8.3.1). The map $\overline{\eta}$ is given by $\overline{\eta}(\overline{m}) = \eta(m)$. It remains to show that $\overline{\eta}$ is R-linear. Suppose given $a \in R$ and $\overline{m} \in M/K$. Then

$$\overline{\eta}(a\overline{m}) = \overline{\eta}(\overline{am}) = \eta(am) = a\eta(m) = a\overline{\eta}(\overline{m}).$$

We have shown that $\overline{\eta}$ is R-linear. The rest of the theorem follows from Theorem 8.3.1.   □

By applying Theorem 18.2.7, we have the following results.

Theorem 18.2.8 (Correspondence Theorem). *Let M be an R-module, N be a submodule of N and let $\pi\colon M \to M/N$ be the canonical epimorphism. Let $\mathscr{A}$ be the family of submodules of M containing N and let $\mathscr{B}$ be the* *family of submodules of M/N. There is an order-preserving one-to-one correspondence between*

$$
\begin{array}{ccc}
\mathscr{A} & \longleftrightarrow & \mathscr{B} \\
K & \longmapsto & \overline{K} \\
\pi^{-1}(\mathfrak{K}) & \longleftarrow & \mathfrak{K}
\end{array}
$$

We leave the proof as an exercise (see Exercise 5).

The following are the *three isomorphism theorems*.

Theorem 18.2.9 (First Isomorphism Theorem). *Let $\eta\colon M \twoheadrightarrow N$ be an epimorphism. Then*

$$
\frac{M}{\ker \eta} \cong N.
$$

Theorem 18.2.10 (Second Isomorphism Theorem). *Let N and K be submodules of the R-module M. Then $N + K$ is a submodule of M, K is a submodule of $N + K$ and $N \cap K$ is a submodule of N. We also have*

$$
\frac{N + K}{K} \cong \frac{N}{N \cap K}.
$$

Theorem 18.2.11 (Third Isomorphism Theorem). *Let $K \subseteq N$ be submodules of the R-module M. Then N/K is also a submodule of M/K and*

$$
\frac{M/K}{N/K} \cong \frac{M}{N}.
$$

We leave the proofs of the three isomorphism theorems to the reader (see Exercise 6).

Let R be a commutative ring. Then as an R-module $R = R1$ is a cyclic module. As ideal I of R may be viewed as a submodule of R. The quotient module $R/I = R\overline{1}$ is also a cyclic module. Next we will see that basically R and R/I are the only cyclic modules over R.

Let M be an R-module and let $m \in M$. We define the **annihilator of m in R** to be

$$
\operatorname{ann}_R m = \{a \in R : am = 0\}.
$$

When R is understood, we may simply write $\operatorname{ann} m$ for $\operatorname{ann}_R m$.

We leave it as an exercise to verify that $\operatorname{ann} m$ is an ideal of R (see Exercise 7).

In the case of abelian groups (as $\mathbb{Z}$-modules), $\operatorname{ann} m$ is generated by the order of m. Hence, $\operatorname{ann} m$ is also called the **order ideal** of m.

Recall that a cyclic R-module is of the form

$$M = Rm = \{am : a \in R\}$$

for some $m \in M$.

Proposition 18.2.12. *Let M be an R-module and let $m \in M$. Then*

$$Rm \cong \frac{R}{\operatorname{ann} m}.$$

Proof. Consider the map $\phi \colon R \to Rm$ sending r to rm. Let $r,\, r' \in R$ (as module elements) and $a \in R$ (as a scalar). We have

$$\phi(r + r') = (r + r')m = rm + r'm = \phi(r) + \phi(r'),$$
$$\phi(ar) = (ar)m = a(rm) = a\phi(r).$$

Thus ϕ is R-linear. It is clear that ϕ is onto and $\ker \phi = \operatorname{ann} m$. The result now follows from the first isomorphism theorem for modules. $\qquad\square$

Example 18.2.13. Consider the $\mathbb{Z}$-module

$$M = \frac{\mathbb{Z} \times \mathbb{Z}}{\langle (2,2),\, (4,-2) \rangle}.$$

Compute $\operatorname{ann} \overline{(1,1)}$ and $\operatorname{ann} \overline{(1,0)}$.

Solution. First, we find $\operatorname{ann} \overline{(1,1)}$. It is an ideal in $\mathbb{Z}$. Clearly 2 is in $\operatorname{ann} \overline{(1,1)}$ since

$$2\overline{(1,1)} = \overline{(2,2)} = 0.$$

If $1 \in \operatorname{ann} \overline{(1,1)}$, then $\overline{(1,1)} = \overline{(0,0)}$. This says that $(1,1) \in \langle (2,2), (4,-2) \rangle$, or equivalently, there are $k,\, \ell \in \mathbb{Z}$ such that

$$(1,1) = k(2,2) + \ell(4,-2),$$

which is impossible. Hence we conclude that $\operatorname{ann} \overline{(1,1)} = (2)$.

Next, let's try to find $\operatorname{ann} \overline{(1,0)}$. Since $(2,2) + (4,-2) = (6,0)$, we have $6\overline{(1,0)} = 0$. This says that $6 \in \operatorname{ann} \overline{(1,0)}$ and the order ideal is nontrivial. Let $\operatorname{ann} \overline{(1,0)} = (n)$ where n is the least positive integer such that $n\overline{(1,0)} = 0$. This says that

$$(n,0) = k(2,2) + \ell(4,-2)$$

for some k, $\ell \in \mathbb{Z}$. We then have $k = \ell$ and $n = 6k$. We now conclude that $\operatorname{ann} \overline{(1,0)} = (6)$.   ◇

Exercises 18.2

1. Prove Proposition 18.2.3.

2. Let $\eta : M \to M'$ be an R-linear map and let N be a submodule of M. Show that the **preimage** of η,

$$\eta^{-1}(N) = \{m \in M : \eta(m) \in N\}$$

   is a submodule of M containing $\ker \eta$.

3. Prove Corollary 18.2.5.

4. Let M be an R-module and let N be a submodule of M. Verify that the quotient module M/N with the given scalar multiplication as in (18.2.2) is indeed an R-module.

5. Prove Theorem 18.2.8.

6. Prove Theorem 18.2.9, Theorem 18.2.10 and Theorem 18.2.11.

7. Let M be an R-module and let $m \in M$. Show that $\operatorname{ann}_R m$ is an ideal of R.

8. Let M and N be additive groups. Show that the $\mathbb{Z}$-linear maps from M into N are the group homomorphisms from M into N. Hence combining the results in Example 18.1.10, Example 18.1.14 and this problem, we see that any problem on abelian groups can be treated as a problem on $\mathbb{Z}$-module.

9. Let $\mathrm{Hom}_R(M, N)$ (or simply $\mathrm{Hom}(M, N)$) denote the set of R-linear maps from the R-module M into the module N. Let $\eta, \zeta \in \mathrm{Hom}(M, N)$ and $a \in R$. Define

$$(\eta + \zeta)(m) = \eta(m) + \zeta(m),$$
$$(a\eta)(m) = a(\eta(m))$$

for all $m \in M$. Show that $\eta + \zeta$ and $a\eta \in \mathrm{Hom}(M, N)$ and that $\mathrm{Hom}(M, N)$ is an R-module.

We often call an R-linear map from an R-module M to itself a **linear endomorphism**, and hence $\mathrm{Hom}_R(M, M)$ is often written as $\mathrm{End}_R(M)$. For $\zeta, \eta \in \mathrm{End}_R(M)$, define

$$(\zeta\eta)(m) = \zeta(\eta(m)).$$

Show that $\zeta\eta \in \mathrm{End}(M)$ and that $\mathrm{End}(M)$ is a (not necessarily commutative) ring with unity.

10. Let n be a positive integer.

   (a) Determine $\mathrm{Hom}_{\mathbb{Z}}(\mathbb{Z}, \mathbb{Z}/(n))$.

   (b) Determine $\mathrm{Hom}_{\mathbb{Z}}(\mathbb{Z}/(n), \mathbb{Z})$.

18.3　Direct Sums of Modules

Let $M_1, \ldots, M_n$ be R-modules. Let M be the set $M_1 \times \cdots \times M_n$. There is a natural R-module structure on M given by

$$(m_1, \ldots, m_n) + (m_1', \ldots, m_n') = (m_1 + m_1', \ldots, m_n + m_n'),$$
$$a(m_1, \ldots, m_n) = (am_1, \ldots, am_n)$$

where $m_i, m_i' \in M_i$ for each i and $a \in R$. We leave it as an exercise to verify that this is indeed an R-module (see Exercise 1). We call this M the **(external) direct sum** of the modules M_i, and denote it by $M_1 \oplus_R \cdots \oplus_R M_n$, or $M_1 \oplus \cdots \oplus M_n$ if R is understood, or simply $\bigoplus_1^n M_i$. In particular, the direct sum of n copies of R is denoted as R^n (*Cf.* Example 18.1.3).

Example 18.3.1. Let M, M_1, ..., M_n be R-modules. Suppose given R-linear maps $\eta_i : M_i \to M$ for each i. This induces a natural map

$$\begin{aligned} \eta: \quad M_1 \oplus \cdots \oplus M_n \quad &\longrightarrow \quad M \\ (m_1, \ldots, m_n) \quad &\longmapsto \quad \sum_1^n \eta_i(m_i). \end{aligned}$$

Let $(m_1, \ldots, m_n)$ and $(m'_1, \ldots, m'_n) \in M_1 \oplus \cdots \oplus M_n$ and $a \in R$. Then

$$\begin{aligned} &\eta\big((m_1, \ldots, m_n) + (m'_1, \ldots, m'_n)\big) \\ =&\eta\big(m_1 + m'_1, \ldots, m_n + m'_n\big) \\ =&\sum_1^n \eta_i(m_i + m'_i) = \sum_1^n \eta_i(m_i) + \eta_i(m'_i) \\ =&\sum_1^n \eta_i(m_i) + \sum_1^n \eta_i(m'_i) \\ =&\eta(m_1, \ldots, m_n) + \eta(m'_1, \ldots, m'_n); \\ &\eta\big(a(m_1, \ldots, m_n)\big) = \eta(am_1, \ldots, am_n) \\ =&\sum_1^n \eta_i(am_i) = \sum_1^n a\eta_i(m_i) = a \sum_1^n \eta_i(m_i) \\ =&a\eta(m_1, \ldots, m_n). \end{aligned}$$

It follows that η is an R-linear map.

Example 18.3.2. Let M, M_1, ..., M_n be R-modules. Suppose given R-linear maps $\eta_i : M \to M_i$ for each i. We have this mapping

$$\begin{aligned} \eta: \quad M \quad &\longrightarrow \quad M_1 \oplus \cdots \oplus M_n \\ m \quad &\longmapsto \quad (\eta_1(m), \ldots, \eta_n(m)). \end{aligned}$$

Let $(m_1, \ldots, m_n)$ and $(m'_1, \ldots, m'_n) \in M_1 \oplus \cdots \oplus M_n$ and $a \in R$. Then

$$\begin{aligned} &\eta\big((m_1, \ldots, m_n) + (m'_1, \ldots, m'_n)\big) \\ =&\eta\big(m_1 + m'_1, \ldots, m_n + m'_n\big) \\ =&\big(\eta_1(m_1 + m'_1), \ldots, \eta_n(m_n + m'_n)\big) \\ =&\big(\eta_1(m_1) + \eta_1(m'_1), \ldots, \eta_n(m_n) + \eta_n(m'_n)\big) \\ =&\big(\eta_1(m_1), \ldots, \eta_n(m_n)\big) + \big(\eta_1(m'_1), \ldots, \eta_n(m'_n)\big) \\ =&\eta(m_1, \ldots, m_n) + \eta(m'_1, \ldots, m'_n); \end{aligned}$$

$$\eta\big(a(m_1,\ldots,m_n)\big) = \eta(am_1,\ldots,am_n)$$
$$=\big(\eta_1(am_1),\ldots,\eta_n(am_n)\big) = \big(a\eta_1(m_1),\ldots,a\eta_n(m_n)\big)$$
$$=a\big(\eta_1(m_1),\ldots,\eta_n(m_n)\big) = a\eta(m_1,\ldots,m_n).$$

It follows that η is an R-linear map.

Definition 18.3.3. Let M be a module and suppose $M_1,\ldots,M_n$ are submodules of M. We say $M_1,\ldots,M_n$ are **independent** submodules of M if whenever $m_1 + \cdots + m_n = 0$ for $m_i \in M_i$, we have $m_1 = \cdots = m_n = 0$.

Let $M_1,\ldots,M_n$ be independent submodules of M. If $\sum_1^n m_i = \sum_1^n m_i'$ where $m_i, m_i' \in M_i$, then $m_i = m_i'$ for all i.

Theorem 18.3.4. *Suppose that M is an R-module with independent submodules $M_1,\ldots,M_n$ such that $M = M_1 + \cdots + M_n$ (that is, M is generated by the M_i's). Then $M \cong M_1 \oplus \cdots \oplus M_n$.*

Conversely, let $M_1,\ldots,M_n$ be R-modules and let $M = M_1 \oplus \cdots \oplus M_n$. Let

$$(*) \qquad M_i' = \{(0,\ldots,0,\underset{i}{m_i},0,\ldots,0) : m_i \in M_i\}.$$

Then M_i' is a submodule of M and $M_i' \cong M_i$. Furthermore, $M_1',\ldots,M_n'$ are independent submodules which generates M.

Proof. For each i we have the inclusion map $\eta_i\, M_i \hookrightarrow M$. We may use the R-linear map in Example 18.3.1 given by

$$\eta: \quad M_1 \oplus \cdots \oplus M_n \quad \longrightarrow \qquad\quad M$$
$$(m_1,\ldots,m_n) \quad \longmapsto \quad m_1 + \cdots + m_n.$$

The map η is surjective since $M = M_1 + \cdots + M_n$. Let $(m_1,\ldots,m_n) \in \ker\eta$, we have that $m_1 + \cdots + m_n = 0$. Thus $m_1 = \cdots = m_n = 0$ since the M_i's are independent. This says that η is injective. We conclude that η is an isomorphism.

Now let's look at the converse. It is quite clear that the map from M_i to M_i' sending m_i to $(0,\ldots,0,\underset{i}{m_i},0,\ldots,0)$ is an isomorphism. An arbitrary element $(m_1,\ldots,m_n)$ in $M_1 \oplus \cdots \oplus M_n$ is the sum

$$\sum_{i=1}^n (0,\ldots,0,\underset{i}{m_i},0,\ldots,0).$$

Hence $M = M'_1 + \cdots M'_i$. On the other hand, let

$$\sum_{i=1}^{n} (0, \ldots, 0, \underset{i}{m_i}, 0, \ldots, 0) = (0, \ldots, 0)$$

where $m_i \in M_i$. This implies that $(m_1, \ldots, m_n) = (0, \ldots, 0)$, or equivalently $m_i = 0$ for $i = 1, \ldots, n$. We have just shown that $M'_1, \ldots, M'_n$ are independent. This completes the proof. $\square$

In Theorem 18.3.4, we have shown that under the given conditions, M is isomorphic to the (external) direct sum of its submodules. The components come from inside the module M. We call the module M the **(internal) direct sum** of the submodules $M_1, \ldots, M_n$. For any $m \in M$, there is a *unique* expression

$$m = m_1 + \cdots + m_n$$

where $m_i \in M_i$ for $i = 1, \ldots, n$.

There are other criteria for decomposing a module into direct sums of submodules. See Exercises 4 and 5.

Example 18.3.5. Let V be a vector space over a field K. The non-zero vectors $v_1, \ldots, v_n$ are linearly independent over K if and only if the subspaces $Kv_1, \ldots Kv_n$ are independent (see Exercises 2). The vectors $v_1, \ldots, v_n$ form a basis if and only if $V = Kv_1 \oplus \cdots \oplus Kv_n$.

Exercises 18.3

In the following exercises, R denotes a commutative ring with unity.

1. Let $M_1, \ldots, M_n$ be R-modules. Show that the external direct sum of $M_1, \ldots, M_n$ is indeed an R-module.

2. Let F be a field. Let V be an F-vector space and let $v_1, \ldots, v_n \in V$. Show that $v_1, \ldots, v_n$ are linearly independent over F if and only if $Fv_1, \ldots, Fv_n$ are independent over F.

3. Let $M_1, \ldots, M_n$ and $N_1, \ldots, N_n$ be R-modules, and let $\eta_i \colon M_i \to N_i$ be R-linear maps for $i = 1, \ldots, n$. Show that the mapping

$$\eta_1 \oplus \cdots \oplus \eta_n \colon \quad M_1 \oplus \cdots \oplus M_n \quad \longrightarrow \quad N_1 \oplus \cdots \oplus N_n$$
$$(m_1, \ldots, m_n) \quad \longmapsto \quad (\eta_1(m_1), \ldots, \eta_n(m_n))$$

is an R-linear map.

4. Let $M_1, \ldots, M_n$ be submodules of an R-module M having the properties:

(a) $M = M_1 + \cdots + M_n$, and

(b) for every i, $1 \leq i \leq n$, we have

$$M_i \cap (M_1 + \cdots + M_{i-1} + M_{i+1} + \cdots + M_n) = 0.$$

Show that $M \cong M_1 \oplus \cdots \oplus M_n$.

5. Under the same conditions as in Exercise 4 except that condition (b) be replaced by

(b') the "triangular" set of conditions

$$M_1 \cap M_2 = 0$$
$$(M_1 + M_2) \cap M_3 = 0$$
$$\vdots$$
$$(M_1 + \cdots + M_{n-1}) \cap M_n = 0.$$

Show that $M \cong M_1 \oplus \cdots \oplus M_n$.

6. Let M_1, M_2 be R-modules and N_1, N_2 be submodules of M_1, M_2 respectively. Show that $N_1 \oplus N_2$ is a submodule of $M_1 \oplus M_2$. Show that

$$\frac{M_1 \oplus M_2}{N_1 \oplus N_2} \cong \frac{M_1}{N_1} \oplus \frac{M_2}{N_2}.$$

7. If M_1 and M_2 are R-submodules of M such that $M = M_1 \oplus M_2$, show that $M/M_1 \cong M_2$ and $M/M_2 \cong M_1$.

8. Let M be as in Example 18.2.13.

(a) Show that $M = \mathbb{Z}\overline{(1,1)} \oplus \mathbb{Z}\overline{(1,0)} \cong \mathbb{Z}_2 \times \mathbb{Z}_6$.

(b) Compute $|M|$. As an additive group, is there an element of order 8 in M?

Review Exercises for Chapter 18

1. We say a *nontrivial* module M is **irreducible** if 0 and M are the only submodules of M.

   (a) Show that a module M is irreducible if and only if $M \neq 0$ and M is generated by any nonzero element in M.

   (b) Show that an R-module M is irreducible if and only if $M \cong R/I$ for some maximal ideal I of R.

2. Let F be a field, V be a vector space over F and T be a linear endomorphism on V. Let λ be an indeterminate over F.

   (a) Show that one can make V into a $F[\lambda]$-module by letting

   $$(a_0 + a_1\lambda + \cdots + a_m\lambda^m)v = a_0v + a_1T(v) + \cdots + a_mT^m(v),$$

   where $a_0, a_1, \ldots, a_m \in F$.

   (b) We say a subspace W of V is **stabilized** by T if $T(W) \subseteq W$. Show that the $F[\lambda]$-submodules of V are those subspaces stabilized by T.

3. Let M and N be R-modules. Let $f: M \to N$ and $g: N \to M$ be R-linear maps such that $fg(n) = n$ for all $n \in N$. Show that $M \cong \ker f \oplus \operatorname{Im} g$.

CHAPTER 19

Free Modules

A free R-module is a module with an R-basis. In this chapter we will investigate the similarity and differences between free modules and vector spaces.

In this chapter we will also give a detailed treatment on the basics of determinant. The reader should pay attention to the subtle differences how matrices and determinant work over commutative rings from how they work over fields.

19.1   Free Modules

To simplify matter, we would like to remind the reader that from this point on R or any other rings always denote a commutative ring with unity unless otherwise noted.

Let M be an R-module. Let $m_1, \ldots, m_n$ be elements in M. We say $\{m_1, \ldots, m_n\}$ is a **linearly independent** set over R if $\sum_1^n a_i m_i = 0$ implies that $a_i = 0$ for all i. We say an infinite set S is **linearly independent** over R if any of its finite subsets is linearly independent over R. If a linearly independent set S generates an R-module M, we say it is a **base** or a **basis** for M over R. We may also say that S is an R-**basis** for M.

Example 19.1.1. In Example 18.1.2 we have seen that R^n is an R-module. Let
$$e_i = (0, \ldots, 0, \underset{i}{1}, 0, \ldots, 0)$$
for $i = 1, \ldots, n$. Note that
$$(a_1, \ldots, a_n) = \sum_1^n a_i e_i.$$
Thus $e_1, \ldots, e_n$ generate R^n. Moreover,
$$\sum_1^n a_i e_i = 0 \implies a_i = 0$$
for $i = 1, \ldots, n$. Thus $\{e_1, \ldots, e_n\}$ is a *base* or *basis* for R^n over R. In fact, we often call $\{e_1, \ldots, e_n\}$ the **standard basis** for R^n over R.

We would like to point out that not all modules possess a basis. To see this, note that the annihilator of a base element must be trivial!

Example 19.1.2. The additive group $\mathbb{Z}_2$ can be seen as a $\mathbb{Z}$-module or a $\mathbb{Z}_2$-module. Neither $\bar{0}$ nor $\bar{1}$ in $\mathbb{Z}_2$ can be part of a basis over $\mathbb{Z}$. However, $\{\bar{1}\}$ form a basis for $\mathbb{Z}_2$ over $\mathbb{Z}_2$.

We say an R-module is a **free** R-module if it has a basis over R.

In Proposition 19.2.13, we will show that when R is commutative with unity and M is a finitely generated free R-module, any two R-bases of M are of the same cardinality. Hence, if M has a basis of m elements, we will say that M is a free module of **rank** m. When $R = F$ is a field, the rank of an F-vector space is called the **dimension**. When a free module possesses a finite base, it is called a free module of **finite rank**. If the underlying ring is a field, we say it is a **finite dimensional** vector space.

Example 19.1.3. The trivial module is a module of rank 0, since the empty set is its basis.

The R-module R^n is free of rank n from our discussion in Example 19.1.1.

However, the notion of *rank* only applies to bases, not to minimal generating sets. This is very different from the situation for vector spaces.

Example 19.1.4. Consider $M = \mathbb{Z}$ as a $\mathbb{Z}$-module. Then $\{1\}$ is a minimal generating set. It is also a basis. On the other hand, the set $\{2, 3\}$ is also a minimal generating set for M. If you drop the 2, or if you drop the 3, it no longer generates M. Two generating sets of a module may not be of the same cardinality, even when it is a free module.

In a free module M with a given basis, any element may be expressed as a linear combination of base elements. This is because the given basis generates M. This expression is unique because the base elements are linearly independent.

Let $(f_1, \ldots, f_n)$ be an ordered basis for M over R. If the element

$$m = a_1 f_1 + \cdots + a_n f_n,$$

we say $(a_1, \ldots, a_n)$ is the **coordinate** of m with respect to the ordered base $(f_1, \ldots, f_n)$.

Theorem 19.1.5. *Let $\{m_i\}_{i \in I}$ be a basis for a free R-module M and let N be another R-module. For any choice of $(n_i)_{i \in I}$ (the n_i's may be repeated), there exists a unique R-linear map from M to N sending m_i to n_i for each $i \in I$.*

Proof. Let $m \in M$. Then we may write

$$m = a_1 m_{i_1} + a_2 m_{i_2} + \cdots + a_r m_{i_r}$$

where $a_k \in R$ and $i_k \in I$. This expression is unique. Hence we have a well-defined map

$$\phi: \quad\quad M \quad\quad \longrightarrow \quad\quad N$$
$$\sum_{k=1}^{r} a_k m_{i_k} \quad \longmapsto \quad \sum_{k=1}^{r} a_k n_{i_k}.$$

It is routine to check that this is indeed R-linear. Let $m = \sum_{k=1}^{r} a_k m_{i_k}$.

If ψ is any R-linear map sending m_i to n_i for all $i \in I$, then

$$\psi(m) = \psi\Big(\sum_{k=1}^{r} a_k m_{i_k}\Big) = \sum_{k=1}^{r} a_k \psi(m_{i_k}) = \sum_{k=1}^{r} a_k n_{i_k} = \phi(m).$$

We have that $\phi = \psi$. $\qquad\qquad\qquad\qquad\qquad\qquad\qquad\qquad\qquad\qquad\square$

Corollary 19.1.6. *Let M and N be R-modules. Suppose they have R-bases of the same cardinality. Then $M \cong N$ as R-modules.*

Proof. Let S and T be bases for M and N respectively. Since they are of the same cardinality, there is a bijective map η from S to T. By Theorem 19.1.5, η can be extended to an R-linear map $\phi: M \to N$. Similarly, η^{-1} can also be extended to an R-linear map $\psi: N \to M$. The composition $\varphi\phi$ maps $s \in S$ to itself. By the uniqueness part of Theorem 19.1.5, $\psi\phi = \mathbf{1}_M$. Similarly, $\phi\psi = \mathbf{1}_N$. Thus ϕ is an isomorphism from M to N. $\qquad\square$

We now have the following immediate result.

Corollary 19.1.7. *If M is a free R-module of rank n, then $M \cong R^n$.*

Proof. This is true because R^n is free of rank n. $\qquad\qquad\qquad\qquad\qquad\square$

Next we study $\operatorname{Hom}_R(R^m, R^n)$ for arbitrary m and n. We will also study the relation between R-linear maps and $m \times n$ matrices over R.

Choose ordered bases $(f_1, \ldots, f_n)$ and $(g_1, \ldots, g_m)$ for R^m and R^n respectively. If η is an R-linear map from R^n to R^m, we tabulate

$$\eta(f_1) = a_{11}g_1 + a_{21}g_2 + \cdots + a_{m1}g_m$$
$$\eta(f_2) = a_{12}g_1 + a_{22}g_2 + \cdots + a_{m2}g_m$$
$$\vdots$$
$$\eta(f_n) = a_{1n}g_1 + a_{2n}g_2 + \cdots + a_{mn}g_m.$$

We may use the matrix

$$A = \begin{bmatrix} a_{11} & a_{12} & \cdots & a_{1n} \\ a_{21} & a_{22} & \cdots & a_{2n} \\ \vdots & \vdots & \ddots & \vdots \\ a_{m1} & a_{m2} & \cdots & a_{mn} \end{bmatrix}$$

to represent the R-linear map η. We call the matrix A the **matrix associated with** η with respect to the ordered bases $(f_1, \ldots, f_n)$ and $(g_1, \ldots, g_m)$.

Suppose $\eta(x) = y$. Let $(x_1, \ldots, x_n)$ and $(y_1, \ldots, y_m)$ be the coordinates of x and y with respect to the ordered bases $(f_1, \ldots, f_n)$ and $(g_1, \ldots, g_m)$ respectively. Then

$$y = \eta(x) = \eta\left(\sum_j x_j f_j\right) = \sum_j x_j \eta\left(f_j\right) = \sum_{i,j} x_j a_{ij} g_i.$$

Hence we obtain that $y_i = \sum_{j=1}^m x_j a_{ij}$ for $i = 1, \ldots, n$. In matrix form we have

$$\eta : \begin{pmatrix} x_1 \\ x_2 \\ \vdots \\ x_m \end{pmatrix} \longmapsto \begin{pmatrix} y_1 \\ y_2 \\ \vdots \\ y_n \end{pmatrix} = A \begin{pmatrix} x_1 \\ x_2 \\ \vdots \\ x_m \end{pmatrix}.$$

One can see that this approach is the same as how people deal with linear transformations in a course on Linear Algebra.

Proposition 19.1.8. *Let $\alpha = (f_1, \ldots, f_n)$ and $\beta = (g_1, \ldots, g_m)$ be ordered bases for R^n and R^m respectively. Let η and ζ be two R-linear maps from R^n into R^m and let A and B be the matrices representing them respectively with respect to the given ordered bases. Let $a \in R$. With respect to the same pair of given bases, the matrices representing $\eta + \zeta$ and $a\eta$ are $A + B$ and aA respectively. Let $\gamma = (h_1, \ldots, h_k)$ be an ordered basis for R^k and let C be a matrix for the R-linear map $\xi : R^m \to R^k$ with respect to the bases β and γ. Then the matrix for $\xi\eta$ with respect to α and γ is CA.*

We leave the proof of this proposition as an easy review exercise (see Exercise 8). From this proposition, we have the following immediate results on isomorphisms.

Corollary 19.1.9. *Let M and N be free R-modules of the same finite rank. Let β and γ be ordered R-bases for M and for N respectively.*

(a) *The matrix of 1_M with respect to β and β is the identity matrix.*

(b) *Let A be the matrix of the R-linear map $\eta\colon M \to N$ with respect to β and γ. Then A is an invertible matrix if η is an isomorphism. The matrix of η^{-1} with respect to γ and β is A^{-1}.*

We leave the proof of this corollary as an exercise (see Exercise 9).

Sometimes it is preferable to use a different basis for a module. Let $\beta = (f_1, \ldots, f_n)$ and $\beta' = (f'_1, \ldots, f'_n)$ be two ordered bases for a free module M over R. Suppose

$$f'_j = \sum_{i=1}^{n} p_{ij} f_i \quad \text{and} \quad f_l = \sum_{k=1}^{n} q_{kl} f'_k.$$

We call $P = (p_{ij})_{n \times n}$ the **base change matrix**, or simply **base change**, from β' to β. Similarly, $Q = (q_{kl})_{n \times n}$ is the base change matrix from β to β'. We have

$$f'_j = \sum_{i=1}^{n} p_{ij} f_i = \sum_{i=1}^{n} \left(p_{ij} \sum_{k=1}^{n} q_{ki} f'_k \right) = \sum_{i,k=1}^{n} q_{ki} p_{ij} f'_k = \sum_{k=1}^{n} \left(\sum_{i=1}^{n} q_{ki} p_{ij} \right) f'_k.$$

This says that

$$\sum_{i,k=1}^{n} q_{ki} p_{ij} = \begin{cases} 1, & \text{if } k = j; \\ 0, & \text{if } k \neq j. \end{cases}$$

Hence we have that $QP = I_n$. Similarly, we have that $PQ = I_n$. We conclude that P is an invertible matrix. We summarize our discussion in the following proposition.

Proposition 19.1.10. *Let P be a base change matrix between two ordered bases of a free R-module of rank n. It is an invertible matrix in $M_n(R)$. Moreover, the base change from an ordered basis β to another ordered basis β' is the inverse of the base change from β' to β.*

The converse of this proposition is also true. See Exercise 10.

The base change from β' to β may be regarded as the matrix of 1_M with respect to β' and β.

Corollary 19.1.11. *Let M and N be free R-modules of finite rank. Let A be the matrix of the R-linear map $\eta : R^n \to R^m$ with respect to the bases $\beta = \{f_1, \ldots, f_n\}$ and $\gamma = \{g_1, \ldots, g_m\}$. Suppose given two other bases $\beta' = \{f'_1, \ldots, f'_n\}$ and $\gamma' = \{g'_1, \ldots, g'_m\}$ for R^n and R^m respectively. Let P be the base change from β' to β, and Q be the base change from γ' to γ. Then the matrix of η with respect to β' and γ' is $Q^{-1}AP$.*

Proof. The matrix P is the matrix of $\mathbf{1}_M$ with respect to β' and β. The matrix A is the matrix of η with respect to β and γ. The matrix Q^{-1} is the matrix of $\mathbf{1}_N$ with respect to γ and γ'. The result follows from Proposition 19.1.8. $\square$

In Corollary 19.1.11, we can see that A and $Q^{-1}AP$ represent the same R-linear map. These two matrices must share certain properties. Therefore, we will say that the matrices A and B are **equivalent** if $B = QAP$ where P and Q are invertible matrices. We leave it to the reader to verify that this is an equivalent relation.

At last, we would like to point out that an invertible matrix may represent:

- an isomorphism, or

- a base change.

This flexibility can be quite a useful tool for many problems.

Exercises 19.1

In the following exercises, R denote a commutative ring with unity.

1. Let M be an R-module and let $m_1, \ldots, m_n \in M$. Suppose that $m_1, \ldots, m_n$ are linearly independent over R. Show that $Rm_1, \ldots, Rm_n$ are independent over R. Give a counterexample to show that the converse is not true in general. (*Cf.* Exercise 2, §19.)

2. Give a minimal generating set of 3 elements for $\mathbb{Z}$ over $\mathbb{Z}$.

3. Let x be an indeterminate over R.

(a) Show that $R[x]$ is a free module over R by giving a basis.

(b) In Exercise 7, §18.1 we have given a submodule $\mathscr{P}_n$ of $R[x]$ over R. Show that $\mathscr{P}_n$ is a free module by giving a basis.

4. Let $I = (2, x)$ be the ideal in $\mathbb{Z}[x]$. Show that I is a not free as a $\mathbb{Z}$-module.

5. Let R be a commutative ring. Show that $M_{m \times n}(R)$ is a free R-module by giving a basis.

6. For this exercise, please refer to Exercise 9, §18.2.

(a) Show that $\operatorname{Hom}_R(R^n, R^m) \cong M_{m \times n}(R)$ as an R-module.

(b) Show that $\operatorname{End}_R(R^n) \cong M_n(R)$ as a ring.

7. Let R be a PID. Show that any ideal of R is a free module.

8. Prove Proposition 19.1.8.

9. Prove Corollary 19.1.9.

10. Let M be a free module over R. Let $\beta = (f_1, \ldots, f_n)$ be an ordered basis for M over R. Let $\beta' = (f_1', \ldots, f_n')$ be such that

$$f_j' = \sum_{i=1}^{n} p_{ij} f_i.$$

Show that β' is an ordered basis if $P = \left[p_{ij} \right]$ is invertible in $M_n(R)$.

11. Let M be a free R-module of finite rank. Let $\eta \in \operatorname{End}_R(M)$. Since the domain and the codomain of an endomorphism are supposed to be the same, it often requires that only one basis be used for both domain and codomain. Let A be the matrix of η with respect to the ordered base β. Let P be the base change from another ordered basis γ to β. Verify that the matrix of η with respect to the base γ is $P^{-1}AP$.

We say that two square matrices A and B are **similar** if $B = P^{-1}AP$ where P is an invertible matrix. Show that similarity between two matrices is an equivalence relation.

19.2   Determinant

In this section, we will give a formal discussion on determinant. Remember that R is a commutative ring with unity.

Definition 19.2.1. Let $A = \begin{bmatrix} a_{ij} \end{bmatrix}_{n \times n}$ be a square matrix over R. Define the **determinant** of A, denoted as $\det A$, to be

$$\det A = \sum_{\sigma \in S_n} (\mathrm{sgn}\,\sigma) a_{1\sigma(1)} \cdots a_{n\sigma(n)}.$$

Note that

$$\det A = \sum_{\sigma \in S_n} (\mathrm{sgn}\,\sigma)\, a_{\sigma^{-1}(1)1} a_{\sigma^{-1}(2)2} \cdots a_{\sigma^{-1}(n)n}$$

$$= \sum_{\sigma \in S_n} (\mathrm{sgn}\,\sigma^{-1})\, a_{\sigma^{-1}(1)1} a_{\sigma^{-1}(2)2} \cdots a_{\sigma^{-1}(n)n}$$

$$= \sum_{\tau \in S_n} (\mathrm{sgn}\,\tau) a_{\tau(1)1} a_{\tau(2)2} \cdots a_{\tau(n)n}.$$

The key point here is that the determinant of a square matrix is the sum of many terms, in which each term is a product of n matrix entries not lying on the same column nor on the same row.

Students are familiar with the determinant of matrices of small size:

- $\det(a_{11}) = a_{11}$;

- $\det \begin{bmatrix} a_{11} & a_{22} \\ a_{21} & a_{22} \end{bmatrix} = a_{11}a_{12} - a_{12}a_{21}.$

In the determinant of the size two matrix, the second term corresponds to the permutation $(1\ 2)$ which is an odd permutation, hence the minus sign.

The determinant of the size three matrix is given by

$$\det \begin{bmatrix} a_{11} & a_{12} & a_{13} \\ a_{21} & a_{22} & a_{23} \\ a_{31} & a_{32} & a_{33} \end{bmatrix}$$

$$\begin{aligned}
= \quad & a_{11}a_{22}a_{33} & \text{the identity permutation} \\
& -a_{12}a_{21}a_{33} - a_{13}a_{22}a_{31} - a_{11}a_{23}a_{32} & \text{the 2-cycles} \\
& +a_{12}a_{23}a_{31} + a_{13}a_{21}a_{32} & \text{the 3-cycles.}
\end{aligned}$$

For the determinant of size four matrices, there are 24 terms to add together. It stops to be efficient to use the definition to find the determinant. We will introduce some tools to deal with determinant in general.

Definition 19.2.2. Let $A = \left[a_{ij}\right] \in M_n(R)$. Define the **transpose** of A, denoted A^{t}, to be the matrix $\left[b_{ij}\right] \in M_n(R)$ where $b_{ij} = a_{ji}$.

Proposition 19.2.3. *Let $A \in M_n(R)$. Then $\det A = \det A^{\mathrm{t}}$.*

Proof. Let $A^{\mathrm{t}} = \left[b_{ij}\right]$. Then

$$
\begin{aligned}
\det A^{\mathrm{t}} &= \sum_{\sigma \in S_n} (\mathrm{sgn}\,\sigma) b_{1\sigma(1)} \cdots b_{n\sigma(n)} \\
&= \sum_{\sigma \in S_n} (\mathrm{sgn}\,\sigma) a_{\sigma(1)1} \cdots a_{\sigma(n)n} = \det A.
\end{aligned}
$$

Hence the result.   □

We will use $\widehat{}$ in the following manner:

$$
a_1 \cdots \widehat{a_i} \cdots a_n = a_1 \cdots a_{i-1} a_{i+1} \cdots a_n.
$$

We use $\widehat{}$ to indicate which element(s) to delete from the product.

Proposition 19.2.4. *Let $\mathbf{r}_i$ be the row vector $(a_{i1}, \ldots, a_{in})$ in R^n for $i = 1, \ldots, n$. Let $\mathbf{r}'_k$ be the row vector $(b_{k1}, \ldots, b_{kn})$. Then*

$$
\det \begin{bmatrix} \mathbf{r}_1 \\ \vdots \\ \mathbf{r}_k + \mathbf{r}'_k \\ \vdots \\ \mathbf{r}_n \end{bmatrix} = \det \begin{bmatrix} \mathbf{r}_1 \\ \vdots \\ \mathbf{r}_k \\ \vdots \\ \mathbf{r}_n \end{bmatrix} + \det \begin{bmatrix} \mathbf{r}_1 \\ \vdots \\ \mathbf{r}'_k \\ \vdots \\ \mathbf{r}_n \end{bmatrix}.
$$

Let $\mathbf{c}_j$ be the column vector $(a_{1j}, \ldots, a_{nj})^{\mathrm{t}}$ in R^n for $j = 1, \ldots, n$. Let $\mathbf{c}'_k$ be the column vector $(b_{1k}, \ldots, b_{nk})^{\mathrm{t}}$. Then

$$
\det \begin{bmatrix} \mathbf{c}_1 & \cdots & \mathbf{c}_k + \mathbf{c}'_k & \cdots & \mathbf{c}_n \end{bmatrix}
$$
$$
= \det \begin{bmatrix} \mathbf{c}_1 & \cdots & \mathbf{c}_k & \cdots & \mathbf{c}_n \end{bmatrix} + \det \begin{bmatrix} \mathbf{c}_1 & \cdots & \mathbf{c}'_k & \cdots & \mathbf{c}_n \end{bmatrix}.
$$

Proof. For the first identity, the determinant on the left-hand side is

$$\sum_{i=1}^{n} \left((a_{ki} + b_{ki}) \sum_{\substack{\sigma \in S_n \\ \sigma(k)=i}} (\operatorname{sgn} \sigma) a_{1\sigma(1)} \cdots \widehat{a}_{ki} \cdots a_{n\sigma(n)} \right)$$

$$= \sum_{i=1}^{n} \left(a_{ki} \sum_{\substack{\sigma \in S_n \\ \sigma(k)=i}} (\operatorname{sgn} \sigma) a_{1\sigma(1)} \cdots \widehat{a}_{ki} \cdots a_{n\sigma(n)} \right)$$

$$+ \sum_{i=1}^{n} \left(b_{ki} \sum_{\substack{\sigma \in S_n \\ \sigma(k)=i}} (\operatorname{sgn} \sigma) a_{1\sigma(1)} \cdots \widehat{a}_{ki} \cdots a_{n\sigma(n)} \right),$$

which is the determinant on the right-hand side.

The second identity follows from Proposition 19.2.3 and the first identity. $\square$

Proposition 19.2.5. *Let $A \in M_n(R)$. The following statements are true.*

(a) *If the entries of one row (or one column) of A are all 0, then $\det A = 0$.*

(b) *Let B be the matrix obtained by multiplying the ith row (or the jth column) of A by r. Then $\det B = r \det A$.*

(c) *If two rows (or two columns) of A are the same, then $\det A = 0$.*

(d) *Let B be the matrix obtained by exchanging two rows (or two columns) of A. Then $\det B = -\det A$.*

(e) *Let B be the matrix obtained by adding a scalar multiple of the ith row (or column) to the jth row (or column, respectively) of A. Then $\det B = \det A$.*

Proof. By Proposition 19.2.3 we only need to check the row version.

For the rest of the proof, we let $A = \left[a_{ij} \right]$ and $B = \left[b_{ij} \right]$.

(a) The result follows from the fact that all the summands are 0 in the definition of determinant.

(b) We have

$$\det B = \sum_{\sigma \in S_n} (\operatorname{sgn} \sigma) a_{1\sigma(1)} \cdots (r a_{i\sigma(i)}) \cdots a_{n\sigma(n)}$$

$$= r \sum_{\sigma \in S_n} (\operatorname{sgn} \sigma) a_{1\sigma(1)} \cdots a_{i\sigma(i)} \cdots a_{n\sigma(n)}$$

$$= r \det A.$$

(c) Suppose $i < j$ and the i-th row and the j-th row of A are the same. Let $\sigma \in S_n$. Suppose $\sigma(i) = r$ and $\sigma(j) = s$. Define $\tau_\sigma = (r\ s)$. Then

$$\tau_\sigma \sigma(i) = s,$$
$$\tau_\sigma \sigma(j) = r,$$
$$\tau_\sigma \sigma(k) = \sigma(k), \quad \text{if } k \neq i, j.$$

The permutations σ and $\tau_\sigma \sigma(k)$ have the same values except at i and j where they switch values. The two permutations σ and $\tau_\sigma \sigma$ are of different signs. In other words, $\operatorname{sgn} \sigma + \operatorname{sgn} \tau_\sigma \sigma = 0$ or $\operatorname{sgn} \tau_\sigma \sigma = -\operatorname{sgn} \sigma$. The determinant of A is

$$\sum_{1 \leq r < s \leq n} \left(\sum_{\substack{\sigma(i)=r \\ \sigma(j)=s}} (\operatorname{sgn} \sigma) a_{1\sigma(1)} \cdots a_{n\sigma(n)} + (\operatorname{sgn} \tau_\sigma \sigma) a_{1\tau_\sigma \sigma(1)} \cdots a_{n\tau_\sigma \sigma(n)} \right)$$

$$= \sum_{1 \leq r < s \leq n} \left(A_{ij} \sum_{\substack{\sigma(i)=r \\ \sigma(j)=s}} (\operatorname{sgn} \sigma) a_{1\sigma(1)} \cdots \widehat{a_{i\sigma(i)}} \cdots \widehat{a_{j\sigma(j)}} \cdots a_{n\sigma(n)} \right)$$

where $A_{ij} = a_{ir} a_{js} - a_{is} a_{jr} = a_{ir} a_{is} - a_{is} a_{ir} = 0$. Thus $\det A = 0$.

(d) Let $\tau = (i\ j)$. We have $b_{st} = a_{\tau(s)t}$ for all s, t. Thus

$$\det B = \sum_{\sigma \in S_n} (\operatorname{sgn} \sigma) b_{1\sigma(1)} \cdots b_{n\sigma(n)}$$

$$= \sum_{\sigma \in S_n} (\operatorname{sgn} \sigma) a_{\tau(1)\sigma(1)} \cdots a_{\tau(n)\sigma(n)}$$

$$= \sum_{\sigma \in S_n} (\operatorname{sgn} \sigma) a_{1\sigma(\tau^{-1}(1))} \cdots a_{\tau(n)\sigma(\tau^{-1}(n))}$$

$$= \sum_{\sigma \in S_n} (\operatorname{sgn} \sigma) a_{1\sigma\tau(1)} \cdots a_{n\sigma\tau(n)}$$

$$= -\sum_{\sigma \in S_n} (\operatorname{sgn} \sigma\tau) a_{1\sigma\tau(1)} \cdots a_{n\sigma\tau(n)}$$

$$= -\sum_{\pi \in S_n} (\operatorname{sgn} \pi) a_{1\pi(1)} \cdots a_{n\pi(n)}, \qquad \text{where } \pi = \sigma\tau$$

$$= -\det A.$$

For part (e), we check the column version instead. Assume $i < j$. Let $A = \begin{bmatrix} \mathbf{c}_1 & \cdots & \mathbf{c}_i & \cdots & \mathbf{c}_j & \cdots & \mathbf{c}_n \end{bmatrix}$. Then

$$\det \begin{bmatrix} \mathbf{c}_1 & \cdots & \mathbf{c}_i & \cdots & \mathbf{c}_j + \mathbf{c}_i & \cdots & \mathbf{c}_n \end{bmatrix}$$

$$= \det \begin{bmatrix} \mathbf{c}_1 & \cdots & \mathbf{c}_i & \cdots & \mathbf{c}_j & \cdots & \mathbf{c}_n \end{bmatrix}$$

$$\quad + \det \begin{bmatrix} \mathbf{c}_1 & \cdots & \mathbf{c}_i & \cdots & \mathbf{c}_i & \cdots & \mathbf{c}_n \end{bmatrix}$$

$$= \det A$$

by part (c) and Proposition 19.2.4. In case $i > j$ the proof is similar and we will not repeat the argument. $\qquad \square$

Let $A = \begin{bmatrix} a_{ij} \end{bmatrix}_{n \times n}$. The (i,j)-**minor** of A, denoted M_{ij}, is the square matrix of size $(n-1)$ obtained by deleting the i-th row and the j-th column from A. If $M_{ij} = \begin{bmatrix} m_{kl} \end{bmatrix}_{(n-1) \times (n-1)}$, then

$$m_{kl} = \begin{cases} a_{kl}, & \text{if } k < i \text{ and } l < j, \\ a_{k+1,l}, & \text{if } k \geq i \text{ and } l < j, \\ a_{k,l+1}, & \text{if } k < i \text{ and } l \geq j, \\ a_{k+1,l+1}, & \text{if } k \geq i \text{ and } l \geq j. \end{cases}$$

For $1 \leq s \leq n$, we use π_s to denote the cycle $(n \; n-1 \; \cdots \; s+1 \; s)$. Then

$$a_{kl} = m_{\pi_i(k)\pi_j(l)}$$

for $k \neq i$ and $l \neq j$.

Let i be such that $1 \leq i \leq n$. The determinant of A is $\sum_{j=1}^n a_{ij} A_{ij}$ where A_{ij} is

$$\sum_{\substack{\sigma \in S_n \\ \sigma(i)=j}} (\operatorname{sgn} \sigma) \, a_{1\sigma(1)} \cdots \widehat{a}_{ij} \cdots a_{n\sigma(n)}$$

$$= \sum_{\substack{\sigma \in S_n \\ \sigma(i)=j}} (\text{sgn}\,\sigma)\, m_{1,\pi_j\sigma(1)} \cdots m_{i-1,\pi_j\sigma(i-1)} m_{i,\pi_j\sigma(i+1)} \cdots m_{n-1,\pi_j\sigma(n)}$$

$$= \sum_{\substack{\sigma \in S_n \\ \sigma(i)=j}} (\text{sgn}\,\sigma)\, m_{1,\pi_j\sigma\pi_i^{-1}(1)} \cdots m_{n-1,\pi_j\sigma\pi_i^{-1}(n-1)}$$

$$= \sum_{\substack{\sigma \in S_n \\ \sigma(i)=j}} ((-1)^{j-1}(-1)^{i-1}\,\text{sgn}\,\pi_j\sigma\pi_i^{-1})\, m_{1,\pi_j\sigma\pi_i^{-1}(1)} \cdots m_{n-1,\pi_j\sigma\pi_i^{-1}(n-1)}.$$

For $\sigma \in S_n$ such that $\sigma(i) = j$, we have

$$\pi_j\sigma\pi_i^{-1}(n) = \pi_j\sigma(i) = n.$$

We have a natural embedding

$$\{\pi_j\sigma\pi_i^{-1} : \sigma \in S_n,\ \sigma(i)=j\} \subseteq S_{n-1}.$$

Since there are $(n-1)!$ such σ, we have

$$\{\pi_j\sigma\pi_i^{-1} : \sigma \in S_n,\ \sigma(i)=j\} = S_{n-1}.$$

Hence

$$A_{ij} = (-1)^{i+j} \det M_{ij}.$$

We call A_{ij} the (i,j)-**cofactor** of A.

Proposition 19.2.6. *Let $A \in M_n(R)$. Then*

$$a_{i1}A_{i1} + a_{i2}A_{i2} + \cdots + a_{in}A_{in} = \det A,$$
$$a_{1j}A_{1j} + a_{2j}A_{2j} + \cdots + a_{nj}A_{nj} = \det A$$

for $1 \le i,\,j \le n$.

Proof. The column version follows from the previous discussion and Proposition 19.2.3.   □

Corollary 19.2.7. *Let $A \in M_n(R)$. Then*

$$a_{i1}A_{j1} + a_{i2}A_{j2} + \cdots + a_{in}A_{jn} = 0,$$
$$a_{1i}A_{1j} + a_{2i}A_{2j} + \cdots + a_{ni}A_{nj} = 0$$

for $1 \le i,\,j \le n$ such that $i \ne j$.

Proof. Let B be the matrix obtained by replacing the j-th row of A by the i-th row. Then

$$a_{i1}A_{j1} + a_{i2}A_{j2} + \cdots + a_{in}A_{jn} = \det B.$$

However, B is a matrix with repeated rows. We have that $\det B = 0$ by Proposition 19.2.5(c).

Similarly we have the column version of this corollary.   □

Definition 19.2.8. Let A be a square matrix of size n. The **adjoint** of A, denoted adj A, is defined to be $\left[A_{ij}\right]_{n\times n}^{t}$.

Corollary 19.2.9. *Let $A \in M_n(R)$. Then*

$$A(\text{adj } A) = (\det A)I_n = (\text{adj } A)A.$$

Proof. The result follows from Proposition 19.2.6 and Corollary 19.2.7.   □

Theorem 19.2.10. *Let A and B be two $n \times n$ matrices over R. Then*

$$\det AB = \det A \cdot \det B.$$

Proof. Case 1. When R is a field, this is an essential result in a course on Linear Algebra. We will just assume the result without proof. We proceed to deal with the general case.

Case 2. Assume that $R = \mathbb{Z}[x_{ij}, y_{ij} : 1 \le i, j \le n]$ where the x_{ij}'s and the y_{ij}'s are variables over $\mathbb{Z}$. Then R is an integral domain by Proposition 16.2.2. Let Q be the quotient field of R. Then R may be considered as a subring of Q. Hence A and B may also be regarded a matrices over Q. From Case 1, we have that $\det AB = \det A \det B$ in Q. It follows that this is true in R as well.

The general case. Let $A = \left[a_{ij}\right]$ and $B = \left[b_{ij}\right]$.

Let $S = \mathbb{Z}[x_{ij}, y_{ij} : 1 \le i, j \le n]$ be the polynomial ring of $2n^2$ variables over $\mathbb{Z}$. There is a canonical ring homomorphism ϕ from $\mathbb{Z}$ into R. From Corollary 16.1.5, ϕ may be extended to a ring homomorphism $\overline{\phi}$ from S into R sending x_{ij} to a_{ij} and y_{ij} to b_{ij}. Let $X = \left[x_{ij}\right]_{n\times n}$ and $Y = \left[y_{ij}\right]_{n\times n}$. Then

$$\det AB = \overline{\phi}(\det XY)$$

$$= \overline{\phi}(\det X \det Y), \qquad \text{by Case 2,}$$
$$= \overline{\phi}(\det X)\overline{\phi}(\det Y)$$
$$= \det A \det B.$$

Hence the result. □

Proposition 19.2.11. *Let $A \in M_n(R)$. The matrix A is an invertible matrix over R if and only if $\det A$ is invertible in R.*

Proof. The "only if" part: Suppose A is invertible. Then there exists $B \in M_n(R)$ such that $AB = I_n$. It follows that

$$\det A \det B = \det AB = \det I_n = 1$$

in R. Hence $\det A$ is invertible in R.

The "if" part: If $\det A$ is invertible in R, then

$$A^{-1} = (\det A)^{-1} \operatorname{adj} A$$

by Corollary 19.2.9. □

Proposition 19.2.12. *Let A and B be two square matrices over a commutative ring with unity. If $AB = I$ then $BA = I$.*

Proof. Since $AB = I$, we have $\det A \det B = \det AB = \det I = 1$. From Proposition 19.2.11, we know that A is invertible. Thus

$$B = IB = A^{-1}AB = A^{-1}I = A^{-1}.$$

It follows that $BA = A^{-1}A = A$. □

When R is non-commutative, there are counterexamples to Proposition 19.2.12, which in turn will give counterexamples to Theorem 19.2.13.

Proposition 19.2.13. *Let R be a commutative ring with unity. If $R^m \cong R^n$ then $m = n$.*

Proof. Suppose this assertion is false. We may assume $m < n$ without loss of generality. Let $\phi \colon R^m \to R^n$ be an isomorphism over R. Let $A_{n \times m}$ be the matrix of ϕ with respect to the standard bases. Let $B_{m \times n}$ be the matrix of ϕ^{-1} with respect to the standard bases. Since $\phi\phi^{-1}$ is the identity matrix, we have that $AB = I_n$. Add $(n - m)$ columns to the right of A to enlarge it into an $n \times n$ matrix A'. Add $(n - m)$ rows below B to enlarge it into an $n \times n$ matrix B' as well. Then

$$A'B' = \begin{bmatrix} A & \mathbf{0}_{n\times(n-m)} \end{bmatrix} \begin{bmatrix} B \\ \mathbf{0}_{(n-m)\times n} \end{bmatrix} = AB = I_n.$$

By Proposition 19.2.12, we should have $B'A' = I_n$ as well. However,

$$B'A' = \begin{bmatrix} B \\ \mathbf{0}_{(n-m)\times n} \end{bmatrix} \begin{bmatrix} A & \mathbf{0}_{n\times(n-m)} \end{bmatrix}$$

$$= \begin{bmatrix} BA & \\ & \mathbf{0}_{(n-m)\times(n-m)} \end{bmatrix} \neq I_n,$$

a contradiction. □

Thanks to this proposition, we now have a well-defined notion of rank for finitely generated free modules.

Exercises 19.2

1. Express explicitly $\det \begin{bmatrix} a_{ij} \end{bmatrix}_{4\times4}$.

2. We say a matrix $A = \begin{bmatrix} a_{ij} \end{bmatrix}$ in $M_n(R)$ is **upper triangular** if $a_{ij} = 0$ for all $i > j$. We say A is **lower triangular** if $a_{ij} = 0$ for all $i < j$. Upper and lower triangular matrices are called **triangular** matrices categorically.

   Show that the determinant of a triangular matrix is the product of the diagonal entries.

3. Let $A = M_n(F)$ where F is a field. Show that $\det \operatorname{adj} A = (\det A)^{n-1}$. Is this still true if $A \in M_n(R)$ where R is just a commutative ring with unity? Prove your assertion or give a counterexample.

4. Show that

$$
\det \begin{bmatrix} 1 & a_1 & a_1^2 & \cdots & a_1^{n-2} & a_1^{n-1} \\ 1 & a_2 & a_2^2 & \cdots & a_2^{n-2} & a_2^{n-1} \\ \cdots\cdots\cdots\cdots\cdots\cdots\cdots\cdots\cdots\cdots\cdots\cdots \\ \cdots\cdots\cdots\cdots\cdots\cdots\cdots\cdots\cdots\cdots\cdots \\ 1 & a_n & a_n^2 & \cdots & a_n^{n-2} & a_n^{n-1} \end{bmatrix} = \prod_{i<j}(a_j - a_i).
$$

This is the so-called **Vandermonde determinant**.

5. Let $n \geq 2$ and let A be the $n \times n$ matrix

$$
\begin{bmatrix} i & 1 & 1 & \cdots & 1 \\ 1 & i & 1 & \cdots & 1 \\ 1 & 1 & i & \cdots & 1 \\ \vdots & \vdots & \vdots & \ddots & \vdots \\ 1 & 1 & 1 & \cdots & i \end{bmatrix}
$$

with complex entries. Find A^{-1}, $\det A$ and $\operatorname{adj} A$.

Review Exercises for Chapter 19

1. Show that $R^2/R(1,\, 1)$ is a free $\mathbb{Z}$-module of rank 1.

2. Let D be a PID and let M be a free R-module of rank n. Show that any submodule of M is free of rank $\leq n$. (Hint: We may assume that $M = R^n$ without loss of generality. When $n = 1$, this is just Exercise 7, §7. In general, let N be a submodule of R^n. Let $N' = N \cap (0 \times R^{n-1})$ and use induction to find a basis for N'. Try to expand this to a basis for N over R.)

3. Let A, B and C be matrices over R. Show that

$$
\det \begin{bmatrix} A_{m\times m} & B_{m\times n} \\ 0_{n\times m} & C_{n\times n} \end{bmatrix}_{(m+n)\times(m+n)} = \det A \det C.
$$

CHAPTER 20

Vector Spaces over Arbitrary Fields

In an introductory course to Linear Algebra, the emphasis is usually on vector spaces and matrices over the field $\mathbb{R}$. However, probably eighty percent of the material is still valid for vector spaces over an arbitrary field. Knowledge on vector spaces in general is essential to all fields of mathematics, and to all disciplines of sciences.

In this chapter, we will describe all the basic results on vector spaces that we will need for the study of fields. Some of the proofs will not be supplied, since you can find proofs that will work almost word for word in any elementary textbook for undergraduate Linear Algebra.

20.1   A Brief Review on Vector Spaces

·Throughout this chapter F always denotes a field. Here we restate the definition for vector spaces in case the instructor or the reader chooses to skip Chapter 18.

Definition 20.1.1. A **vector space** V over F or an F-vector space is an abelian group V with the **vector addition** "+" and the **scalar product** from $F \times V$ to V such that for all α, $\beta \in F$ and $v, w \in V$ we have

(V1) $\alpha(v + w) = \alpha v + \alpha w$,

(V2) $(\alpha + \beta)v = \alpha v + \beta v$,

(V3) $\alpha(\beta v) = (\alpha\beta)v$, and

(V4) $1v = v$.

Elements in V are called **vectors** while we refer to elements in F as **scalars**. A **subspace** W of V is an additive subgroup of V which is closed under the scalar product. A subspace is itself a vector space over F.

We say that F is a **subfield** of the field E if F is a subring of E and F is itself a field. Below are some comments on subfields.

(1) To show that F is a subfield of E is equivalent to showing that 1, $a + b$, ab, $-a$, $a^{-1} \in F$ for all $a, b \in F$ and $a \neq 0$.

(2) The unity in E is the same as the unity in any of its subfields F.

(3) If V is a vector space over E and F is a subfield of E, then it is also a vector space over F.

(4) Let F be a subfield of E and E be a subfield of K. Then F is still a subfield of K. Furthermore, K is a vector space over E as well as over F. On the other hand, E is a subspace of K over F.

   Moreover, if a field F is a subring of the ring R, R is a vector space over F. If I is an ideal of R, then I is a subspace of R over F.

Let $v_1, \ldots, v_m$ be vectors in the vector space V over F. A vector of the form

$$\alpha_1 v_1 + \alpha_2 v_2 + \cdots + \alpha_m v_m, \qquad \alpha_i \in F,$$

is called a **linear combination** of $v_1, \ldots, v_m$ over F. The set of all linear combinations of $v_1, \ldots, v_m$ over F forms the smallest subspace in V containing $v_1, \ldots, v_m$. This subspace is usually denoted as

$$\langle v_1, \ldots, v_m \rangle \qquad \text{or} \qquad F v_1 + \cdots + F v_m$$

and is called the **linear span** of $v_1, \ldots, v_m$ over F. If $V = \langle v_1, \ldots, v_m \rangle$, we also say V is **spanned** by $v_1, \ldots, v_m$ over F.

Even if S is an infinite set, we will still use $\langle S \rangle$ to denote the smallest subspace in V containing S. In fact, $\langle S \rangle$ is the subspace containing all linear combinations of finitely many vectors in S.

We say that the vectors $v_1, \ldots, v_m$ in the F-vector space V are **linearly independent** over F if for $\alpha_1, \ldots, \alpha_m \in F$,

$$\alpha_1 v_1 + \alpha_2 v_2 + \cdots + \alpha_m v_m = 0 \quad \Longrightarrow \quad \alpha_1 = \alpha_2 = \cdots = \alpha_m = 0.$$

When $v_1, \ldots, v_m$ are linearly independent over F, the expression of a vector as a linear combination of $v_1, \ldots, v_m$ is unique. To be more precise,

$$\sum_{i=1}^{m} \alpha_i v_i = \sum_{i=1}^{m} \beta_i v_i \quad \Longrightarrow \quad \alpha_i = \beta_i \qquad \text{for } i = 1, 2, \ldots, m.$$

We say that $v_1, \ldots, v_m$ form a **basis** or a **base** for V over F if

(i) $v_1, \ldots, v_m$ are linearly independent over F, and

(ii) V is spanned by $v_1, \ldots, v_m$ over F.

This means that every element of V can be expressed as a linear combination of $v_1, \ldots, v_m$ in a unique way.

By a **finite dimensional** vector space V over a field F, we mean that V is spanned by a finite subset of vectors $v_1, v_2, \ldots, v_m$. Now we explain how to construct a basis from a set of generators for V. If $v_1, \ldots, v_m$ are linearly independent over F, we have a basis and we are done. Suppose not. Let $\alpha_1 v_1 + \alpha_2 v_2 + \cdots + \alpha_m v_m = 0$ where $\alpha_i \in F$ are not all zero. We might as well assume that $\alpha_m \neq 0$. Then

$$v_m = -\alpha_m^{-1} (\alpha_1 v_1 + \alpha_2 v_2 + \cdots + \alpha_{m-1} v_{m-1}).$$

Hence any subspace containing $v_1, \ldots, v_{m-1}$ will also contain v_m. We have shown that $\langle v_1, \ldots, v_{m-1} \rangle = \langle v_1, \ldots, v_m \rangle$. Continue with this process, eventually we will reach a subset, say $\{v_1, \ldots, v_d\}$, which is linearly independent over F while spanning V at the same time. We have demonstrated that every finite dimensional vector space has a basis. This result is true even for infinite dimensional vector spaces.

Theorem 20.1.2. *Every vector space has a basis.*[1]

As you can see, the proof of the theorem above is exactly the same as the one for the same theorem dealing with only vector spaces over $\mathbb{R}$. Next, we are going to list more results that we will be needing without giving the actual proofs. We will leave proving these results as exercises.

Theorem 20.1.3. *Any two bases of a vector space have the same cardinality.*

We define the **dimension** of V over F, denoted $\dim_F V$ or simply $\dim V$, to be the cardinality of any basis of V over F.

Proposition 20.1.4. *Let V be an n-dimensional vector space over F. Then the following conditions are equivalent.*

(i) *The set $\{v_1, \ldots, v_n\}$ is a basis for V over F.*

(ii) *The set $\{v_1, \ldots, v_n\}$ is a linearly independent set over F.*

(iii) *The set $\{v_1, \ldots, v_n\}$ spans V over F.*

In fact, in the previous discussion we have also proved the part (a) of the following important result for finite dimensional vector spaces.

Proposition 20.1.5. *Let V be a vector space over F.*

(a) *Any set of generators for V over F can be reduced to a basis for V over F.*

(b) *Any linearly independent set over F in V can be expanded to a basis for V over F.*

Corollary 20.1.6. *Let W be a subspace of an F-vector space. Then $\dim W \leq \dim V$.*

---

[1] For a rigorous proof of this theorem one needs to assume Zorn's Lemma.

Example 20.1.7. The field $\mathbb{C}$ can be viewed as a vector space over $\mathbb{R}$ or over $\mathbb{C}$. It is easy to see that $\dim_{\mathbb{R}} \mathbb{C} = 2$ and $\dim_{\mathbb{C}} \mathbb{C} = 1$. The set $\{1, i\}$ is a basis for $\mathbb{C}$ over $\mathbb{R}$ and the set $\{1\}$ is a basis for $\mathbb{C}$ over $\mathbb{C}$.

If W is an F-subspace of V, then as additive groups we have the quotient group V/W. Since $\alpha W \subseteq W$, one can see that it is also natural to have a scalar product on V/W given by

$$\alpha(v + W) = \alpha v + W, \qquad \text{for any } v \in V.$$

Thus, V/W is also a vector space over F.

Proposition 20.1.8. *Let V be a finite dimensional vector space over F and let W be a subspace. Then $\dim V/W = \dim V - \dim W$.*

Proof. Find a basis $\{w_1, \ldots, w_m\}$ for W over F. Since this is also a linearly independent set in V over F, by Theorem 20.1.5(b) it can be expanded to a basis $\{w_1, \ldots, w_m, v_1, \ldots, v_n\}$ for V over F. It follows that

$$\frac{V}{W} = \langle \overline{w}_1, \ldots, \overline{w}_m, \overline{v}_1, \ldots, \overline{v}_n \rangle = \langle \overline{v}_1, \ldots, \overline{v}_n \rangle.$$

Suppose that there are $\alpha_1, \ldots, \alpha_n \in F$ such that $\alpha_1 \overline{v}_1 + \cdots + \alpha_n \overline{v}_n = 0$. Then $\alpha_1 v_1 + \cdots + \alpha_n v_n \in W$. We can thus find $\beta_1, \ldots, \beta_m \in F$ such that $\alpha_1 v_1 + \cdots + \alpha_n v_n = \beta_1 w_1 + \cdots + \beta_m w_m$. Hence,

$$-\beta_1 w_1 - \cdots - \beta_m w_m + \alpha_1 v_1 + \cdots + \alpha_n v_n = 0.$$

Since $\{w_i, v_j\}_{i,j}$ is linearly independent over F, we have that all the α_i's are 0. We have shown that $\{\overline{v}_1, \ldots, \overline{v}_n\}$ is a basis for V/W over F. It follows that $\dim V = n + m = \dim V/W + \dim W$. $\qquad\square$

Exercises 20.1

In the following exercises, F always denotes a field and V, W always denote vector spaces over F.

1. Let the field F be a subring of the commutative ring R and let I be any ideal of R. Show that I is a vector space over F (this includes the improper ideal R itself).

2. Prove Theorem 20.1.3 and Proposition 20.1.5(b) for finite dimensional vector spaces.

3. Prove Proposition 20.1.4.

4. Let W be a subspace of a finite dimensional vector space V over F. Show that $\dim W = \dim V$ if and only if $W = V$.

5. Let x be an indeterminate over F and let $f(x)$ be a nonzero polynomial of degree n in $F[x]$. Show that $\dim_F F[x]/(f(x)) = n$.

6. Let U and W be subspaces of a vector space V over a field F. Define

$$U + W = \{\, u + w \in V : u \in U, w \in W \,\}.$$

   (a) Show that $U + W$ is a subspace of V.

   (b) Show that $U \cap W$ is a subspace of V.

   (c) If U and W are finite dimensional over F, show that

$$\dim_F(U + W) = \dim_F U + \dim_F W - \dim_F(U \cap W).$$

7. Let n be a positive integer and p a positive prime integer. If V is an n-dimensional vector space over $\mathbb{Z}_p$, show that V has p^n elements.

8. Determine whether the given set of vectors is linearly independent over $\mathbb{Z}_5$.

   (a) $\{\, (2,1,0), (0,-2,-3), (1,4,3) \,\}$

   (b) $\{\, (4,1,3), (1,4,2) \,\}$

   (c) $\{\, (1,2,3), (2,4,2), (3,1,4) \,\}$

   (d) $\{\, (2,3,2), (3,2,3) \,\}$

9. Let n be a positive integer and p be a positive prime integer.

   (a) Show that $|GL_n(\mathbb{Z}_p)| = (p^n - 1)(p^n - p) \cdots (p^n - p^{n-1})$.

   (b) Show that $|SL_n(\mathbb{Z}_p)| = (\dfrac{1}{p-1})(p^n - 1)(p^n - p) \cdots (p^n - p^{n-1})$.

20.2   A Brief Review on Linear Transformations

Again, in this section F always denotes a field. This section is optional reading.

Let V and W be vector spaces over F. A mapping T from V to W is called a **linear transformation** over F if

$$T(u + v) = T(u) + T(v) \quad \text{and} \quad T(\alpha u) = \alpha T(u)$$

for u, $v \in V$ and $\alpha \in F$. A bijective linear transformation is called an **isomorphism** as well. We say V is **isomorphic to** W, denoted $V \cong W$, if there exists an isomorphism from V to W. The relation $\cong$ is an equivalence relation. If T is a homomorphism, then its image $T(V)$ is a subspace of W and its kernel $\ker T = \{\, v \in V : T(v) = 0 \,\}$ is a subspace of V. We have

$$T(V) \cong \frac{V}{\ker T}$$

as vector spaces over F.

The following is a most important theorem which tells us how to construct linear transformations between vector spaces.

Theorem 20.2.1. *Let V and W be vector spaces over F. Let $v_1, \ldots, v_n$ form a basis for the vector space V over F. Then for any choice of $w_1, \ldots, w_n$ in W (which may be repeated), there exists a unique F-linear transformation $T: V \to W$ such that $T(v_i) = w_i$ for each i.*

Proof. Since $v_1, \ldots, v_n$ generate V, a linear transformation is completely determined by the images of v_i. Hence T is unique if it exists. Define T by letting $T\left(\sum_{i=1}^{n} \alpha_i v_i\right) = \sum_{i=1}^{n} \alpha_i w_i$. Since the expression of any vector in V as a linear combination of the v_i's is unique, T is well-defined. It is now easy to verify that T is the required linear transformation over F. □

Corollary 20.2.2. *Let V and W be finite dimensional vector spaces over the field F. Then $V \cong W$ if and only if $\dim_F V = \dim_F W$.*

Proof. The "only if" part is clear for an isomorphism carries a basis onto a basis. To show the "if" part, let $\{\, v_1, \ldots, v_n \,\}$ and $\{\, w_1, \ldots, w_n \,\}$ be bases for V and W over F respectively. Let $S\colon V \to W$ be the F-linear transformation sending v_i to w_i and let $T\colon W \to V$ be the F-linear transformation sending w_i to v_i. Then S and T are inverse to each other. Thus S is an isomorphism. $\square$

Corollary 20.2.3. *Let n be a positive integer and V be an n-dimensional vector space over a field F. Then $V \cong F^n$ as vector spaces.*

Let $T\colon V \to W$ be a linear transformation over F. Let $\mathscr{B} = \{\, v_1, \ldots, v_n \,\}$ and $\mathscr{C} = \{\, w_1, \ldots, w_m \,\}$ be bases for V and W respectively. If

$$T(v_j) = \sum_{i=1}^{n} a_{ij} w_i, \qquad j = 1, \ldots, n,$$

where $a_{ij} \in F$, we can use the matrix $A = \left[a_{ij} \right]_{m \times n}$ to represent the linear transformation T with respect to the ordered bases $\mathscr{B}$ and $\mathscr{C}$. Let $B = \left[b_{ij} \right]_{\ell \times m}$ be the matrix representing $S\colon W \to U$ with respect to the ordered bases $\mathscr{C}$ and $\mathscr{D} = (x_1, \ldots, x_\ell)$. From Proposition 19.1.8 and Corollary 19.1.9 we have the following properties regarding matrix representations.

(1) The matrix representing ST with respect to ordered bases $\mathscr{B}$ and $\mathscr{D}$ is BA.

(2) The matrix representing the identity function on V with respect to $\mathscr{B}$ and $\mathscr{B}$ is the identity matrix.

(3) If T is an isomorphism, then with respect to $\mathscr{C}$ and $\mathscr{B}$ the isomorphism T^{-1} is represented by the matrix A^{-1}.

Next, we will concentrate on the F-**linear endomorphisms** on V. A linear endomorphism is a linear transformations from V into itself. We prefer to use one fixed ordered basis for V when we use a matrix to represent a linear endomorphism.

Let the matrix $A = \left[a_{ij} \right] \in M_n(F)$ be the matrix representing the linear endomorphism T with respect to the ordered basis $\mathscr{B} = (u_1, u_2, \ldots, u_n)$ for V over F. Now assume that $\mathscr{B}' = (v_1, v_2, \ldots, v_n)$ is another ordered basis for V over F where

$$v_j = \sum_{i=1}^{n} p_{ij} u_i, \quad i = 1, 2, \ldots, n.$$

Then $P = \left[p_{ij} \right]_{n \times n}$ is called the **base change** matrix from $\mathscr{B}'$ to $\mathscr{B}$. From Proposition 19.1.10 and Corollary 19.1.11, we have the following properties regarding base change.

(1) The matrix $P^{-1} = \left[q_{ij} \right]_{n \times n}$ is the base change matrix from $\mathscr{B}$ to $\mathscr{B}'$ and $u_j = \sum_{j=1}^{n} q_{ij} v_i$ for $j = 1, 2, \ldots, n$.

(2) The matrix representing T with respect to the ordered basis $\mathscr{B}'$ is $P^{-1} A P$.

With a suitable base change, one can represent a linear endomorphism with a particularly simple matrix.

We say two $n \times n$ matrices are **similar** if there is an invertible matrix P such that $B = P^{-1} A P$. Hence a linear endomorphism on an n-dimensional vector space into itself is represented by similar matrices with respect to different ordered bases. From now on we will talk about the matrices and the linear endomorphisms interchangeably.

Let V be an n-dimensional vector space over the field F and T a linear endomorphism on V. Let A be the matrix representing T with respect to the ordered basis $\mathscr{B} = (v_1, \ldots, v_n)$. Suppose V also possesses an ordered basis $\mathscr{B}' = (u_1, u_2, \ldots, u_n)$ consisting of eigenvectors with corresponding eigenvalues $\lambda_1, \lambda_2, \ldots, \lambda_n$ in F. Let P be the base change matrix from $\mathscr{B}'$ to $\mathscr{B}$. Then

$$P^{-1} A P = \operatorname{diag}[\lambda_1, \lambda_2, \ldots, \lambda_n].$$

In this case, we say that A is a **diagonalizable** matrix. A diagonalizable matrix is certainly easier to analyze after it is diagonalized.

Symmetric matrices over $\mathbb{R}$ are diagonalizable. A matrix U in $M_n(\mathbb{C})$ is **unitary** if $U^{-1} = \overline{U}^t$, the **conjugate transpose** of U. Unitary matrices are diagonalizable.

In general, not all matrices are diagonalizable. The next best thing is to find its canonical form. A general discussion of canonical forms is beyond the scope of this textbook. Here, we simply give a description of the most basic canonical form: the *rational form*. No proofs will be provided for the results stated below.

For the rest of this section, let x be an indeterminate over F. For any $A = \left[a_{ij}\right] \in M_n(F)$, the polynomial

$$f(x) = \det(xI_n - A)$$

is a monic polynomial called the **characteristic polynomial** of A.

Theorem 20.2.4 (Cayley-Hamilton). *Let $A = \left[a_{ij}\right] \in M_n(F)$. There is a monic polynomial $g(x)$ in $F[x]$, called the **minimal polynomial** of A, such that A is the lowest degree monic polynomial with $g(A) = 0$. The minimal polynomial $g(x)$ is a factor of the characteristic polynomial of A. Moreover, if $h(x) \in F[x]$ is such that $h(A) = 0$ then $g(x)$ is a factor of $h(x)$.*

Let $A \in F[x]$ be a matrix such that its minimal polynomial equals its characteristic polynomial $f(x) = x^n + a_{n-1}x^{n-1} + a_{n-2}x^{n-2} + \cdots + a_1 x + a_0$. Let T be the linear endomorphism on F^n represented by A with respect to the standard basis

$$\{\, e_1 = (1, 0, \ldots, 0),\ e_2 = (0, 1, \ldots, 0),\ \ldots,\ e_n = (0, 0, \ldots, 1)\,\}.$$

It is possible to find $v \in F^n$ such that

$$\mathscr{B} = \{\, v,\ T(v),\ T^2(v),\ \ldots,\ T^{n-1}(v)\,\}$$

form a basis for F^n over F. Since

$$T^n(v) = -a_{n-1}T^{n-1}(v) - a_{n-2}T^{n-2}(v) - \cdots - a_1 T(v) - a_0 v,$$

the matrix representing T with respect to $\mathscr{B}$ is

(20.2.1)
$$R = \begin{bmatrix} 0 & 0 & 0 & \cdots & 0 & -a_0 \\ 1 & 0 & 0 & \cdots & 0 & -a_1 \\ 0 & 1 & 0 & \cdots & 0 & -a_2 \\ \vdots & \vdots & \vdots & \ddots & \vdots \\ 0 & 0 & 0 & \cdots & 0 & -a_{n-2} \\ 0 & 0 & 0 & \cdots & 1 & -a_{n-1} \end{bmatrix}.$$

This is called the **rational canonical form** of T. In general, a matrix of the form in (20.2.1) is called the **rational block** of $f(x)$. For an arbitrary linear endomorphism T on a finite dimensional vector space over F, we can decompose V into a direct sum of subspaces

$$V = W_1 \oplus \cdots \oplus W_s$$

such that the characteristic polynomial and the minimal polynomial are the same for each $T|_{W_i}$. Let $f_i(x)$ be the minimal polynomial of $T|_{W_i}$ for each i. Then with respect to a suitably chosen basis for V, the matrix representation of T is

$$\begin{bmatrix} R_1 & & & \\ & R_2 & & \\ & & \ddots & \\ & & & R_s \end{bmatrix},$$

where B_i is the rational block of $f_i(x)$ and $f_i|f_{i+1}$ for $i = 1, \ldots, s-1$. This is called the **rational canonical form** of (any matrix representing) T.

Exercises 20.2

In the following problems, F always denotes a field and V, W always denote vector spaces over F.

1. Show that "$\cong$" is an equivalence relation for vector spaces over F.

2. Show that the relation "similar" is an equivalence relation on $M_n(F)$.

3. Let $T: V \to W$ be a linear transformation.

   (a) Show that $T(V)$ is a subspace of W.

   (b) Show that $T(V) \cong V/\ker T$ as vector spaces over F.

   (c) If V is finite dimensional over F, show that

   $$\dim_F T(V) = \dim_F V - \dim_F \ker T.$$

4. Find the characteristic polynomial of the matrix R in (20.2.1) by actually finding the determinant of $xI_n - R$.

Review Exercises for Chapter 20

In all the following problems, F denotes a field.

1. Let V be a vector space over F and let U, W be its subspace.

    (a) Show that $(U + W)/W \cong U/(U \cap W)$.

    (b) If $U \subseteq W$, show that $\dfrac{V/U}{W/U} \cong \dfrac{V}{W}$.

2. Show that λ is an eigenvalue of the matrix $A \in M_n(F)$ if and only if it is a zero of the characteristic polynomial of A.

3. Let F be a finite field of characteristic p. Show that $|F| = p^n$ for some integer n.

4. Let $v_i = (a_{i1}, \ldots, a_{in}) \in F^n$ for $i = 1, \ldots, n$. Show that $v_1, \ldots, v_n$ form a basis for F^n over F if and only if the matrix $[a_{ij}] \in M_n(F)$ is invertible.

5. Let V be a finite dimensional vector space over $\mathbb{C}$. Show that

$$\dim_{\mathbb{C}} V = 2 \dim_{\mathbb{R}} V.$$

CHAPTER 21

Field Extensions

We can start with the prime fields, and then adjoin new elements to make field extensions. The new elements can be divided into two categories, the algebraic and the transcendental. We will also study basic properties on field extensions and extension degrees.

Since ancient Greece, people have been intrigued by geometric constructions. Especially, the geometric construction which can be achieved using only a straightedge without markings and a compass. However, the question whether it is possible to achieve the following constructions continued to perplex mathematicians for over 2000 years:

- Duplicating a cube

- Squaring a circle

- Trisecting an arbitrary angle

This question was finally answered in the negative in the nineteenth century when these problems of geometry were finally translated into problems of algebra.

21.1 Algebraic or Transcendental?

Let E and F be fields. We call E a **field extension** of F if E contains F as a subfield. We also use E/F to denote the field extension $F \subseteq E$. Let S be a subset of E. We use $F(S)$ to denote the smallest subfield in E containing both F and S. It is also called the field **generated** by S over F. If $S = \{\alpha_1, \ldots, \alpha_n\}$, we write $F(\alpha_1, \ldots, \alpha_n)$ for $F(S)$. If $E = F(\alpha_1, \ldots, \alpha_n)$ we say that E is **finitely generated** over F as a field. Below are two important properties.

(1) The field extension $F(S)$ is the quotient field of $F[S]$.

(2) The field $F(\alpha_1, \ldots, \alpha_n) = F(\alpha_1)(\alpha_2) \cdots (\alpha_n)$.

If $E = F(\alpha)$, we say E/F is a **simple extension**, and α is called a **primitive** element of E over F. Thanks to the second property above, to study a finitely generated field extension, it is feasible to break it into several steps and study the simple extensions. Now we will focus on simple extensions.

Let $F \subseteq E$ be a field extension. We say an element $\alpha \in E$ is **algebraic** over F if there is a nonzero polynomial $f(x)$ over F such that $f(\alpha) = 0$. Otherwise, we say that α is **transcendental** over F.

Traditionally, a number α in $\mathbb{C}$ is called an **algebraic number** if it is algebraic over $\mathbb{Q}$. Otherwise, it is called a **transcendental number**. For example, $\sqrt[3]{2}$ and $\sqrt[5]{4 + \sqrt{2}}$ are algebraic numbers since they are zeros of $x^3 - 2$ and $(x^5 - 4)^2 - 2$ respectively. The two numbers e and π are the two most well-known examples of transcendental numbers.[1]

Consider the ring homomorphism from $F[x]$ to E sending x to α. If α is transcendental over F, then

$$\ker \varphi = \{g(x) \in F[x] : g(\alpha) = 0\} = (0).$$

Thus, $F[x] \cong F[\alpha]$ and α can be treated exactly like an indeterminate over F. Hence $F(\alpha)$ is the quotient field of $F[\alpha]$ which is isomorphic to the rational field $F(x)$.

Next, we consider the case where α is algebraic over F. Then $\ker \varphi$ is a nonzero principal ideal in $F[x]$. Find a monic polynomial $f(x)$ such

---

[1] You can find a proof in some of the textbooks for Advanced Calculus.

that $\ker \varphi = (f(x))$. Since $F[x]/\ker \varphi \cong F[\alpha]$ is a subring of E, it is an integral domain. Hence $\ker \varphi$ is a prime ideal. By Theorem 17.1.10, $f(x)$ is irreducible over F and $\ker \varphi$ is a maximal ideal. Hence $F[\alpha]$ is a field and $F(\alpha) = F[\alpha]$. The monic polynomial $f(x)$ is called the **minimal polynomial** of α over F. The degree of the minimal polynomial of α over F is called the **degree** of α over F. Furthermore, $\dim_F F[\alpha]$ is equal to the degree of α over F (see Exercise 5, §20.1).

Proposition 21.1.1. *Let α be algebraic over F. Then the following conditions are equivalent for a polynomial $f(x) \in F[x]$:*

 (i) *$f(x)$ is the minimal polynomial of α over F;*

 (ii) *$f(x)$ is the lowest degree monic polynomial such that $f(\alpha) = 0$;*

 (iii) *$f(x)$ is monic irreducible over F and $f(\alpha) = 0$.*

Proof. (i) $\Leftrightarrow$ (ii): In the previous discussion, the minimal polynomial $f(x)$ is the monic generator of the ideal

$$\ker \varphi = \{\, g(x) \in F[x] : g(\alpha) = 0 \,\}.$$

From Corollary 16.3.5 we have that the minimal polynomial is the lowest degree monic nonzero polynomial in $\ker \varphi$.

  (i) $\Leftrightarrow$ (iii): Since $\ker \varphi$ is a maximal ideal, the generator is an irreducible polynomial over F by Corollary 17.1.10. Conversely, $f(x)$ generates a maximal ideal contained in the proper ideal $\ker \varphi$. Hence $\ker \varphi = (f(x))$. $\square$

Let α be an algebraic number over F with $f(x)$ as the minimal polynomial. We point out some useful properties on α and $f(x)$ from the discussions above.

 (1) If $\deg f(x) = n$, then elements in $F(\alpha)$ can be expressed uniquely as

$$a_0 + a_1\alpha + \cdots + a_{n-1}\alpha^{n-1}, \qquad \text{for } a_i \in F$$

   (see Exercise 3, §16.2).

 (2) For any $g(x) \in F[x]$ such that $g(\alpha) = 0$, $f(x)$ is a divisor of $g(x)$ since $g(x) \in f(x)F[x]$.

(3) Let E/F be a field extension and let $\alpha \in E$. Then $\alpha \in F$ if and only if the minimal polynomial of α over F is $x - \alpha$.

Example 21.1.2. The minimal polynomial of $\sqrt{2}$ over $\mathbb{Q}$ is $x^2 - 2$ since this monic polynomial is irreducible over $\mathbb{Q}$ by Eisenstein criterion. The minimal polynomial of i over $\mathbb{Q}$ is $x^2 + 1$ since $x^2 + 1$ has no zeros in $\mathbb{Q}$ and thus is irreducible over $\mathbb{Q}$. The minimal polynomial of $e^{2\pi i/5}$ over $\mathbb{Q}$ is $\Phi_5(x) = x^4 + x^3 + x^2 + x + 1$ by Corollary 17.3.13. Similarly, the minimal polynomial of $\sqrt{2}$ over $\mathbb{R}$ is $x - \sqrt{2}$. The minimal polynomial of i over $\mathbb{R}$ is still $x^2 + 1$.

Example 21.1.3. Let $\alpha \in \mathbb{C}$ be a zero of $f(x) = x^4 - 4x^3 + 2$. This is an irreducible polynomial over $\mathbb{Q}$ by Eisenstein criterion. Remember that $\mathbb{Q}[\alpha] = \mathbb{Q}(\alpha)$. Then how do we find the inverse of a polynomial in α, say $\alpha^3 - 1$?

First note that the degree of $g(x) = x^3 - 1$ is less than the degree of $f(x)$. Thus $f(x)$ and $g(x)$ are relatively prime. We make a calculation to find

$$-f(x)(x^2 + 2x + 4) + g(x)(x^3 - 2x^2 - 4x - 15) = 7.$$

Substitute in α and we obtain

$$g(\alpha)(\alpha^3 - 2\alpha^2 - 4\alpha - 15) = 7.$$

Thus $(\alpha^3 - 1)^{-1} = (\alpha^3 - 2\alpha^2 - 4\alpha - 15)/7$.

Example 21.1.4. Consider the ring of Gaussian integers $\mathbb{Z}[i]$. Let p be a positive prime integer. Since $\mathbb{Z}[i]$ is a PID, (p) is a maximal ideal if and only if p is irreducible in $\mathbb{Z}[i]$ if and only if $p \equiv 3 \pmod 4$ (see Theorem 17.1.10 and Exercise 11 on P. 276). When $p \equiv 3 \pmod 4$, $\mathbb{Z}[i]/(p) \cong \mathbb{Z}_p[i]$ is a field. The element $\bar{i}$ is a zero of $x^2 + 1$. Since -1 is a quadratic nonresidue in $\mathbb{Z}_p$ in this case, $x^2 + 1$ has no zero in $\mathbb{Z}_p$ and is thus irreducible over $\mathbb{Z}_p$. Hence, the minimal polynomial of $\bar{i}$ over $\mathbb{Z}_p$ is $x^2 + 1$ and the degree of $\bar{i}$ over $\mathbb{Z}_p$ is 2. We conclude that $\mathbb{Z}[i]/(p)$ is a field of p^2 elements (see Exercise 7, §20.1).

On the other hand, if $p \equiv 1 \pmod 4$, then $p = (a + bi)(a - bi)$ for some $a, b \in \mathbb{Z}$. Since $N(a + bi) = p$, we know that $a + bi$ is irreducible in $\mathbb{Z}[i]$ by Proposition 17.1.5(b). Now we know that $(a + bi)$ is a maximal ideal in $\mathbb{Z}[i]$. Since $p \in (a + bi)$, the maximal ideal $p\mathbb{Z}$ of $\mathbb{Z}$ is contained in the kernel of $\varphi: \mathbb{Z} \to \mathbb{Z}[i]/(a + bi)$. Hence $\ker \varphi = p\mathbb{Z}$. It follows that $\mathbb{Z}_p$ is a subfield of $\mathbb{Z}[i]/(a + bi)$. We have that $\mathbb{Z}[i]/(a + bi) = \mathbb{Z}_p[\bar{i}]$. Note that $p \nmid b$. In this case, $\bar{i} = -\bar{b}^{-1}\bar{a} \in \mathbb{Z}_p$. We conclude that $\mathbb{Z}[i]/(a + bi) \cong \mathbb{Z}_p$.

Exercises 21.1

1. Find a basis for the indicated vector space over the field $\mathbb{Q}$.

    (a) $\mathbb{Q}(\sqrt{3})$

    (b) $\mathbb{Q}(\sqrt[4]{2}i)$

    (c) $\mathbb{Q}(e^{2\pi i/p})$ where p is a positive prime integer

2. Let α be a root of the equation $x^3 + 3x + 2 = 0$ in $\mathbb{C}$. Rewrite the following products as polynomials in α of degree ≤ 2.

    (a) $(\alpha^2 + 3)(\alpha^2 + 2\alpha - 3)$

    (b) $(\alpha - 1/\alpha)(\alpha^2 + 1/\alpha^2)$

    (c) $(\alpha^2 + \alpha + 1)/(3\alpha^2 + 2\alpha + 1)$

3. Let $\alpha = \sqrt{2 - \sqrt{3}} \in \mathbb{R}$.

    (a) Find the minimal polynomial of α over $\mathbb{Q}$. Show that α is of degree 4 over $\mathbb{Q}$.

    (b) Rewrite $(\alpha^2 + 2)(\alpha^2 + 2\alpha + 1)$ and $(\alpha^2 + \alpha + 1)/(\alpha^2 - \alpha + 1)$ as polynomials in α of degree ≤ 3.

4. Show that $\pm\sqrt{3/2}$ and $\pm\sqrt{3} \notin \mathbb{Q}$.

5. Show that $\pi^2 - 2$ and $\sqrt{\pi}$ are transcendental numbers assuming that π is a transcendental number.

21.2   Finite and Algebraic Extensions

Remember that for any field extension E/F, E can be viewed as a vector space over F. We say that the field extension E/F is a **finite extension** or simply E is **finite** over F, if E is a finite-dimensional vector space over F. The **extension degree** of E over F, denoted $[E : F]$, is defined to be $\dim_F E$.

In the previous section, we have shown that if α is algebraic of degree n over F, then $[F(\alpha) : F] = n$. On the other hand, if α is transcendental over F, then $F(\alpha) \cong F(x)$, the rational field of one variable over F. The rational field $F(x)$ contains $F[x]$ as a subspace and $F[x]$ is infinite-dimensional over F. Thus, in this case, $[F(\alpha) : F] = \infty$. We conclude with the following proposition.

Proposition 21.2.1. *Let E/F be a field extension and $\alpha \in E$. Then α is algebraic over F if and only if $F(\alpha)/F$ is a finite extension.*

In particular, if α is algebraic of degree n over F, then $[F[\alpha] : F] = n$.

At the outset, $F[\alpha]$ is an integral domain. However, when it is also finite-dimensional over F, it becomes a field. We can generalize this result as follows.

Theorem 21.2.2. *Let D be an integral domain containing the field F. If D is a finite-dimensional vector space over F, then D is a field.*

Proof. Let $\alpha \in D$ and $\alpha \neq 0$. The set $\{1, \alpha, \ldots, \alpha^m, \ldots\}$ cannot be linearly independent over F since D is finite-dimensional over F. There must exist some n and $a_0, a_1, \ldots, a_n \in F$ not all zero such that

$$a_0 + a_1 \alpha + \cdots + a_n \alpha^n = 0.$$

Thus, α is a zero of some nonzero polynomial over F. We now know that α is algebraic over F. It then follows that $F(a) = F[a] \subseteq D$. Hence $a^{-1} \in D$. This shows that D is a field. $\square$

Theorem 21.2.3. *Let $F \subseteq K \subseteq L$ be three fields. Then L/F is a finite extension if and only if L/K and K/F are both finite extensions. In particular, when L/F is a finite extension, we have that*

(21.2.1) $$L : F] = [L : K][K : F].$$

Proof. Suppose that $[K : F] = \infty$. Then $[L : F] \geq [K : F]$ since K is a subspace of L over F. Hence L/F is not a finite extension.

Now assume that L/K is not a finite extension. If L/F is a finite extension, then any (finite) base for L over F will also generate L over K. This implies that L/K is a finite extension, a contradiction. Hence L/F is not a finite extension.

It remains to show that L/F is a finite extension if both L/K and K/F are finite extensions. We can prove this by proving (21.2.1).

Let $\{v_1, v_2, \ldots, v_m\}$ be a basis for L over K and $\{w_1, w_2, \ldots, w_n\}$ be a basis for K over F. We shall prove that

$$\mathscr{B} = \{v_i w_j \in L : 1 \leq i \leq m, \ 1 \leq j \leq n\}$$

is a basis for L over F. First we prove that these elements span L over F. For any $v \in L$, one has

$$v = \sum_{i=1}^{m} a_i v_i, \qquad a_i \in K.$$

For each i, one also has

$$a_i = \sum_{j=1}^{n} b_{ij} w_j, \qquad b_{ij} \in F.$$

This implies that

$$v = \sum_{i=1}^{m} a_i v_i = \sum_{i=1}^{m} \left(\sum_{j=1}^{n} b_{ij} w_j \right) v_i = \sum_{\substack{1 \leq i \leq m \\ 1 \leq j \leq n}} b_{ij} v_i w_j, \qquad b_{ij} \in F,$$

is spanned by the $v_i w_j$'s. Next, we prove that the $v_i w_j$'s are linearly independent over F. Suppose that $\sum_{i,j} c_{ij} v_i w_j = 0$ where $c_{ij} \in F$. Since

$$\sum_{\substack{1 \leq i \leq m \\ 1 \leq j \leq n}} c_{ij} v_i w_j = \sum_{i=1}^{m} \left(\sum_{j=1}^{n} c_{ij} w_j \right) v_i = 0$$

where $\sum_{j=1}^{n} c_{ij} w_j \in K$, we have that

$$\sum_{j=1}^{n} c_{ij} w_j = 0 \qquad \text{for all } i$$

by the linear independence of $v_1, v_2, \ldots, v_m$ over K. However, $w_1, w_2, \ldots, w_n$ are also linearly independent over F. It implies that $c_{ij} = 0$ for all i and j. Hence we conclude that $\mathcal{B}$ is indeed a basis for L over F, and that $[L : F] = mn = [L : K][K : F]$. $\qquad\qquad\qquad\qquad\qquad\qquad\qquad\qquad\qquad\qquad\qquad\quad\square$

By induction on n we have the following result.

Corollary 21.2.4. *Let $F_0 \subseteq F_1 \subseteq \cdots \subseteq F_n$ be a series of finite extensions. Then F_n/E is a finite extension and*

$$[F_n : F_0] = [F_n : F_{n-1}][F_{n-1} : F_{n-2}] \cdots [F_1 : F_0].$$

Example 21.2.5.   (1) Compute $[\mathbb{Q}[\sqrt{2}, \sqrt{3}] : \mathbb{Q}]$.

(2) Compute $[\mathbb{Q}[\sqrt{2} + \sqrt{3}] : \mathbb{Q}]$ and find the minimal polynomial of $\sqrt{2} + \sqrt{3}$ over $\mathbb{Q}$.

(3) Conclude that $\mathbb{Q}[\sqrt{2}, \sqrt{3}] = \mathbb{Q}[\sqrt{2} + \sqrt{3}]$.

Solution. To solve this problem we have to obtain the minimal polynomial of $\sqrt{2} + \sqrt{3}$ over $\mathbb{Q}$ as well as some of the other relevant minimal polynomials.

(1) First, we know that the minimal polynomial of $\sqrt{2}$ over $\mathbb{Q}$ is $x^2 - 2$. Hence, $[\mathbb{Q}[\sqrt{2}] : \mathbb{Q}] = 2$. Next, note that since $\sqrt{3}$ is a zero of $x^2 - 3$ over $\mathbb{Q}[\sqrt{2}]$, we have that $[\mathbb{Q}[\sqrt{2}, \sqrt{3}] : \mathbb{Q}[\sqrt{2}]] \leq 2$. If the extension degree is indeed 1, then $\sqrt{3} \in \mathbb{Q}[\sqrt{2}]$. It means that $\sqrt{3} = a + b\sqrt{2}$ for some $a, b \in \mathbb{Q}$. It follows that

$$3 = (a + b\sqrt{2})^2 = a^2 + 2b^2 + 2ab\sqrt{2}.$$

This gives the relation

$$(a^2 + 2b^2 - 3) + 2ab\sqrt{2} = 0.$$

Since 1 and $\sqrt{2}$ form a basis for $\mathbb{Q}[\sqrt{2}]$ over $\mathbb{Q}$, we have that

$$(a^2 + 2b^2 - 3) = 2ab = 0.$$

If $a = 0$ then $\pm\sqrt{3/2} \in \mathbb{Q}$. If $b = 0$, then $\pm\sqrt{3} \in \mathbb{Q}$. Both are impossible by Exercise 4, §21.1. Thus,

$$[\mathbb{Q}[\sqrt{2}, \sqrt{3}] : \mathbb{Q}[\sqrt{2}]] = 2 \quad \text{and} \quad [\mathbb{Q}[\sqrt{2}, \sqrt{3}] : \mathbb{Q}] = 4.$$

Remember that $\{1, \sqrt{2}\}$ is a basis for $\mathbb{Q}[\sqrt{2}]$ over $\mathbb{Q}$ and $\{1, \sqrt{3}\}$ is a basis for $\mathbb{Q}[\sqrt{2}, \sqrt{3}]$ over $\mathbb{Q}[\sqrt{2}]$. Hence $\{1, \sqrt{2}, \sqrt{3}, \sqrt{6}\}$ is a basis for $\mathbb{Q}[\sqrt{2}, \sqrt{3}]$ over $\mathbb{Q}$ by the proof of Theorem 21.2.3.

(2) We have the relation $\mathbb{Q} \subseteq \mathbb{Q}[\sqrt{2} + \sqrt{3}] \subseteq \mathbb{Q}[\sqrt{2}, \sqrt{3}]$. Thus, we have that

$$[\mathbb{Q}[\sqrt{2} + \sqrt{3}] : \mathbb{Q}] \mid [\mathbb{Q}[\sqrt{2}, \sqrt{3}] : \mathbb{Q}] = 4.$$

Since

$$\sqrt{6} = \frac{(\sqrt{2} + \sqrt{3})^2 - 5}{2} \in \mathbb{Q}[\sqrt{2} + \sqrt{3}],$$

We have that $[\mathbb{Q}[\sqrt{2} + \sqrt{3}] : \mathbb{Q}] \geq [\mathbb{Q}[\sqrt{6}] : \mathbb{Q}] = 2$. If $[\mathbb{Q}[\sqrt{2} + \sqrt{3}] : \mathbb{Q}] = 2$, we have $\sqrt{2} + \sqrt{3} \in \mathbb{Q}[\sqrt{6}]$. There exist a and $b \in \mathbb{Q}$ such that

$$\sqrt{2} + \sqrt{3} = a + b\sqrt{6}.$$

This means that $1, \sqrt{2}, \sqrt{3}$ and $\sqrt{6}$ are linearly dependent over $\mathbb{Q}$, a contradiction. We conclude that

$$[\mathbb{Q}[\sqrt{2} + \sqrt{3}] : \mathbb{Q}] = 4.$$

We leave it to the reader to verify that $\alpha = \sqrt{2} + \sqrt{3}$ is a zero of $f(x) = x^4 - 10x^2 + 1$. Since the degree of α over $\mathbb{Q}$ is 4, we have that $f(x)$ is the minimal polynomial of α over $\mathbb{Q}$.

(3) From (1) and (2) we have that $[\mathbb{Q}[\sqrt{2}, \sqrt{3}] : \mathbb{Q}[\sqrt{2} + \sqrt{3}]] = 1$. Hence $\mathbb{Q}[\sqrt{2}, \sqrt{3}] = \mathbb{Q}[\sqrt{2} + \sqrt{3}]$. ◇

Let E/F be a field extension. We say that E is an **algebraic extension** over F, or simply E is **algebraic** over F, if every element in E is algebraic over F.

Corollary 21.2.6. *Any finite extension of F is an algebraic extension of F.*

Proof. Let E/F be a finite extension and let $\alpha \in E$. Since $F \subseteq F(\alpha) \subseteq E$, we have that $F(\alpha)/F$ is a finite extension by Theorem 21.2.3. Now by Proposition 21.2.1, α is algebraic over F. □

Proposition 21.2.7. *Let E/F be a field extension. Then*

$$\overline{F}_E = \{\alpha \in E : \alpha \text{ is algebraic over } F\}$$

is a subfield of E containing F.

Proof. It suffices to show that $\overline{F}_E$ is a field. Let α and β in E be algebraic over F. We must prove that $\alpha \pm \beta$, $\alpha\beta$ and α/β, if $\beta \neq 0$, are algebraic over F. First note that β is also algebraic over $F(\alpha)$. Hence, by Proposition 21.2.1, both $F(\alpha)/F$ and $F(\alpha, \beta)/F(\alpha)$ are finite extensions. Now by Theorem 21.2.3, $F(\alpha, \beta)/F$ is a finite extension. Since $\alpha + \beta \in F(\alpha, \beta)$, we have that $F \subseteq F(\alpha + \beta) \subseteq F(\alpha, \beta)$ and $F(\alpha+\beta)/F$ is a finite extension again by Theorem 21.2.3. Thus, $\alpha + \beta$ is algebraic over F. Exactly the same argument shows that $\alpha - \beta$, $\alpha\beta$ and α/β are algebraic over F. $\square$

The field $\overline{F}_E$ is called the **algebraic closure** of F in E. It is an algebraic extension of F.

Example 21.2.8. We usually use $\overline{\mathbb{Q}}$ to denote $\overline{\mathbb{Q}}_{\mathbb{C}}$ and call it the **field of algebraic numbers**. Although $\overline{\mathbb{Q}}$ is an algebraic extension of $\mathbb{Q}$, it is not a finite extension of $\mathbb{Q}$. For any positive integer n, $x^n - 2$ is the minimal polynomial of $\sqrt[n]{2}$ over $\mathbb{Q}$. Hence $[\overline{\mathbb{Q}} : \mathbb{Q}] \geq [\mathbb{Q}[\sqrt[n]{2}] : \mathbb{Q}] = n$ for any n. It follows that $[\overline{\mathbb{Q}} : \mathbb{Q}] = \infty$. However, the field $\overline{\mathbb{Q}}$ is still relatively small in $\mathbb{C}$. In fact, $\overline{\mathbb{Q}}$ is countable while $\mathbb{C}$ is uncountable.

Proposition 21.2.9. *Let $F \subseteq K \subseteq L$ be field extensions. Then L/F is an algebraic extension if and only if L/K and K/F are both algebraic extensions.*

Proof. If L/F is an algebraic extension, then obviously K/F is an algebraic extension since K is a subfield of L. An element algebraic over F is definitely algebraic over K. Thus L/K is also algebraic.

Conversely, let L/K and K/F be algebraic extensions. Take any $\alpha \in L$ and assume

$$f(x) = x^n + a_{n-1}x^{n-1} + \cdots + a_1 x + a_0$$

is the minimal polynomial of α over K. Then α is also algebraic over $F(a_0, a_1, \ldots, a_{n-1})$. It follows that $F(a_0, a_1, \ldots, a_{n-1}, \alpha)$ is a finite extension of $F(a_0, a_1, \ldots, a_{n-1})$. Since for each i, a_i is algebraic over F, a_i is algebraic over $F(a_1, \ldots, a_{i-1})$. We have $F(a_1, \ldots, a_i)$ is a finite extension of $F(a_1, \ldots, a_{i-1})$ for all i. By Corollary 21.2.4, $F(a_0, a_1, \ldots, a_{n-1}, \alpha)$ is finite over F. It follows that $F(\alpha)/F$ is a finite extension and α is algebraic over F. $\square$

Corollary 21.2.10. *If $\alpha \in E$ is algebraic over $\overline{F}_E$, then $\alpha \in \overline{F}_E$.*

Proof. Since $\overline{F}_E[\alpha]$ is an algebraic extension of $\overline{F}_E$ and $\overline{F}_E$ is an algebraic extension of F, $\overline{F}_E[\alpha]$ is an algebraic extension of F. Hence α is algebraic over F and $\alpha \in \overline{F}_E$. $\square$

Exercises 21.2

1. For the following elements, find their minimal polynomials over $\mathbb{Q}$.

    (a) $\alpha = 3 - 2\sqrt{2}$

    (b) $\alpha = \sqrt{2} + \sqrt{5}$

    (c) $\alpha = \sqrt{2 + \sqrt{5}}$

    (d) $\alpha = i + \sqrt[3]{2}$

2. Show that $\mathbb{Q}(\sqrt{3} + \sqrt{5}) = \mathbb{Q}(\sqrt{3}, \sqrt{5})$.

3. Find the extension degree and a basis for the following field extensions.

    (a) $\mathbb{Q}(\sqrt{2}, \sqrt{5})$ over $\mathbb{Q}(\sqrt{5})$

    (b) $\mathbb{Q}(\sqrt{2}, \sqrt{3})$ over $\mathbb{Q}(\sqrt{6})$

    (c) $\mathbb{Q}(\sqrt{2}, \sqrt[3]{3})$ over $\mathbb{Q}$

4. If α is algebraic of odd degree over F, show that α^2 is algebraic of the same degree and $F(\alpha) = F(\alpha^2)$.

5. Let E be a finite extension of a field F with $[E : F]$ a prime integer. Show that $E = F(\alpha)$ for any $\alpha \in E \setminus F$. Hence there are no proper intermediate fields between E and F.

6. Let E/F be a finite extension. Suppose $p(x) \in F[x]$ is irreducible over F and $\deg p(x)$ is not a divisor of $[E : F]$. Show that $p(x)$ has no zeros in E.

21.3   Construction with Straightedge and Compass

In this section we want to use the basic results on fields that we have learnt so far to answer some very classical problems regarding *Euclidian constructions* (what geometric constructions can be achieved using only straight-edge and compass). Note that it is required that the construction be completed in a finite number of steps and be repeatable.

The concept of Euclidean construction can be precisely formulated in the following definition.

Definition 21.3.1. Given a finite set of points $S = \{P_1, P_2, \ldots, P_n\}$ in a plane ω, define the subsets S_m, $m = 1, 2, \ldots$, of ω inductively by letting $S_1 = S$, and S_{r+1} be the union S_r and

(i) the set of points of intersections of pairs of lines connecting distinct points of S_r,

(ii) the set of points of intersections of lines described in (i) with circles having centers in S_r and radii equal to segments having end points in S_r, and

(iii) the set of points of intersections of pairs of circles described in (ii).

Define

$$C(P_1, \ldots, P_n) = \bigcup_1^\infty S_i.$$

We shall say a point P of ω is a **constructible point** (by straight-edge and compass) from $P_1, \ldots, P_n$ if and only if $P \in C(P_1, \ldots, P_n)$.

Obviously, the purpose of straight-edge and compass is to make lines and circles through already constructed points. Intersections of these lines and/or circles give us new constructed points. The compass can also be used to produce the length between two constructed points. Hence, all problems regarding Euclidean construction can be translated as problems of constructing certain points. For example, let's look at the problem of trisecting the angle of $60°$. This is equivalent to asking whether

$$(\cos 20°,\ \sin 20°) \in C\big(O,\ (1,0),\ (\cos 60°,\ \sin 60°)\big)?$$

However, we remark that in Euclidean constructions we sometimes encounter an instruction to use an "arbitrary" point or length which might or might not be restricted by a condition that the point be contained in a certain region or that the length satisfy a certain inequality. In either case, we are instructed to choose points in designated (nonempty) open subsets of the plane, and no matter which points we actually choose it should not affect the outcome, for otherwise the construction will not be repeatable. We shall see very soon in Lemma 21.3.5 that the set $C(P_1, P_2, \ldots, P_n)$ is *dense* in the plane as long as $n \geq 2$, which is the case since we should have at least P_1 and P_2 whose distance gives the unit length. Hence any instruction involving choosing a point in an nonempty open set can be fulfilled by choosing some point in $C(P_1, P_2, \ldots, P_n)$. Consequently, our notion of constructible points is equivalent to what have been intended in the problem of Euclidean constructions.

We shall now reformulate our definition algebraically. Without loss of generality we shall assume that $n \geq 2$. We choose a Cartesian coordinate system so that $P_1 = (0,0)$ and $P_2 = (1,0)$. We associate with the point $P = (x, y)$ the complex number $z = x + iy$. In this way the plane ω is identified with the complex plane or the field $\mathbb{C}$. For the rest of this section we will assume that the given points $P_1, \ldots, P_n$ correspond to the complex numbers $z_1, \ldots, z_n$ where $z_1 = 0$ and $z_1 = 1$.

Definition 21.3.2. The set $C(z_1, \ldots, z_n)$ of complex numbers which corresponds to the set $C(P_1, P_2, \ldots, P_n)$ of points of ω are called **constructible numbers** (by straight-edge and compass) from $z_1, \ldots, z_n$.

How does one determine whether a number is constructible? We now propose a few approaches. But, before we start, we first review what we have been taught to do with straightedge and compass in high school algebra.

- To reproduce a length and to freely move an angle around the plane.

- To bisect a line segment (a length) by constructing the perpendicular bisector.

- To construct a line perpendicular to a given line and passing through a given point.

- To construct a line parallel to a given line and passing through a point outside the line.

- To bisect an angle.

With these techniques in mind, we are also able to achieve the following two things.

- To reproduce a length anywhere on the plane.

- To reproduce an angle anywhere on the plane.

We will call the distance between any two constructible points (or numbers) a **constructible length**. Many of the classical problems regarding Euclidean construction involves verifying whether some length is constructible. A numbers in $C(z_1, \ldots, z_n) \cap \mathbb{R}$ is called a **constructible real**, it is easy to see that a real number r is a constructible real if and only if $|r|$ is a constructible length. Let r be a constructible real, then obviously $|r|$ is the distance between r and 0 and hence a constructible length. On the other hand, suppose $|r|$ is a constructible length. Draw a circle with 0 as the center and $|r|$ as the radius using straight edge and compass. One of the intersections of this circle and the real line is r.

Lemma 21.3.3. *Let* $z = x + yi$ *where* x, $y \in \mathbb{R}$. *Then the following conditions are equivalent:*

(i) $z \in C(z_1, \ldots, z_n)$;

(ii) x *and* y *are constructible reals;*

(iii) $|x|$ *and* $|y|$ *are constructible lengths.*

We leave the proof as an easy exercise (see Exercise 1).

On the other hand, a complex number $z = x + yi$ can be written in the form

$$z = re^{i\theta} = r(\cos\theta + i\sin\theta),$$

where $r = |z| = \sqrt{x^2 + y^2}$ and θ is the angle between the positive x-axis (the initial ray) and z viewed as a position vector (the terminal ray, see Figure 21.1). If z is a constructible number, the length r being the distance between 0 and z, is a constructible length. It is also reasonable to think of the angle θ determined by the positive x-axis and the constructible number z (or just any three constructible numbers) as a **constructible angle**.

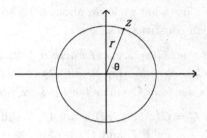

Figure 21.1: $z = re^{i\theta}$

It is possible to use straight-edge and compass to freely move an angle around in the plane ω. In particular, we may move the angle so that the initial ray of the angle falls on the positive x-axis. If θ is a constructible angle, the number $z = \cos\theta + i\sin\theta$ is the intersection of the unit circle and the terminal ray of a constructible angle θ and is thus a constructible number. By Lemma 21.3.3, $\cos\theta$ and $\sin\theta$ are constructible reals. Conversely, when $\cos\theta$ and $\sin\theta$ are constructible reals, it will make $z = \cos\theta + i\sin\theta$ a constructible number which will determine θ as a constructible angle. However, assume we only have that $\cos\theta$ is a constructible real. Mark $\cos\theta$ on the real axis. Draw a vertical line passing through $\cos\theta$. One of the intersections of this line with the unit circle is $z = \cos\theta + i\sin\theta$. Hence we have the following lemma.

Lemma 21.3.4. *Let $\theta \in \mathbb{R}$. Then θ is a constructible angle if and only if $\cos\theta$ is a constructible real.*

Moreover, let $z = re^{i\theta}$, where r and $\theta \in \mathbb{R}$. Then the following conditions are equivalent:

(i) $z \in C(z_1, \ldots, z_n)$;

(ii) $|r|$ *is a constructible length and θ is a constructible angle;*

(iii) r *and $\cos\theta$ are constructible reals;*

(iv) $|r|$ *and* $|\cos\theta|$ *are constructible lengths.*

Now we have several methods to choose from to determine whether a complex number is constructible. Our next step is to show that $C(z_1,\ldots,z_n)$ is a field. We can then use what we know about fields to attack the classical problems on Euclidean constructions.

Lemma 21.3.5. *The set* $C(z_1,\ldots,z_n)$ *where* $n \geq 2$, $z_1 = 0$ *and* $z_2 = 1$, *is a subfield of* $\mathbb{C}$. *Consequently, the field* $C(z_1,\ldots,z_n)$ *contains the field* $\mathbb{Q}[i] = \mathbb{Q} \oplus \mathbb{Q}i$ *and is dense in* $\mathbb{C}$ *under the Euclidean topology.*

Proof. To show that $C = C(z_1,\ldots,z_n)$ is a field, it suffices to show that if $z, z' \in C$ then $z + z'$, $-z$, $zz' \in \mathbb{C}$ and $1/z \in C$ if $z \neq 0$.

Draw a line through z and parallel to the line through O and z'. Draw another line through z' and parallel to the line through O and z. The intersection of these two lines is the number $z + z'$. Make a circle with the origin as the center and $|z|$ as the radius. The intersections of this circle and the line through O and z are z and $-z$ (see Figure 21.2).

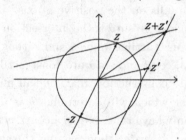

Figure 21.2: How to construct $z + z'$ and $-z$ from z and z'

Next, we show how to construct zz' if z and z' are already constructed. Let $z = re^{i\theta}$ and $z' = r'e^{i\theta'}$. Then $zz' = rr'e^{i(\theta+\theta')}$. It is sufficient to construct rr' and the angle $\theta + \theta'$. The angle $\theta + \theta'$ is easy to construct since we can move an angle freely around the plane ω using straightedge and compass. The main point is to construct the length rr'. Make a line through $(r, 0)$ and parallel to the line through $(1, 0)$ and $(0, r')$. The intersection of this line with the y-axis will give us the length rr' (see Figure 21.3).

Similarly, the number $1/z = (1/r)e^{i(-\theta)}$. It suffices to construct the length $1/r$ and the angle $-\theta$. The angle $-\theta$ is easy to construct using

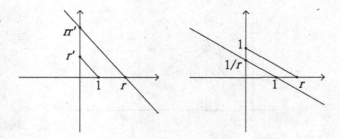

Figure 21.3: How to construct the lengths rr' and $1/r$ from the length r and r'

straightedge and compass. To construct the length $1/r$, make a line through $(1,0)$ and parallel to the line through $(r,0)$ and $(0,1)$. The intersection of this line with the y-axis will give us the length $1/r$ (see Figure 21.3).

To conclude, we have that C is a field. Since C is a subfield of $\mathbb{C}$, it is of characteristic 0. Thus $\mathbb{Q}$ is the prime field of C. It follows that all the rational numbers are constructible reals. From Lemma 21.3.3, $\mathbb{Q}i$ are also contained in C. We have shown that $\mathbb{Q}[i] = \mathbb{Q} \oplus \mathbb{Q}i \subseteq C$. $\qquad\square$

There are extra properties regarding $C(z_1, \ldots, z_n)$.

Lemma 21.3.6. *Let $C = C(z_1, \ldots, z_n)$.*

(a) *The field C is closed under* conjugation, *that is, $\overline{z} \in C$ if $z \in C$.*

(b) *The field C is closed under* square roots, *that is, $z \in C$ if $z^2 \in C$.*

Proof. (a) Let $z = x + yi$ where $x, y \in \mathbb{R}$. Since z is constructible, x and y are constructible reals. Now by Lemma 21.3.3, $\overline{z} = x - yi \in C$.

(b) Suppose $z^2 = re^{i\theta} \in C$, where $r > 0$. Then r is a constructible length and θ is a constructible angle. Note that $z = \pm\sqrt{r}e^{i\theta/2}$. It is sufficient to construct the length $\sqrt{r}$ and the angle $\theta/2$. Since bisecting an angle is possible with straightedge and compass, $\theta/2$ is a constructible angle. It suffices to construct the length $\sqrt{r}$.

Bisect the line segment connecting 0 and $1 + r$ and use the midpoint as the center to construct a circle C_1 with $(1 + r)/2$ as its radius. Make a line L through 1 and perpendicular to the x-axis. We claim that the

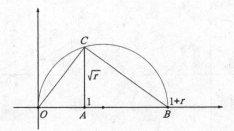

Figure 21.4: How to construct the length $\sqrt{r}$ from the length r

intersection of L and the upper half of the circle C_1 will give us the length $\sqrt{r}$ (see Figure 21.4). Let A be the point at 1, B the point at $1+r$ and C be the new intersection. Since the angle $\angle OCB$ is a right angle, we have that $\angle COA = \angle BCA$. Hence the triangles $\triangle COA \sim \triangle BCA$. Let x be the length of the line segment $\overline{AC}$. Then

$$\frac{\overline{AC}}{\overline{OA}} = \frac{x}{1} = \frac{\overline{AB}}{\overline{AC}} = \frac{r}{x} \qquad \Longrightarrow \qquad x = \sqrt{r}.$$

We have now completed the proof.   □

Proposition 21.3.7. *The set* $C(z_1, \ldots, z_n)$ *is the smallest subfield of* $\mathbb{C}$ *containing* $z_1, \ldots, z_n$ *and closed under conjugation and square roots.*

Proof. Let $C = C(z_1, \ldots, z_n)$ and let F be any subfield of $\mathbb{C}$ containing $z_1, \ldots, z_n$ and closed under conjugation and square roots. By the previous lemma, it suffices to show that $C \subseteq F$. We will do this by showing that $S_i \subseteq F$ for all i.

Clearly, $S_1 = \{z_1, \ldots, z_n\} \subseteq F$. Now assume that $S_i \subseteq F$. Note that any line through two points in S_i (and thus in F) or the circles obtained by using a point in S_i as the center and the distance between two points in S_i as the radius all have equations with coefficients in F. Numbers of S_{i+1} are simultaneous solutions of such a line/circle with another such line/circle. In solving this type of systems of equations, it involves taking sums, differences, products, quotients and square roots of the coefficients. Hence numbers in S_{i+1} are still in F. Now we are done with the induction and we conclude that $C \subseteq F$.   □

A field of the form $K(u_1, u_2, \ldots, u_r)$, where $u_i^2 \in K(u_1, \ldots, u_{i-1})$ for $1 \le i \le r$, is called a **square root tower** over K. In other words, a square root tower is a series of simple extensions

$$K \subseteq K(u_1) \subseteq K(u_1, u_2) \subseteq \cdots \subseteq K(u_1, \ldots, u_{r-1}) \subseteq K(u_1, \ldots, u_{r-1}, u_r)$$

such that in each step a square root is adjoined.

We now have the following criterion to determine whether a complex number is constructible.

Theorem 21.3.8 (Characterization of $C(z_1, \ldots, z_n)$). *Let $z_1 = 0$, $z_2 = 1$, $z_3, \ldots, z_n \in \mathbb{C}$ and let $K = \mathbb{Q}(z_1, \ldots, z_n, \overline{z}_1, \ldots, \overline{z}_n)$. Then a complex number z is constructible from $z_1, \ldots, z_n$ if and only if z belongs in a square root tower over K.*

Proof. Before starting the proof, we first comment that K is closed under conjugation. Let $\alpha \in K$. Then

$$\alpha = \sum_{\substack{I=(i_1,\ldots,i_n) \\ J=(j_1,\ldots,j_n)}} a_{I,J} z_1^{i_1} \cdots z_n^{i_n} \overline{z}_1^{j_1} \cdots \overline{z}_n^{j_n}, \qquad a_{I,J} \in \mathbb{Q}$$

where all the i_k's and j_k's are nonnegative integers. It follows that

$$\overline{\alpha} = \sum_{\substack{I=(i_1,\ldots,i_n) \\ J=(j_1,\ldots,j_n)}} a_{I,J} \overline{z}_1^{i_1} \cdots \overline{z}_n^{i_n} z_1^{j_1} \cdots z_n^{j_n} \in K.$$

Let F be the union of all square root towers over K. Next we show that F is the smallest subfield of $\mathbb{C}$ containing $z_1, \ldots, z_n$ and closed under conjugation and square roots. Then by the previous proposition we have that $C(z_1, \ldots, z_n) = F$ and we are done.

Let z and $z' \in F$. Find a square root tower $K(u_1, \ldots, u_r)$ over K in which z belongs. Then find a square root tower $K(v_1, \ldots, v_s)$ in which z' belongs. Then z and z' both belong in the square root tower $T = K(u_1, \ldots, u_r, v_1, \ldots, v_s)$ over K. Since T is a field, we have that $z + z'$, $-z$, zz' and $1/z$ with $z \neq 0$, are all in T and hence all in F. This shows that F is a field containing $z_1, \ldots, z_n$. Furthermore, let

$$z = \sum_{I=(i_1,\ldots,i_r)} \alpha_I u_1^{i_1} \cdots u_r^{i_r}, \qquad \alpha_I \in K$$

where all the i_k's are nonnegative integers. It follows that

$$\overline{z} = \sum_{I=(i_1,\ldots,i_r)} \overline{\alpha}_I \overline{u}_1^{i_1} \cdots \overline{u}_r^{i_r} \in K(\overline{u}_1,\ldots,\overline{u}_r).$$

It is clear that $\overline{u}_i^2 \in K(\overline{u}_1,\ldots,\overline{u}_{i-1})$ for each i. Hence $K(\overline{u}_1,\ldots,\overline{u}_r)$ is also a square root tower over K. This implies that $\overline{z} \in F$ and F is closed under conjugation. On the other hand, if $z^2 \in F$, then z^2 belongs in some square root tower over K, say $K(u_1,\ldots,u_r)$. Then z belongs in the square root tower $K(u_1,\ldots,u_r,z)$ over K. Hence $z \in F$ and F is closed under square root.

Clearly, K is contained in any field containing $z_1,\ldots,z_n$ and closed under conjugation. A square root tower over K is obtained by adjoining consecutive square roots to K. Hence any square root tower over K is also contained in a field containing K and closes under square roots. Thus, F, being the union of all square root towers over K, is the smallest subfield of $\mathbb{C}$ containing $z_1,\ldots,z_n$ and closed under conjugation and square roots. □

Corollary 21.3.9. *Let $K = \mathbb{Q}(z_1,\ldots,z_n,\overline{z}_1,\ldots,\overline{z}_n)$. If z is constructible from $z_1,\ldots,z_n$, then z is algebraic of degree 2^r over K for some r.*

Proof. Let z belongs in a square root tower over K, say $K(u_1,\ldots,u_s)$. Then $[K(u_1,\ldots,u_i) : K(u_1,\ldots,u_{i-1})] = 1$ or 2. Hence $[K(u_1,\ldots,u_s) : K] = 2^t$ for some t. Since $K(z)$ is a subfield of $K(u_1,\ldots,u_s)$, $[K(z) : K]$ is a divisor of 2^t. □

We now use this criterion to solve the three ancient Greek problems regarding constructions using only straightedge and compass.

(1) *Is it possible to construct a square whose area is the same as that of the unit circle?*

   This is the same as asking whether $\sqrt{\pi} \in C(0,1)$? In this case $K = \mathbb{Q}$. Since $\sqrt{\pi}$ is a transcendental number, the answer is NO by Corollary 21.3.9!

(2) *Is it possible to construct a cube whose volume is 2?*

   This is equivalent to asking whether $\sqrt[3]{2} \in C(0,1)$. The minimal polynomial of $\sqrt[3]{2}$ over $\mathbb{Q}$ is $x^3 - 2$. Hence $\sqrt[3]{2}$ is algebraic of degree 3 over $\mathbb{Q}$. The answer is again NO by Corollary 21.3.9!

(3) *Is it possible to trisect an arbitrary angle?*

Let's try to trisect $60°$. We are asking whether $\cos 20° \in C(0, 1, \cos 60°)$. In this case $K = \mathbb{Q}$. Using the formula

$$\cos 3\theta = 4\cos^3\theta - 3\cos\theta,$$

we have that $\cos 20°$ is a zero of

$$f(x) = x^3 - \frac{3}{4}x - \frac{1}{8}.$$

We can use Proposition 17.2.13 to verify that $f(x)$ has no zero in $\mathbb{Q}$. Hence $f(x)$ is the minimal polynomial of $\cos 20°$ over $\mathbb{Q}$. The number $\cos 20°$ is algebraic of degree 3 over $\mathbb{Q}$. The answer is still NO!

Exercises 21.3

1. Prove Lemma 21.3.3.

2. Find the minimal polynomials of $\cos 10°$ and $\sin 10°$ over $\mathbb{Q}$ respectively.

3. Find the minimal polynomial of $\cos 72°$ over $\mathbb{Q}$.

4. Let $K_0 \subseteq K_1 \subseteq \cdots \subseteq K_n$ be a series of field extensions such that $[K_i : K_{i-1}] = 2$ for $1 \le i \le n$. Show that this is a square root tower over K_0 when Char $K_0 \ne 2$.

5. In this problem we prove that a regular pentagon is constructible with straightedge and compass using Theorem 21.3.8.

   Recall that the minimal polynomial of $\alpha = e^{2\pi i/5}$ over $\mathbb{Q}$ is $\Phi_5(x) = x^4 + x^3 + x^2 + x + 1$. Hence α is algebraic of degree 4 over $\mathbb{Q}$.

   (a) Show that $\Phi_4(x) = (x - \alpha)(x - \alpha^2)(x - \alpha^3)(x - \alpha^4)$.

   (b) Show that $\mathbb{Q} \subseteq \mathbb{Q}(\alpha + \alpha^4) \subseteq \mathbb{Q}(\alpha)$ is a square root tower over $\mathbb{Q}$.

   (c) Conclude that the regular pentagon is constructible with straightedge and compass.

6. Let p be a positive odd prime integer.

    (a) Show that if the regular p-gon is constructible with straightedge and compass, $p = 2^s + 1$ for some integer s.

    (b) Show that the $s = 2^t$ for some integer t.

A prime integer is called a **Fermat's prime** if it is of the form $2^{2^t} + 1$. So far, 3, 5, 17, 257 and 65537 are all the Fermat's primes mathematicians have discovered.

Review Exercises for Chapter 21

1. If $\alpha \in \mathbb{C}$ is a zero of the polynomial $p(x) = x^5 + \sqrt{2}x^3 + \sqrt{3}x^2 + \sqrt{5}x + \sqrt{7}$, show that α is algebraic over $\mathbb{Q}$ of degree at most 80.

2. Let m and n be relatively prime integers. Suppose that a and b are algebraic over $\mathbb{Q}$ of degrees m and n respectively. Show that $[\mathbb{Q}[a, b] : \mathbb{Q}] = mn$.

3. Let F be a field and $u = x^3/(x+1)$ in $F(x)$. Show that x is algebraic over $F(u)$. Determine $[F(x) : F(u)]$.

4. Let F be a field. Show that every element in $F(x) \backslash F$ is transcendental over F.

5. Show that the regular 15-gon is constructible.

6. Show that $\arccos(11/16)$ can be trisected with straight-edge and compass.

7. Show that the regular 9-gon cannot be constructed with straight-edge and compass.

CHAPTER 22

All About Roots

In $\mathbb{Q}[x]$ there are irreducible polynomials of any given degree. Hence there are numerous polynomials over $\mathbb{Q}$ which have no zeroes in $\mathbb{Q}$. There is a method to slowly extend $\mathbb{Q}$ to include a zero, and eventually all zeroes, of any of these polynomials. Of course, this method applies to any arbitrary field, not just to $\mathbb{Q}$.

But one thing at a time. We will start with one single polynomial. We will construct a superfield in which a given polynomial would split. More precisely, we will describe how to construct the *splitting field* of a given polynomial over a given field.

351

22.1   Zeros of Polynomials

We say that a nonconstant polynomial **splits** over F or splits in $F[x]$ if

$$f(x) = u(x - a_1)(x - a_2) \cdots (x - a_n), \qquad \text{where } u \text{ and } a_i \in F \text{ for all } i.$$

A field E is said to be a **splitting field** for $f(x)$ over F if $E = F[\alpha_1, \ldots, \alpha_n]$ where $\alpha_1, \ldots, \alpha_n$ are all the zeroes of $f(x)$. In other words, a splitting field of $f(x)$ over F is a minimal field extension of F over which $f(x)$ splits.

Before we discuss how to construct a splitting field for a given polynomial over some field F, we first look at some examples.

Example 22.1.1. The field $\mathbb{Q}(\sqrt{2})$ is a splitting field of $x^2 - 2$ over $\mathbb{Q}$. Indeed, one has $x^2 - 2 = (x - \sqrt{2})(x + \sqrt{2})$.

Example 22.1.2. We claim that the field $F = \mathbb{Q}(\sqrt[3]{2}, \omega)$, where $\omega = e^{2\pi i/3}$, is a splitting field of $x^3 - 2$ over $\mathbb{Q}$. Since

$$x^3 - 2 = (x - \sqrt[3]{2})(x - \sqrt[3]{2}\omega)(x - \sqrt[3]{2}\omega^2),$$

the splitting field of $x^3 - 2$ over $\mathbb{Q}$ is actually $F' = \mathbb{Q}(\sqrt[3]{2}, \sqrt[3]{2}\omega, \sqrt[3]{2}\omega^2)$. Clearly $F \supseteq F'$. However, since $\omega = \sqrt[3]{2}\omega/\sqrt[3]{2}$, we also have $F \subseteq F'$.

Now we describe how to construct splitting fields.

Proposition 22.1.3 (Kronecker). *Let F be a field and $f(x) \in F[x]$ be a polynomial of degree $n > 0$. There exists a finite extension K of F such that*

(i) *$f(x)$ has a zero α in K, and*

(ii) *$[K : F] \leq n$.*

Proof. Let $p(x)$ be an irreducible divisor of $f(x)$ in $F[x]$. Then any zero of $p(x)$ is also a zero of $f(x)$. It suffices to extend F so that $p(x)$ has a zero. Consider the quotient ring $K = F[x]/(p(x))$. This is a field since $p(x)$ is irreducible over F. It is also clear that K contains F as a subfield since we have the natural embedding $F \to F[x]/(p(x))$. Hence K/F is a field extension. Let $\alpha = x + (p(x))$. Then $p(\alpha) = \overline{p(x)} = 0$. Hence α is a zero of $p(x)$ in K. Clearly, $[K : F] = \deg p(x) \leq \deg f(x) = n$. $\qquad\square$

Example 22.1.4. We usually view the field $\mathbb{R}[x]/(x^2+1)$ as the field $\mathbb{C}$ of complex numbers. We may think of $\overline{x}$ as i. Note that

$$\mathbb{R}[\overline{x}] = \{a + b\overline{x} : a, b \in \mathbb{R}\} = \mathbb{R} \oplus \mathbb{R}\overline{x}.$$

Inside this ring

$$(a + b\overline{x})(c + d\overline{x}) = ac + ad\overline{x} + bc\overline{x} + bd\overline{x}^2 = (ac - bd) + (ad + bc)\overline{x}$$

where $a, b, c, d \in \mathbb{R}$. The element $\overline{x}$ works exactly like i.

Example 22.1.5. Let $f(x) = x^2 + 2x + 2 \in \mathbb{Q}[x]$. Find a splitting field of $f(x)$ over $\mathbb{Q}$. Factor $f(x)$ over the splitting field.

Solution. By Eisenstein's criterion, $f(x)$ is irreducible over $\mathbb{Q}$. Thus, $\mathbb{Q}[x]/(f(x))$ is a field extension of $\mathbb{Q}$ in which $\alpha = \overline{x}$ is a zero of $f(x)$. Do the long division and we obtain that

$$x^2 + 2x + 2 = (x - \alpha)(x + (2 + \alpha)) + \alpha^2 + 2\alpha + 2 = (x - \alpha)(x + (2 + \alpha))$$

in $\mathbb{Q}[\alpha]$. The splitting field of $f(x)$ over $\mathbb{Q}$ is $\mathbb{Q}[\alpha, -2 - \alpha] = \mathbb{Q}[\alpha]$. ⬦

Proposition 22.1.6. *Let $f(x) \in F[x]$ be a polynomial of degree $n > 0$. There exists a splitting field K of $f(x)$ over F such that $[K : F] \leq n!$.*

Proof. We prove this by induction on n. It is true for $n = 1$. By Proposition 22.1.3, there exists a field extension K of F such that $f(x)$ has a zero α in K and $[K : F] \leq n$. Thus, in $K[x]$, we have $f(x) = (x - \alpha)q(x)$ with $q(x) \in K[x]$ and $\deg q(x) = n - 1$. By the induction hypothesis, there exists an extension field L of K such that $q(x)$ splits in $L[x]$ and $[L : K] \leq (n - 1)!$. Consequently, $f(x)$ splits in $L[x]$ and

$$[L : F] = [L : K][K : F] \leq n!.$$

Now let E be the field generated by the zeros of $f(x)$ in L over F. Then E is a splitting field of $f(x)$ over F and $[E : F] \leq [L : F] \leq n!$. ☐

Example 22.1.7. The polynomial $x^4 - 2$ is irreducible over $\mathbb{Q}$ by Eisenstein criterion. We would like to find a splitting field of $x^4 - 2$ over $\mathbb{Q}$. Let

$$K = \frac{\mathbb{Q}[x]}{(x^4 - 2)} = \mathbb{Q}[\alpha]$$

where $\alpha = \bar{x}$. Then

$$K = \mathbb{Q} \oplus \mathbb{Q}\alpha \oplus \mathbb{Q}\alpha^2 \oplus \mathbb{Q}\alpha^3.$$

Since $(-\alpha)^4 = \alpha^4 = 2$, both α and $-\alpha$ are zeroes of $x^4 - 2$. We have that

$$x^4 - 2 = (x - \alpha)(x + \alpha)(x^2 + \alpha^2).$$

Thus $K = \mathbb{Q}[\alpha] = \mathbb{Q}[\alpha, -\alpha]$ already has two roots of $x^4 - 2 = 0$. If K is a splitting field of $x^4 - 2$ then K must also contain the zeroes of $x^2 + \alpha^2$. Let's check if this is true. If K contains a zero of $x^2 + \alpha^2$, then there exist rational numbers a, b, c, d not all zero such that

$$(a + b\alpha + c\alpha^2 + d\alpha^3)^2 + \alpha^2$$
$$= (a^2 + 2c^2 + 4bd) + (2ab + 4cd)\alpha$$
$$+ (b^2 + 2d^2 + 2ac + 1)\alpha^2 + (2ad + 2bc)\alpha^3 = 0.$$

This implies that

$$a^2 + 2c^2 + 4bd = ab + 2cd = b^2 + 2d^2 + 2ac + 1 = ad + bc = 0.$$

From $a^2 + 2c^2 + 4bd = 0$ we have that $bd \leq 0$ and from $b^2 + 2d^2 + 2ac + 1 = 0$ we have that $ac < 0$.

(1) If $b = 0$ or $d = 0$, then $a^2 + 2c^2 = 0$ and thus $a = c = 0$. This contradicts to the fact that $ac < 0$.

(2) If b and $d \neq 0$, then b and d are of opposite signs and a and c are of opposite signs. Hence ab and cd are of the same sign. This implies that $ab + 2cd \neq 0$, a contradiction.

From our discussion above we see that $x^2 + \alpha^2$ is irreducible over K. Let $L = K[y]/(y^2 + \alpha^2) = K[\beta]$ where $\beta = \bar{y}$. Similarly we have that $(-\beta)^2 = \beta^2 = -\alpha^2$. Hence

$$x^2 + \alpha^2 = (x - \beta)(x + \beta).$$

Over $L = K[\beta] = \mathbb{Q}[\alpha, \beta]$ we have that

$$x^4 - 2 = (x - \alpha)(x + \alpha)(x - \beta)(x + \beta).$$

The field L is a splitting field of $x^4 - 2$ over $\mathbb{Q}$ and

$$[L : \mathbb{Q}] = [L : K][K : \mathbb{Q}] = 2 \cdot 4 = 8.$$

In the next section we will establish the uniqueness of splitting fields.

Exercises 22.1

1. Find the extension degrees of the splitting fields of the following poly-nomials over $\mathbb{Q}$.

   (a) $f(x) = (x^2 - 2)(x^2 - 5)$
   (b) $f(x) = x^4 + x^2 + 1$
   (c) $f(x) = x^6 + x^3 + 1$

2. Remember that $f(x) = x^3 + x + 1$ is irreducible over $\mathbb{Z}_2$. Find a splitting field of $f(x)$ over $\mathbb{Z}_2$. Factor $f(x)$ over that splitting field.

3. Let p be a positive prime integer. Let F be a splitting field of $\Phi_p(x)$ over $\mathbb{Q}$. Show that $[F : \mathbb{Q}] = p - 1$ and factor $\Phi_p(x)$ over F.

4. If $f(x)$ is irreducible of degree 2 over a field F, show that $F[x]/(f(x))$ is a splitting field of $f(x)$ over F.

5. Let E and F be fields such that $[E : F] = 2$. Prove that E is a splitting field of a quadratic polynomial over F.

6. Find an irreducible polynomial of degree 3 over $\mathbb{Q}$ so that a splitting field of this polynomial over $\mathbb{Q}$ has extension degree 6.

7. Let E/F be a field extension and let $f(x) \in F[x]$. If $f(x)$ splits over E, show that E contains a splitting field of $f(x)$ over F.

8. Let F be a field of characteristic $p > 0$. For $a \in F$, show that $f(x) = x^p - a$ is either irreducible over F or $f(x) = (x - c)^p$ for some $c \in F$.

22.2 Uniqueness of Splitting Fields

Some of you might deduce that $L' = \mathbb{Q}[\sqrt[4]{2}, i\sqrt[4]{2}] = \mathbb{Q}[\sqrt[4]{2}, i] \subseteq \mathbb{C}$ is a splitting field of $x^4 - 2$ over $\mathbb{Q}$ since

$$x^4 - 2 = (x - \sqrt[4]{2})(x + \sqrt[4]{2})(x - i\sqrt[4]{2})(x + i\sqrt[4]{2}).$$

This looks very different from the splitting field constructed in Example 22.1.7. Are these two fields truly different from each other? Fortunately we can show that any two splitting fields for the same polynomial over the same field are isomorphic.

Let η be a ring homomorphism from the field F *into* another field F'. Let $f(x) = \sum_i a_i x^i \in F[x]$. We use $f_\eta(x)$ to denote the polynomial $\sum_i \eta(a_i)x^i$ in $F'[x]$.

Theorem 22.2.1. *Let η be a ring homomorphism from the field F to another field E. Let u be algebraic over F with the minimal polynomial $f(x)$. Then η can be extended to a ring homomorphism from $F[u]$ to E if and only if $f_\eta(x)$ has a zero in E. The number of such extensions is the number of distinct zeroes of $f_\eta(x)$ in E. In fact, for each zero $v \in E$ of f_η there is a unique ring homomorphism $\widetilde{\eta} : F[u] \to E$ such that $\widetilde{\eta}|_F = \eta$ and $\widetilde{\eta}(u) = v$.*

This is a very important result and a key theorem in the developing of Galois pairing in Chapter 23 and also in establishing the uniqueness of splitting fields.

Proof. By Theorem 16.1.4, there is a ring homomorphism η' from $F[x]$ to E extending η and sending x to any element v in E. However, in order for this homomorphism to factor through a ring homomorphism $\widetilde{\eta}$ from $F[u] = F[x]/(f(x))$ to E, one needs to check that $f(x)$ belongs in $\ker \eta'$. This happens if and only if $\eta'(f(x)) = f_\eta(v) = 0$. $\square$

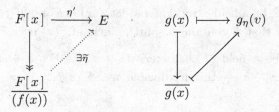

If $\eta : K \to L$ is a ring homomorphism between two fields, then $\operatorname{Char} K = \operatorname{Char} L$. When $\operatorname{Char} K$ is a prime p, both K and L contain the prime field $\mathbb{Z}_p$, and we have that $\eta|_{\mathbb{Z}_p} = \mathbf{1}_{\mathbb{Z}_p}$. When $\operatorname{Char} K = 0$, both K and L contains the prime field $\mathbb{Q}$ and $\eta|_{\mathbb{Q}} = \mathbf{1}_{\mathbb{Q}}$.

Let E/F be a field extension and $u, v \in E$. We say u and v are **conjugates** over F if they have the same minimal polynomial over F. When we are dealing with the case where $\eta\colon F \hookrightarrow E$ such that $f_\eta(x) = f(x)$, to extend η to a ring homomorphism $K[u]$ to E is to look for conjugates of u over F. This is especially true for the case where $F = \mathbb{Q}$.

Example 22.2.2. The elements $\sqrt[4]{2}$, $-\sqrt[4]{2}$, $i\sqrt[4]{2}$ and $-i\sqrt[4]{2}$ are conjugates over $\mathbb{Q}$. Over $\mathbb{Q}[\sqrt[4]{2}]$, $\sqrt[4]{2}$ has no other conjugates than itself. Over $\mathbb{Q}[\sqrt[4]{2}]$ the only conjugates of $i\sqrt[4]{2}$ are $-i\sqrt[4]{2}$ and itself.

A ring homomorphism from a ring R to itself is called a ring **endomorphism** of R.

Example 22.2.3. How many ring endomorphisms of $\mathbb{Q}[\sqrt{2}]$ are there?

Solution. Any ring endomorphism η of $F = \mathbb{Q}[\sqrt{2}]$ restricted to $\mathbb{Q}$ is $1_\mathbb{Q}$. That is where we start. The minimal polynomial of $\sqrt{2}$ over $\mathbb{Q}$ is $x^2 - 2$. By Theorem 22.2.1, u can only be mapped to any root of $x^2 - 2 = 0$ in F. Thus, there are exactly two ring endomorphisms of F. One sends $\sqrt{2}$ to $\sqrt{2}$ and the other sends $\sqrt{2}$ to $-\sqrt{2}$. In other words, one is the identity map of F and the other one sends $a + b\sqrt{2}$ to $a - b\sqrt{2}$. The number of endomorphisms of F is equal to $[F : \mathbb{Q}] = 2$.   ◇

Example 22.2.4. How many ring endomorphisms of $F = \mathbb{Q}[\sqrt[4]{2}]$ are there?

Solution. The minimal polynomial of $\sqrt[4]{2}$ over $\mathbb{Q}$ is $x^4 - 2$. The zeros of $x^4 - 2$ in F are $\pm\sqrt[4]{2}$. Hence there are exactly two endomorphisms of F. One is the identity map of F and the other one sends $\sqrt[4]{2}$ to $-\sqrt[4]{2}$. To be more precise, the second endomorphism sends $a + b\sqrt[4]{2} + c\sqrt[4]{4} + d\sqrt[4]{8}$ to $a - b\sqrt[4]{2} + c\sqrt[4]{4} - d\sqrt[4]{8}$ for $a, b, c, d \in \mathbb{Q}$. In this case the number of endomorphisms of F is less than $[F : \mathbb{Q}] = 4$.   ◇

Lemma 22.2.5. *Let η be an isomorphism from F onto F'. If $f(x) = g(x)h(x)$ in $F[x]$, then $f_\eta(x) = g_\eta(x)h_\eta(x)$ in $F'[x]$. In particular, f is irreducible over F if and only if f_η is irreducible over F'.*

Proof. The first part follows from the identity (16.2.1). If $f(x)$ is reducible over F, then $f(x) = g(x)h(x)$ for f and $h \in F[x]$ with $\deg g$, $\deg h < \deg f$. Hence, $f_\eta(x) = g_\eta(x)h_\eta(x)$ is reducible over F'. Similarly, if f_η is reducible over F', then $f = (f_\eta)_{\eta^{-1}}$ is reducible over F.   □

Lemma 22.2.6. *Let E be a splitting field of $f(x)$ over F and let K be a field with $F \subseteq K \subseteq E$. Then E is a splitting field of $f(x)$ over K.*

Proof. Let $f(x) = u(x - \alpha_1)(x - \alpha_2) \cdots (x - \alpha_n)$ over E and $u \in F^*$. Then $E = F(a_1, \ldots, a_n)$. It follows that $E = K(a_1, \ldots, a_n)$ since $K \subseteq E$. Hence E is also a splitting field of $f(x)$ over K. □

Theorem 22.2.7. *Let η be an isomorphism from F onto F'. Let $f(x)$ be a polynomial over F. Let E and E' be splitting fields of $f(x)$ and $f_\eta(x)$ over F and over F' respectively. Then η can be extended to an isomorphism from E onto E'.*

Proof. We prove this theorem by induction on $[E : F]$. If $[E : F] = 1$, then $f(x)$ splits over F. By Lemma 22.2.5, f_η also splits over F'. Thus $E' = F'$ and we are done. We now assume that $[E : F] > 1$.

Find $\alpha \in E \setminus F$ such that α is a zero of $f(x)$. Let $g(x)$ be the minimal polynomial of α over F. Then $g(x)$ is a divisor of $f(x)$. By Lemma 22.2.5, g_η is a factor of f_η in $F'[x]$. We can extend η to a ring homomorphism $\eta' : F[\alpha] \to$ to E' sending α to a zero β of $g_\eta(x)$ in E'. Then by restricting the codomain of η' to $F'[\beta]$, we can view $\eta' : F[\alpha] \to F'[\beta]$ as an isomorphism. By Lemma 22.2.6, E and E' are splitting fields of $f(x)$ and $f_\eta(x)$ over $F[\alpha]$ and $F'[\beta]$ respectively. Since $[E : F[\alpha]] = [E : F]/[F[\alpha] : F] < [E : F]$, we have that $E \cong E'$ by induction. □

Corollary 22.2.8. *Let E and E' both be splitting fields of $f(x)$ over F. Then $E \simeq E'$. Hence the splitting field of $f(x)$ over F is unique up to isomorphism.*

Proof. Choose η to be $\mathbf{1}_F$ and use the previous theorem. □

Example 22.2.9. There is an isomorphism between the two splitting fields $L = \mathbb{Q}[\alpha, \beta]$ in Example 22.1.7 and $L' = \mathbb{Q}[\sqrt[4]{2}, i]$ for $x^4 - 2$ over $\mathbb{Q}$ (See Exercise 1). If fact, we can identify α with any one of its conjugates, $\pm\sqrt[4]{2}$ or $\pm i\sqrt[4]{2}$. If α is identified as $\sqrt[4]{2}$, then β can be identified as one of the zeros of $x^2 + \sqrt[4]{2}^2 = x^2 + \sqrt{2}$. If α is identified as $-i\sqrt[4]{2}$, then β is identified as one of the zeros of $x^2 + (-i\sqrt[4]{2})^2 = x^2 - \sqrt{2}$. The following table gives all identifications of α and β in $\mathbb{Q}[\sqrt[4]{2}, i]$.

$\alpha \mapsto$	$\sqrt[4]{2}$	$\sqrt[4]{2}$	$-\sqrt[4]{2}$	$-\sqrt[4]{2}$	$i\sqrt[4]{2}$	$i\sqrt[4]{2}$	$-i\sqrt[4]{2}$	$-i\sqrt[4]{2}$
$\beta \mapsto$	$i\sqrt[4]{2}$	$-i\sqrt[4]{2}$	$i\sqrt[4]{2}$	$-i\sqrt[4]{2}$	$\sqrt[4]{2}$	$-\sqrt[4]{2}$	$\sqrt[4]{2}$	$-\sqrt[4]{2}$

Exercises 22.2

1. Show that $\mathbb{Q}[\sqrt[4]{2},\, i]$ is a splitting field of $x^4 - 2$ over $\mathbb{Q}$. How many ring endomorphisms of $\mathbb{Q}[\sqrt[4]{2},\, i]$ are there?

2. Let F be the splitting field of $p(x) = (x^2 - 2)(x^2 - 3)$ over $\mathbb{Q}$ and E be the splitting field of $q(x) = (x^2 - 6)(x^2 - 12)$ over $\mathbb{Q}$. Show that F is isomorphic to E.

3. Let $f(x) = p_1(x)^{d_1} p_2(x)^{d_2} \cdots p_s(x)^{d_s}$ where the p_i's are distinct irreducible polynomials over F. Show that the splitting field of $f(x)$ over F is the splitting field of $g(x) = p_1(x)p_2(x)\cdots p_s(x)$ over F.

4. Find the conjugates of $a + b\sqrt{2}$ over $\mathbb{Q}$ if $a, b \in \mathbb{Q}$. Find the conjugates of $a + bi$ over $\mathbb{R}$ if $a, b \in \mathbb{R}$.

22.3   Algebraically Closed Fields

A field F is **algebraically closed** if every nonconstant polynomial over F has a zero in F.

Example 22.3.1. The well-known *Fundamental Theorem of Algebra* due to Gauss (1777–1855) asserted that $\mathbb{C}$ is algebraically closed. On the other hand, fields such as $\mathbb{R}$ and $\mathbb{Q}$ are not algebraically closed.

Example 22.3.2. Let p be a positive prime integer. The field $\mathbb{Z}_p$ is not algebraically closed. The polynomial

$$f(x) = x(x-1)(x-2)\cdots(x-(p-1)) + 1 \in \mathbb{Z}_p[x].$$

is a nonzero polynomial of degree p with no zeros in $\mathbb{Z}_p$.

Proposition 22.3.3. *Let F be a field. The following conditions are equivalent.*

(i) *The field F is algebraically closed.*

(ii) *Let $f(x)$ be a nonconstant polynomial in $F[x]$. Then $f(x)$ splits into a product of linear factors in $F[x]$.*

(iii) *Any irreducible polynomial in $F[x]$ is of degree 1.*

(iv) *If E/F is any algebraic extension, then $E = F$. In other words, F has no proper algebraic extensions.*

Proof. (i) $\Rightarrow$ (ii): Let $f(x) \in F[x]$ be a polynomial of positive degree n. By assumption, $f(x)$ has a zero, say a, in F. By the factor theorem, $f(x) = (x - a)g(x)$ where $\deg g(x) = n - 1$. If $g(x) \in F$, we are done. Otherwise, use induction on n, $g(x)$ splits over F. It follows that $f(x)$ also splits over F.

(ii) $\Rightarrow$ (iii): By assumption, any polynomial of degree greater than 1 will factor into polynomials of lower degrees. Thus, the only irreducible polynomials are those of degree 1.

(iii) $\Rightarrow$ (iv): Let $\alpha \in E$. Then α is algebraic over F. Since the minimal polynomial of α over F is irreducible over F, it is of degree 1. Hence $\alpha \in F$. We have shown that $E \subseteq F$.

(iv) $\Rightarrow$ (i): Let $f(x)$ be a nonconstant polynomial in $F[x]$ and let E be a splitting field of $f(x)$ over F. Then E is algebraic over F. It follows that $E = F$ by assumption. Thus, the zeros of $f(x)$ are in F. $\qquad\square$

Definition 22.3.4. We say that E is an **algebraic closure** of F if

(i) E is an algebraic extension of F, and

(ii) E is algebraically closed.

For any field, there exists a unique algebraic closure up to isomorphism. However, its proof is beyond the scope of this book. Nonetheless, we will denote the algebraic closure of a field F by $\overline{F}$.

Exercises 22.3

1. Find the algebraic closure of $\mathbb{R}$.

2. Show that $\overline{\mathbb{Q}} = \overline{\mathbb{Q}}_{\mathbb{C}}$, the field of algebraic numbers, is indeed an algebraic closure of $\mathbb{Q}$.

3. Suppose E is an algebraic extension of F. Show that $\overline{E} = \overline{F}$.

4. Let E be an algebraically closed field containing F. Show that $\overline{F} = \overline{F}_E$.

5. Let x be an indeterminate over $\mathbb{C}$. Show that $\mathbb{C}(x)$ is not algebraically closed.

22.4 Multiplicity of Roots

Remember that a polynomial $f(x)$ of degree n over a field F has at most n roots. In a splitting field of $f(x)$ it should have exactly n roots counting *multiplicity*. Let's take a formal look at the concept of multiplicity.

Definition 22.4.1. Let F be a field. If $f(x) \in F[x]$ and c is a root of $f(x) = 0$ in some field extension $E \supseteq F$, then the **multiplicity** of c is the largest positive integer m such that $(x - c)^m \mid f(x)$ over E (see Exercise 4).

Example 22.4.2. Let F be a field of characteristic $p > 0$ and let $f(x) = x^p - a \in F[x]$. If c is a root of $f(x) = 0$ then c is a root of multiplicity p (see Exercise 8, §22.1). Furthermore, if $g(x) = x^{p^n} - a$ and d is a root of $g(x) = 0$, then $g(x) = (x - d)^{p^n}$. Hence, d is a root of multiplicity p^n.

We say c is a **simple** root of $f(x) = 0$ if the multiplicity of c is 1. We say c is a **multiple** root if the multiplicity of c is > 1.

Next we will develop a criterion to determine whether a root is multiple.

Let F be a field and let $f(x) \in F[x]$. Let h be an indeterminate over $F[x]$. Then $f(x+h) - f(x) \in F[x][h]$. Note that $f(x+h) - f(x)\big|_{h=0} = 0$. Hence h is a divisor of $f(x+h) - f(x)$ in $F[x][h]$ and $\big(f(x+h) - f(x)\big)/h \in F[x][h]$. We define the **formal derivative** of $f(x)$, denoted $f'(x)$, to be

$$f'(x) = \frac{f(x + h) - f(x)}{h}\bigg|_{h=0}.$$

Lemma 22.4.3. *Let F be a field. Then the following statements are true for all $f(x), g(x) \in F[x]$ and $a \in F$:*

(a) $a' = 0$ *and* $x' = 1$;

(b) $(f(x) + g(x))' = f'(x) + g'(x)$;

(c) $(f(x)g(x))' = f(x)g'(x) + f'(x)g(x)$;

(d) $(af(x))' = af'(x)$;

(e) *if* $f(x) = a_n x^n + a_{n-1}x^{n-1} + \cdots + a_1 x + a_0$, *then*

$$f'(x) = na_n x^{n-1} + (n-1)a_{n-1}x^{n-2} + \cdots + 2a_2 x + a_1.$$

The proof is extremely similar to those in Calculus. For example,

$$x' = \left.\frac{(x+h) - x}{h}\right|_{h=0} = 1|_{h=0} = 1.$$

We will leave the rest of the proof as an exercise (see Exercise 2).

Proposition 22.4.4. *Let F be a field and let $f(x) \in F[x]$. Then c is a multiple root of $f(x)$ in some field extension E of F if and only if $f(c) = f'(c) = 0$.*

Proof. If c is a root of $f(x)$, then $f(x) = (x-c)^m g(x)$ in $E[x]$ for some $m > 0$. We may assume that $g(c) \neq 0$. Then the multiplicity of c is m. Using Lemma 22.4.3, we have

$$f'(x) = m(x-c)^{m-1}g(x) + (x-c)^m g'(x).$$

If $m = 1$, then $f'(c) = g(c) \neq 0$. If $m \geq 2$ then $f'(c) = 0$. □

Let $p(x)$ be an irreducible polynomial over F. Then $p(x)$ is called a **separable** polynomial over F if it has only simple roots in a splitting field of $p(x)$ over F; otherwise $p(x)$ is called an **inseparable** polynomial over F. An inseparable irreducible polynomial has multiple roots.

An arbitrary polynomial in $F[x]$ is said to be **separable** over F if each of its irreducible factors is separable over F; otherwise it is called **inseparable** over F.

If c is an element in an algebraic extension E of F, then c is said to be **separable** (or **separable algebraic**) over F if its minimal polynomial over F is separable. Otherwise, c is called **inseparable** over F. If every element in E is separable over F, E is called a **separable** (or **separable algebraic**) extension of F; otherwise, E is called an **inseparable** extension of F.

Example 22.4.5. Let F be a field of characteristic $p > 0$ and let $f(x) = x^p - a \in F[x]$. Let c be a root of $f(x) = 0$. If $c \in F$ then $f(x) = (x - c)^p$ in $F[x]$. Both $f(x)$ and c are separable over F. If $c \notin F$ then $f(x)$ is the minimal polynomial of c over F. In this case both $f(x)$ and c are inseparable over F.

Proposition 22.4.6. *Let F be a field. An irreducible polynomial $f(x)$ in $F[x]$ is separable over F if and only if $f'(x)$ is a nonzero polynomial.*

Proof. If $f'(x)$ is the zero polynomial, then any root c of $f(x) = 0$ is multiple by Proposition 22.4.4. Hence $f(x)$ is inseparable over F.

Suppose $f'(x)$ is a nonzero polynomial. Since $\deg f'(x) < \deg f(x)$, $f(x)$ and $f'(x)$ are relatively prime in $F[x]$. There exist $a(x)$ and $b(x)$ in $F[x]$ such that

$$a(x)f(x) + b(x)f'(x) = 1.$$

Then for any root c of $f(x) = 0$, $b(c)f'(c) = 1$. Thus $f'(c) \neq 0$. This shows that all roots of $f(x) = 0$ are simple. Hence $f(x)$ is separable over F. $\square$

In the characteristic 0 case, the formal derivative of any polynomial of degree $n \geq 1$ is of degree $n - 1 \geq 0$. Hence the formal derivative of an irreducible polynomial is never 0. We have the following corollary.

Corollary 22.4.7. *If F is a field of characteristic 0 then any polynomial in $F[x]$ is separable over F.*

We say a field F is **perfect** if every polynomial over F is separable over F. The pervious corollary tells us that a field of characteristic 0 is always perfect.

Lemma 22.4.8. *Let F be a field of characteristic $p > 0$. Then an irreducible polynomial $f(x)$ is inseparable over F if and only if $f(x) = g(x^p)$ for some $g(x) \in F[x]$.*

Proof. By Proposition 22.4.6, $f(x)$ is inseparable over F if and only if $f'(x) = 0$. This implies that the only non-vanishing terms are those of degree pk. Thus,

$$f(x) = a_d x^{pd} + a_{d-1} x^{f(d-1)} + \cdots + a_1 x^p + a_0, \qquad a_i \in F.$$

It follows that $f(x) = g(x^p)$ where $g(x) = \sum_{i=0}^{d} a_i x^i$. $\qquad\square$

Lemma 22.4.9. *Let F be a field of characteristic $p > 0$. Let $f(x)$ be an irreducible polynomial over F. Then there is a separable polynomial $g(x) \in F[x]$ and a nonnegative integer e such that $f(x) = g(x^{p^e})$.*

Proof. If $f(x)$ is separable over F, just choose $g(x) = f(x)$ and $e = 0$ and we are done. Otherwise, find $q_1(x) \in F[x]$ such that $f(x) = q_1(x^p)$. Remember that $\deg q_1 < \deg p$. If $q_1(x)$ is separable over F, we can choose $g(x) = q_1(x)$ and $e = 1$. Otherwise, find $q_2(x) \in F[x]$ such that $q_1(x) = q_2(x^p)$. Note that $\deg q_2(x) < \deg q_1(x)$ and $f(x) = q_2(x^{p^2})$. This process cannot go on indefinitely. Eventually, we will reach a $q_e(x) \in F[x]$ such that $f(x) = q_e(x^{p^e})$ and $q_e(x)$ is separable over F. $\qquad\square$

Theorem 22.4.10. *Let F be a field of characteristic $p > 0$. Let $f(x)$ be an irreducible monic polynomial over F. Then there is a nonnegative integer e and distinct elements $\alpha_1, \dots, \alpha_d$ in a splitting field of $f(x)$ such that*

$$f(x) = (x - \alpha_1)^{p^e} \cdots (x - \alpha_d)^{p^e}, \qquad \text{where } d = \frac{\deg f(x)}{p^e}.$$

Hence, all zeros of $f(x)$ are of the same multiplicity p^e.

Proof. Find $g(x) \in F[x]$ such that $f(x) = g(x^{p^e})$ and $g(x)$ is separable over F. Clearly, $d = \deg g(x) = (\deg f(x))/p^e$. Let E be a splitting field of $g(x)$ over F and let $g(x) = (x - \beta_1)(x - \beta_2) \cdots (x - \beta_d)$ for distinct elements β_i in E. Then

$$f(x) = (x^{p^e} - \beta_1)(x^{p^e} - \beta_2) \cdots (x^{p^e} - \beta_d).$$

Let L be a splitting field of $f(x)$ over E. For each $1 \le i \le d$, find α_i in L such that $\alpha_i^{p^e} = \beta_i$. Note that the α_i's are distinct. Then

$$f(x) = (x^{p^e} - \alpha_1^{p^e})(x^{p^e} - \alpha_2^{p^e}) \cdots (x^{p^e} - \alpha_d^{p^e})$$

$$= (x - \alpha_1)^{p^e} (x - \alpha_2)^{p^e} \cdots (x - \alpha_d)^{p^e}.$$

Thus the α_i's are the zeros of $f(x)$ and they are all of multiplicity p^e. $\quad\square$

Example 22.4.11. Let F be a field of characteristic $p > 0$ and let t be an indeterminate over F. Let $f(x) = x^{2p^2} + tx^{p^2} + t \in F(t)[x]$. Since t is irreducible in $F[t]$, $f(x)$ is irreducible over $F(t)$ by Eisenstein criterion. Let $g(x) = x^2 + tx + t$. Then $f(x) = g(x^{p^2})$. Note that $g(x)$ is also irreducible over $F(t)$ and $g'(x) = 2x + t$ is nonzero. By Proposition 22.4.6, $g(x)$ is separable over $F(t)$. Hence $f(x)$ has two distinct zeros of multiplicity p^2.

Exercises 22.4

1. Determine whether each of the following polynomials is separable over the given field.

   (a) $x^5 - 6x^3 + 15$ over $\mathbb{Q}$;

   (b) $x^4 + x^2 + 1$ over $\mathbb{Z}_2$;

   (c) $x^5 - t$ over $\mathbb{Z}_5(t)$ where t is an indeterminate over $\mathbb{Z}_5$.

2. Prove Lemma 22.4.3.

3. Let $f(x) = g^m(x)$ in $F[x]$. Show that $f'(x) = mg^{m-1}(x)g'(x)$.

4. (a) Let $E \subseteq E'$ both be field extensions of F. Let $f(x) \in F[x]$ and let $c \in E$ be a zero of $f(x)$. Show that the multiplicity of c is the same no matter you consider it over E or over E'.

   (b) Let E/F and E'/F' both be field extensions. Suppose that $F \xrightarrow{\eta} F'$ and $E \xrightarrow{\tilde{\eta}} E'$, which is an extension of η, are both isomorphisms. Show that if c is a zero of $f(x) \in F[x]$ of multiplicity m in E, then so is $\tilde{\eta}(c)$ in E'.

   (c) Show that the concept of multiplicity is well-defined.

5. Let $f(x) \in F[x]$. Show that $f(x)$ has no multiple roots in any field extension of F if and only if f and f' are relatively prime in $F[x]$.

6. Let η be a field isomorphism from F onto F'. Show that a nonconstant polynomial $f(x)$ is separable over F if and only if $f_\eta(x)$ is separable over F'.

22.5 Finite Fields

Throughout this section p denotes a positive prime integer.

A finite field F is of some prime characteristic p and contains the field $\mathbb{Z}_p$ as its prime field. One can consider F as a finite dimensional vector space over $\mathbb{Z}_p$. If $[F : \mathbb{Z}_p] = n$, then $F \cong \mathbb{Z}_p^n$ and $|F| = p^n$. Mathematicians like to use q to denote a number of the form p^n.

Example 22.5.1. Remember that $x^2 + x + 1 \in \mathbb{Z}_2[x]$ is irreducible over $\mathbb{Z}_2$ since it has no zero in $\mathbb{Z}_2$. Hence $F = \dfrac{\mathbb{Z}_2[x]}{(x^2 + x + 1)}$ is a finite field of 2^2 elements. Let $\alpha = \overline{x}$. Then $F = \mathbb{Z}_2 \oplus \mathbb{Z}_2\alpha$. How does the multiplication function in F? For example, let's compute the product $(1 + \alpha)^2$. Note that $\alpha^2 + \alpha + 1 = 0$. Hence $(1 + \alpha)^2 = 1 + \alpha^2 = -\alpha = \alpha$. The complete multiplication table of F is given below:

	0	1	α	$1+\alpha$
0	0	0	0	0
1	0	1	α	$1+\alpha$
α	0	α	$1+\alpha$	1
$1+\alpha$	0	$1+\alpha$	1	α

Example 22.5.2. Both $x^3 + x + 1$ and $x^3 + x^2 + 1$ are irreducible over $\mathbb{Z}_2$. Hence both $F_1 = \dfrac{\mathbb{Z}_2[x]}{(x^3 + x + 1)}$ and $F_2 = \dfrac{\mathbb{Z}_2[x]}{(x^3 + x^2 + 1)}$ are finite fields of 2^3 elements.

Proposition 22.5.3. *Let $q = p^n$. Then the following statements are true.*

(a) *Let F be a splitting field of $x^q - x$ over $\mathbb{Z}_p$. Then F is a finite field of q elements.*

(b) *If F is a finite field of order q, then F is the splitting field of $x^q - x$ over $\mathbb{Z}_p$.*

(c) *For any n, there is a finite field of order $q = p^n$.*

(d) *Let F_1 and F_2 be two finite fields. If $|F_1| = |F_2|$, then $F_1 \simeq F_2$.*

Proof. (a) Let $f(x) = x^q - x$. Since $f'(x) = -1 \neq 0$ in $\mathbb{Z}_p$, $x^q - x$ has q distinct roots over $\mathbb{Z}_p$, we have the following set of q elements:

$$K = \{\alpha \in F : \alpha^q - \alpha = 0\}.$$

We first check that K is a subfield of F. Let α and $\beta \in K$. Then the following statements are true:

(1) $(\alpha + \beta)^q = \alpha^q + \beta^q = \alpha + \beta$ (see Exercise 6, §15.3);

(2) $(-\alpha)^q = (-1)^q \alpha^q = -\alpha$, $((-1)^q = -1$ is true even when $p = 2$ since in that case $1 = -1$);

(3) $(\alpha\beta)^q = \alpha^q \beta^q = \alpha\beta$;

(4) $(\alpha^{-1})^q = (\alpha^q)^{-1} = \alpha^{-1}$ when $\alpha \neq 0$.

This shows that $\alpha + \beta$, $-\alpha$, $\alpha\beta \in K$ and $\alpha^{-1} \in K$ for $\alpha \neq 0$. Since K is a field consisting of all the zeros of $x^q - x$, it is the splitting field of $x^q - x$ over $\mathbb{Z}_p$. Hence $F = K$ is a field of q elements.

(b) Since $(F^*, \cdot)$ is a group of order $q - 1$, we have that $\alpha^{q-1} = 1$ for any $\alpha \in F^*$ by Fermat's little theorem. It then implies that $\alpha^p = \alpha$ for all $\alpha \in F$. Thus, we have found q distinct zeros for the polynomial $x^q - x$ in F. By comparing the leading coefficients on both sides of the following relation, we have that

$$x^q - x = \prod_{\alpha \in F} (x - \alpha)$$

splits in F. Thus, F is the splitting field of $x^q - x$ over $\mathbb{Z}_p$.

Part (c) is a consequence of (a). Part (d) follows from (b) and Corollary 22.2.8. □

We can use $\mathbb{F}_q$ to denote the finite field of q elements. Even though it is not obvious, the two fields in Example 22.5.2 are isomorphic and both are called $\mathbb{F}_8$ (see Exercise 3). Every element in $\mathbb{F}_q$ is separable over $\mathbb{Z}_p$ since its minimal polynomial is a factor of $x^q - x$, and $x^q - x = 0$ has only simple roots.

Corollary 22.5.4. *The multiplicative group $\mathbb{F}_q^*$ is a cyclic group for any* $q = p^n$.

Proof. Let F be a finite field of order $q = p^n$. For any divisor d of $q - 1$, the equation $x^d = 1$ has at most d solutions in $\mathbb{F}^*$ by Corollary 16.2.8. Now $\mathbb{F}^*$ is cyclic by Corollary 4.2.8. $\qquad\square$

Proposition 22.5.5. *The finite field $\mathbb{F}_{p^m}$ can be embedded into $\mathbb{F}_{p^n}$ if and only if* $m \mid n$.

Proof. If $F = \mathbb{F}_{p^m}$ is a subfield of $E = \mathbb{F}_{p^n}$, then E can be viewed as a vector space over F. Hence $|E| = |F|^d$ where $d = [E : F]$. Thus $p^n = (p^m)^d = p^{md}$ and $n = md$.

Conversely, assume $n = md$ for some positive integer d. Then

$$F = \left\{ a \in \mathbb{F}_{p^n} : a^{p^m} = a \right\}$$

is a subfield of at most p^m elements in F^{p^n} for the same reason that K is a subfield in the proof of Proposition 22.5.3(a). By Corollary 22.5.4, $\mathbb{F}_{p^n}^*$ is a cyclic group of order $p^n - 1$ generated by say, g. Then since $p^m - 1$ is a divisor of $p^n - 1 = p^{md} - 1$, we see that $H = \langle g^{(p^n - 1)/(p^m - 1)} \rangle$ is a subgroup of order $p^m - 1$. Thus $F = H \cup \{0\}$ is a field of p^m elements. $\quad\square$

Example 22.5.6. In the characteristic 2 case, the elements of $\mathbb{F}_4$ are the zeros of $x^4 - x$. There are two elements in $\mathbb{F}_4 \setminus \mathbb{F}_2$. These two elements are the zeros of one irreducible polynomial of degree 2 over $\mathbb{F}_2$. Hence

$$x^4 - x = x(x - 1)(x^2 + x + 1)$$

is a factorization into irreducible polynomials over $\mathbb{F}_2$. The elements of $\mathbb{F}_8$ are the zeros of $x^8 - x$ which consist of 2 elements in $\mathbb{F}_2$ and 6 elements of $\mathbb{F}_8 \setminus \mathbb{F}_2$. Since each element of $\mathbb{F}_8 \setminus \mathbb{F}_2$ is one of the 3 distinct zeros of an irreducible polynomial of degree 3, there are altogether 2 irreducible polynomials of degree 3 over $\mathbb{F}_2$. Hence

$$x^8 - x = x(x + 1)(x^3 + x + 1)(x^3 + x^2 + 1)$$

is a factorization into irreducible polynomials over $\mathbb{F}_2$. The only proper subfields of $\mathbb{F}_{16}$ are $\mathbb{F}_4$ and $\mathbb{F}_2$ by Proposition 22.5.5. There are $16 - 4 = 12$

elements in $\mathbb{F}_{16} \setminus \mathbb{F}_4$ each of which is one of the 4 zeros of an irreducible polynomial of degree 4 over $\mathbb{F}_2$. Hence there are 3 irreducible polynomials of degree 4 over $\mathbb{F}$. Since these polynomials have no zeros in $\mathbb{F}_2$, they must be among the following polynomials:

$$x^4 + x^3 + x^2 + x + 1, \quad x^4 + x^3 + 1, \quad x^4 + x^2 + 1, \quad x^4 + x + 1.$$

Among these possibilities, $x^4 + x^2 + 1 = (x^2 + x + 1)^2$. Hence

$$x^{16} - x = x(x+1)(x^2+x+1)(x^4+x^3+x^2+x+1)(x^4+x^3+1)(x^4+x+1).$$

Exercises 22.5

In the following problems p denotes a positive prime integer.

1. Let $q = p^n$. If α is a generator of the multiplicative group $\mathbb{F}_q^*$, show that $\mathbb{F}_q = \mathbb{F}_p[\alpha]$. Hence $\mathbb{F}_q$ is a simple extension of $\mathbb{F}_p$. What is the degree of α over $\mathbb{F}_p$?

2. Let F be a finite field. Show that there are irreducible polynomials of any positive degree over F.

3. Let α be a zero of $x^3 + x^2 + 1$ and β be a zero of $x^3 + x + 1$. Find an actual isomorphism from $\mathbb{Z}_2[\alpha]$ to $\mathbb{Z}_2[\beta]$.

4. Factor $x^9 - x$ and $x^{27} - x$ over $\mathbb{F}_3$.

5. Construct a finite field with 125 elements.

6. Let $f(x) = x^3 - 2$ and $g(x) = x^3 + 2$ be polynomials in $\mathbb{Z}_7[x]$. Show that $f(x)$ and $g(x)$ are irreducible over $\mathbb{Z}_7$ and

$$\mathbb{Z}_7[x]/(f(x)) \cong \mathbb{Z}_7[x]/(g(x)).$$

7. How many irreducible polynomials of degree 12 over $\mathbb{F}_3$ are there?

8. Let F be a finite field and let $f(x)$ be an irreducible polynomial of degree n over F. Let E be the splitting field of $f(x)$ over F. Show that $x^q - x$ where $q = p^n$ splits over E.

9. Let $f(x)$ be an irreducible polynomial of degree m over $\mathbb{F}_p$ and let $q = p^n$. Show that $f(x)$ is a factor of $x^q - x$ in $\mathbb{F}_p[x]$ if and only if $m \mid n$.

10. Show that any finite field is perfect.

11. A ring endomorphism of a field F which is also an isomorphism is called a field **automorphism**. Let $\text{Aut}(F)$ be the set of all field automorphisms of F.

    (a) Show that $\text{Aut}(F)$ is a group under composition of functions.

    (b) Show that the **Frobenius map** $\sigma \colon \mathbb{F}_{p^n} \to \mathbb{F}_{p^n}$ sending a to a^p is an automorphism of $\mathbb{F}_{p^n}$. Show that $\sigma^m(a) = a^{p^m}$ for all $a \in \mathbb{F}_{p^n}$.

    (c) Show that the Frobenius map is of order n in $\text{Aut}(\mathbb{F}_{p^n})$.

Review Exercises for Chapter 22

1. Find a splitting field E of $x^4 - 10x^2 + 21$ over $\mathbb{Q}$. Find $[E : \mathbb{Q}]$.

2. Find the minimal polynomial of i over $\mathbb{Q}[\sqrt[4]{2}]$.

3. Show that any algebraically closed field is infinite.

4. Let F be a field of characteristic $p > 0$ and $a \in F$. Show that $f(x) = x^p - x - a$ is reducible over F if and only if $f(x)$ splits over F if and only if $f(x)$ has a zero in F.

5. The **higher formal derivatives** can be defined inductively by letting $f^{(m)}(x) = (f^{(m-1)})'(x)$ for $m \geq 2$. In the characteristic 0 case, show that the multiplicity of c in $f(x)$ is m if and only if $f(c) = f'(c) = \cdots = f^{(m-1)}(c) = 0$ and $f^{(m)}(c) \neq 0$.

6. Let $F \subseteq K \subseteq E$ be fields. Show that E/F is a separable algebraic extension if and only if E/K and K/F are both separable algebraic extensions.

CHAPTER 23

Galois Pairing

When E is a field extension of F, it is difficult to see how many intermediate fields sit between E and F. However, if E is the splitting field of a separable polynomial over a subfield F, it is possible to use Galois pairing to establish a one-to-one correspondence between the subfields of E containing F and the subgroups of the automorphism group of E fixing F.

23.1   Galois Groups

Let E be a field extension of F. The **Galois group** of E over F, denoted $\mathrm{Gal}\,(E/F)$, is defined to be

$$\{\, \sigma \in \mathrm{Aut}\,(E) : \sigma(x) = x \text{ for all } x \in F \,\}$$

(see Exercise 11, §22.5 and Exercise 2). We say that any automorphism σ in $\mathrm{Gal}\,(E/F)$ is an automorphism of E fixing F. We also say that F is fixed by σ.

First, we review some basic properties regarding field automorphisms.

(1) Any automorphism of a field also fixes its prime field. Suppose F_0 is the prime field of the field E and let σ be a field automorphism of E. For any $m \in \mathbb{Z}$, $\sigma(m) = m$ where m also denotes the image of m in F_0. If E is a field of characteristic $p > 0$, then this is saying that F_0 is fixed by σ. On the other hand, if E is of characteristic 0, then $F_0 = \mathbb{Q}$. In this case $\mathbb{Z}$ is fixed by σ. Furthermore, for all rational number $r = n/m$ with $m, n \in \mathbb{Z}$ and $m \neq 0$, we have

$$\sigma(r) = \sigma\left(\frac{n}{m}\right) = \frac{\sigma(n)}{\sigma(m)} = \frac{n}{m}.$$

It follows that $\mathbb{Q} = F_0$ is also fixed by σ. We conclude that

$$\mathrm{Gal}\,(E/F_0) = \mathrm{Aut}\,(E).$$

(2) Let $\sigma \in \mathrm{Gal}\,(E/F)$. Theorem 22.2.1 tells us that for any $a \in E$, $\sigma(a)$ *is* and can be any of the conjugates of a over F in E. The elements a and $\sigma(a)$ share the same minimal polynomial over F.

Example 23.1.1. Let p be a prime integer ≥ 3 and let $\zeta = e^{2\pi i/p}$. We claim that $G = \mathrm{Gal}\,(K/\mathbb{Q}) = \mathrm{Aut}\,(\mathbb{Q}[\zeta])$ is isomorphic to the multiplicative group $\mathbb{Z}_p^*$.

Let σ be an automorphism of $\mathbb{Q}[\zeta]$. Then σ is completely determined by $\sigma(\zeta)$ and $\sigma(\zeta)$ must be one of the conjugates of ζ over $\mathbb{Q}$. Since the minimal polynomial of ζ over $\mathbb{Q}$ is $\Phi_p(x)$, $\sigma(\zeta) = \zeta^i$ for some $1 \leq i \leq p-1$. It follows that $|\,\mathrm{Gal}\,(\mathbb{Q}[\zeta]/\mathbb{Q})\,| = p-1 = [\mathbb{Q}[\zeta] : \mathbb{Q}]$.

Define a map $\varphi\colon \mathbb{Z}_p^* \to G$ sending $\bar{i}$ to σ_i where $\sigma_i \in G$ is such that $\sigma_i(\zeta) = \zeta^i$. Since for all $kp + i$ we have that $\zeta^{kp+i} = \zeta^i$, this map is well-defined. It is also clearly bijective. It remains to show that φ is a group homomorphism. For any $\bar{i}$ and $\bar{j} \in \mathbb{Z}_p^*$,

$$(\varphi(\bar{i})\varphi(\bar{j}))(\zeta) = \sigma_i(\sigma_j(\zeta)) = \sigma_i(\zeta^j) = (\zeta^i)^j = \zeta^{ij} = \varphi(\overline{ij})(\zeta).$$

Thus, $\varphi(\bar{i})\varphi(\bar{j}) = \varphi(\overline{ij})$. We conclude that φ is a group isomorphism.

Example 23.1.2. Since $\mathrm{Gal}\left(\mathbb{Q}[\sqrt{2}]/\mathbb{Q}\right) = \mathrm{Aut}\left(\mathbb{Q}[\sqrt{2}]\right)$, Example 22.2.3 tells us that $\mathrm{Gal}\left(\mathbb{Q}[\sqrt{2}]/\mathbb{Q}\right)$ consists of two automorphisms, $1_{\mathbb{Q}[\sqrt{2}]}$ and σ which sends $a + b\sqrt{2}$ to $a - b\sqrt{2}$. Remember that $\mathbb{Q}[\sqrt{2}]$ is the splitting field of $x^2 - 2$ over $\mathbb{Q}$. In this case we have $|\,\mathrm{Gal}\left(\mathbb{Q}[\sqrt{2}]/\mathbb{Q}\right)| = 2 = [\mathbb{Q}[\sqrt{2}] : \mathbb{Q}]$.

On the other hand, Example 22.2.4 tells us that $\mathrm{Gal}\left(\mathbb{Q}[\sqrt[4]{2}]/\mathbb{Q}\right)$ also consists of two elements, $1_{\mathbb{Q}[\sqrt[4]{2}]}$ and τ which sends $a + b\sqrt[4]{2} + c\sqrt[4]{4} + d\sqrt[4]{8}$ to $a - b\sqrt[4]{2} + c\sqrt[4]{4} - d\sqrt[4]{8}$ for $a, b, c, d \in \mathbb{Q}$. Note that in this case $\mathbb{Q}[\sqrt[4]{2}]$ is not the splitting field of $x^4 - 2$ over $\mathbb{Q}$ and $|\,\mathrm{Gal}\left(\mathbb{Q}[\sqrt[4]{2}]/\mathbb{Q}\right)| = 2 < [\mathbb{Q}[\sqrt[4]{2}] : \mathbb{Q}]$.

Example 23.1.3. Remember that $E = \mathbb{Q}[\alpha, i]$, where $\alpha = \sqrt[4]{2}$, is the splitting field of $x^4 - 2$ over $\mathbb{Q}$. In Example 22.1.7, we have computed that $[E : \mathbb{Q}] = 8$ and $[E : \mathbb{Q}[\alpha]] = 2$. Hence the minimal polynomial of i over $\mathbb{Q}[\alpha]$ is $x^2 + 1$.

To find $\sigma \in \mathrm{Gal}\left(\mathbb{Q}[\alpha, i]/\mathbb{Q}\right)$, we see that σ fixes $\mathbb{Q}$ and can send α to any of its conjugate over $\mathbb{Q}$ in E, that is, to $\pm\alpha$ or to $\pm i\alpha$. The coefficients of $x^2 + 1$ remain fixed under any of these four maps. Hence, σ maps i to any zero of $x^2 + 1$ in E, that is, to i or to $-i$. Thus, there are altogether 8 elements in $\mathrm{Gal}\left(E/\mathbb{Q}\right)$ as listed below.

	ρ_0	ρ_1	ρ_2	ρ_3	μ_1	δ_1	μ_2	δ_2
$\alpha \mapsto$	α	$i\alpha$	$-\alpha$	$-i\alpha$	α	$i\alpha$	$-\alpha$	$-i\alpha$
$i \mapsto$	i	i	i	i	$-i$	$-i$	$-i$	$-i$

(*Cf.* Example 22.2.9). Let's take δ_1 as an example to show how it really functions. Any element in E can be expressed uniquely as

$$z = a + b\alpha + c\alpha^2 + d\alpha^3 + ei + fi\alpha + gi\alpha^2 + hi\alpha^3,$$

for $a, b, c, d, e, f, g, h \in \mathbb{Q}$. Then

$$\delta_1(z) = a + bi\alpha - c\alpha^2 - di\alpha^3 - ei + f\alpha + gi\alpha^2 - h\alpha^3.$$

In this example we have $|\operatorname{Gal}(E/\mathbb{Q})| = 8 = [E : \mathbb{Q}]$.

Example 23.1.4. Let p be a positive prime integer. Let t be an inde-terminate over $\mathbb{Z}_p$ and let $F = \mathbb{Z}_p(t)$. Then $f(x) = x^p - t$ is irreducible over $\mathbb{Z}_p$. Let E be the splitting field of $f(x)$ over F. Let α be a zero of $f(x)$. Then $f(x) = (x - \alpha)^p$. It is conventional to denote α by $\sqrt[p]{t}$, and $E = F[\sqrt[p]{t}] = \mathbb{Z}_p(\sqrt[p]{t})$.

If $\sigma \in \operatorname{Gal}(E/F)$, then σ can only map $\sqrt[p]{t}$ to the sole conjugate of $\sqrt[p]{t}$ over F, which is itself. Hence, $\operatorname{Gal}(E/F)$ is the trivial group. In this case, $|\operatorname{Gal}(E/F)| = 1 < [E : F] = p$. The reason here obviously is that $f(x)$ is inseparable over F.

Proposition 23.1.5. *If E is the splitting field of a polynomial $f(x)$ over F and $f(x)$ has n distinct zeros in E, then $\operatorname{Gal}(E/F)$ is isomorphic to a subgroup of S_n.*

Proof. Suppose that $\alpha_1, \alpha_2, \ldots, \alpha_n$ are n distinct zeros of $f(x)$. Since $E = F(\alpha_1, \alpha_2, \ldots, \alpha_n)$, any automorphism σ of E fixing F is determined by $\sigma(\alpha_i)$, $i = 1, 2, \ldots, n$. Theorem 22.2.1 tells us that σ induces a permutation on $A = \{\alpha_1, \alpha_2, \ldots, \alpha_n\}$. Hence, there is an injective map from $\operatorname{Gal}(E/F)$ to S_n by sending σ to the induced permutation. We leave it as an exercise to check that this is a group homomorphism (see Exercise 8). $\square$

Next we discuss the relation between $\operatorname{Gal}(E/F)$ and $[E : F]$.

Lemma 23.1.6. *Let E/F be a field extension and η be any ring monomor-phism from F to E. Suppose E is the splitting field of both $f(x)$ and $f_\eta(x)$ over F and $\eta(F)$ respectively. Then the number of extensions of η to E is the same as the number of extensions of the inclusion map $F \stackrel{i}{\hookrightarrow} E$ to E.*

Proof. By Theorem 22.2.7, we can find $\eta' : E \to E$ which is an extension of η. Let

$$\begin{cases} S_1 = \{\tau \in \operatorname{Aut}(E) : \tau|_F = \eta\}; \\ S_2 = \{\sigma \in \operatorname{Aut}(E) : \sigma|_F \text{ is the inclusion map } F \hookrightarrow E\}. \end{cases}$$

Let $\Phi_1 : S_2 \to S_1$ be the map sending σ to $\eta'\sigma$. For $a \in F$, $\eta'\sigma(a) = \eta'(\sigma(a)) = \eta(a)$. Hence $\eta'\sigma \in S_1$. On the other hand, let $\Psi : S_1 \to S_2$ be the map sending τ to $(\eta')^{-1}\tau$. If $a \in F$ then $\tau(a) = \eta(a)$. Thus for all

$a \in F$, $(\eta')^{-1}\tau(a) = (\eta')^{-1}(\eta(a)) = a$. It follows that $(\eta')^{-1}\tau \in S_2$. We have shown that both Φ and Ψ are well-defined. It is clear that Φ and Ψ are inverse to each other. Hence $|S_1| = |S_2|$. $\square$

Theorem 23.1.7. *Let E be the splitting field of $f(x)$ over F. Then $|\operatorname{Gal}(E/F)|$ is a divisor of $[E:F]$. Furthermore, $|\operatorname{Gal}(E/F)| = [E:F]$ if and only if f is a separable polynomial over F.*

We simply prove Theorem 22.2.7 again, but more carefully this time.

Proof. We prove this theorem by induction on $[E:F]$. If $[E:F] = 1$, there is nothing to prove. We now assume that $[E:F] > 1$.

Find $\alpha \in E \setminus F$ such that α is a zero of $f(x)$. In case $f(x)$ is inseparable over F, we choose α to be a multiple root of its minimal polynomial $g(x)$ over F. In either case $g(x)$ is a divisor of $f(x)$ in $F[x]$. We can extend η to a ring homomorphism $\eta' : F[\alpha] \to E$ sending α to any zero β of $g(x)$ in E. By Theorem 22.2.1, if k is the number of distinct zeros of g in E, then η has exactly k extensions to $F[\alpha]$. By Theorem 22.4.10, k is a divisor of $\deg g = [F[\alpha] : F]$, and $k = [F[\alpha] : F]$ exactly when g is separable over F. Next we are going to extend η' to a ring homomorphism from E to E. Any such extension is injective. Since E is a finite dimensional vector space over F, any monomorphism from E to E is also onto, and hence an automorphism.

Note that $f_{\eta'}(x) = f(x)$ and E remains the splitting field of $f(x)$ over both $F[\alpha]$ and $\eta'(F[\alpha]) = F[\beta]$. There are as many further extensions for η' to E as there are for $F[\alpha] \hookrightarrow E$ by Lemma 23.1.6. Thus,

$$(23.1.1) \qquad |\operatorname{Gal}(E/F)| = k |\operatorname{Gal}(E/F[\alpha])|.$$

Hence, by induction, $|\operatorname{Gal}(E/F)|$ divides $[E:F]$. When E is inseparable over F, we have chosen α so that $|\operatorname{Gal}(F[\alpha]/F)|$ is less than $[F[\alpha]:F]$. In this case $|\operatorname{Gal}(E/F)|$ is a proper divisor of $[E:F]$. If $f(x)$ is separable over F, then E is also separable over $F[\alpha]$. By induction, $|\operatorname{Gal}(E/F[\alpha])| = [E:F[\alpha]]$. Thus, $|\operatorname{Gal}(E/F)| = [E:F]$ by (23.1.1). $\square$

Example 23.1.8. Let $q = p^n$ where p is a positive prime integer. The finite field $\mathbb{F}_q$ is the splitting field of the separable polynomial $x^q - x$ over $\mathbb{Z}_p$. Thus, $|\operatorname{Gal}(\mathbb{F}_q/\mathbb{F}_p)| = [\mathbb{F}_q : \mathbb{F}_p] = n$. However, we also know that the

Galois group $\mathrm{Gal}\,(\mathbb{F}_q/\mathbb{F}_p)$ contains the Frobenius map σ which generates a cyclic group of order n (see Exercise 11, §22.5). Thus, $\mathrm{Gal}\,(\mathbb{F}_q/\mathbb{F}_p) = \langle\,\sigma\,\rangle$.

Exercises 23.1

1. Find all conjugates in $\mathbb{C}$ of the following given complex numbers over $\mathbb{Q}$.

   (a) $\sqrt{2} + \sqrt{3}$

   (b) $\sqrt[3]{2}$

   (c) $\sqrt{3} + i$

   (d) $\sqrt{1 + \sqrt{3}}$

   (e) $\sqrt[3]{2 - \sqrt{3}}$

2. Let E/F be a field extension. Show that $\mathrm{Gal}\,(E/F)$ is a subgroup of $\mathrm{Aut}\,(E)$.

3. Show that the only automorphism of $E = \mathbb{Q}(\sqrt[3]{2})$ is the identity automorphism.

4. Let E be the splitting field of $x^2 + 4$ over $\mathbb{Q}$. Compute $\mathrm{Gal}\,(E/\mathbb{Q})$.

5. Find all subgroups of $\mathrm{Gal}\,(\mathbb{Q}[\sqrt{2}, \sqrt{3}\,]/\mathbb{Q})$.

6. Let E be the splitting field of $x^3 - 2$ over $\mathbb{Q}$ and F be the splitting field of $x^2 + x + 1$ over $\mathbb{Q}$. Show that F can be embedded into E and find the Galois group of E over F.

7. Verify that $\mathrm{Gal}\,(\mathbb{Q}[\sqrt[4]{2}, i\,]/\mathbb{Q}) \cong D_4$ in Example 23.1.3. Find all its subgroups.

8. Complete the proof of Proposition 23.1.5.

9. Let E be the splitting field of $f(x)$ over F and $p(x)$ be an irreducible factor of $f(x)$ in $F[x]$. If $\alpha_1, \alpha_2, \ldots, \alpha_r$ are zeros of $p(x)$, show that there exists $\sigma_i \in \mathrm{Gal}\,(E/F)$ such that $\sigma_i(\alpha_1) = \alpha_i$ for each i.

23.2   The Fixed Subfields of a Galois Group

When we say that H is a **group of automorphisms** of a field E, we mean that H is a subgroup of $\mathrm{Aut}\,(E)$. For any H which is a group of automorphisms of E, define

$$E^H = \{\, a \in E : \sigma(a) = a \text{ for all } \sigma \in H \,\}.$$

If $a,\, b \in E^H$, then $\sigma(a) = a$ and $\sigma(b) = b$ for all $\sigma \in H$. It is routine to check that $a+b$, $a-b$, ab and a/b if $b \neq 0$, all lie in E^H. Hence E^H is indeed a subfield of E. We call E^H the subfield of H-**invariants** or the H-**fixed subfield** of E. It is clear that H is also a subgroup of $\mathrm{Gal}\,(E/E^H)$.

Example 23.2.1. Let's find the fixed subfields in Example 23.1.2. Let $E_1 = \mathbb{Q}[\sqrt{2}\,]$ and $G_1 = \mathrm{Gal}\,(E_1/\mathbb{Q})$. Let $a, b \in \mathbb{Q}$. Then $a + b\sqrt{2} \in E_1^{G_1}$ if and only if

$$\sigma(a + b\sqrt{2}) = a - b\sqrt{2} = a + b\sqrt{2}.$$

It follows that $b = 0$. Hence $E_1^{G_1} = \mathbb{Q}$.

Let $E_2 = \mathbb{Q}[\sqrt[4]{2}\,]$ and $G_2 = \mathrm{Gal}\,(E_2/\mathbb{Q})$. Let a, b, c, $d \in \mathbb{Q}$. Then $a + b\sqrt[4]{2} + c\sqrt[4]{4} + d\sqrt[4]{8} \in E_2^{G_2}$ if and only if

$$\tau(a + b\sqrt[4]{2} + c\sqrt[4]{4} + d\sqrt[4]{8}) = a - b\sqrt[4]{2} + c\sqrt[4]{4} - d\sqrt[4]{8} = a + b\sqrt[4]{2} + c\sqrt[4]{4} + d\sqrt[4]{8}.$$

It follows that $b = d = 0$. Hence $E_2^{G_2} = \mathbb{Q} \oplus \mathbb{Q}\sqrt[4]{4} = \mathbb{Q}[\sqrt{2}\,]$.

Let E be a field. Then we have a map from the family $\mathfrak{F}$ of subfields of E to the family $\mathfrak{G}$ of groups of automorphisms of E:

$$F \longmapsto \mathrm{Gal}\,(E/F).$$

Similarly, we have a map from $\mathfrak{G}$ to $\mathfrak{F}$:

$$G \longmapsto E^G.$$

Lemma 23.2.2. *Let F, F' be subfields of E and let G and G' be groups of automorphisms of E. Then the following basic properties hold:*

(a) $G \subseteq G' \Rightarrow E^G \supseteq E^{G'}$;

(b) $F \subseteq F' \Rightarrow \mathrm{Gal}\,(E/F) \supseteq \mathrm{Gal}\,(E/F')$;

(c) $E^{(\text{Gal}(E/F))} \supseteq F$;

(d) $\text{Gal}\left(E/E^G\right) \supseteq G$.

We leave proving this lemma as an exercise (see Exercise 1). Next we state the only non-trivial result in this section.

Theorem 23.2.3 (Artin's Lemma). *Let G be a finite group of automorphisms of a field E and let $F = E^G$. Then*

$$[E : F] \leq |G|.$$

Proof. Let $G = \{\eta_1 = 1, \eta_2, \ldots, \eta_n\}$. Let $u_1, \ldots, u_m$ be any m elements of L where $m > n$. It suffices to show that $u_1, \ldots, u_m$ are linearly dependent over F. Consider the system of n linear equations in m unknowns with coefficients in L:

(23.2.1)
$$\sum_{j=1}^{m} \eta_i(u_j)x_j = 0, \qquad 1 \leq i \leq n.$$

Since there are more unknowns than the equations, we know that there are non-trivial solutions for this system of equations. Among such solutions we choose one $(a_1, \ldots, a_m) \neq (0, \ldots, 0)$ with the least number of non-zero a_j's. Without loss of generality we may assume that $a_1 = 1$. Thus we have

(23.2.2) $\qquad \eta_i(u_1) + \eta_i(u_2)a_2 + \cdots + \eta_i(u_m)a_m = 0, \qquad i = 1, \ldots, n.$

We claim that $a_1, \ldots, a_m \in F$. If not, without loss of generality we may assume $a_2 \notin F$. Then there exists k such that $\eta_k(a_2) \neq a_2$. Apply η_k and we obtain

$$\sum_{j=1}^{m} \eta_k \eta_i(u_j)\eta_k(a_j) = 0, \qquad 1 \leq i \leq n.$$

Since $G = \{\eta_k \eta_i : i = 1, \ldots, n\}$, we have the following relations

(23.2.3) $\eta_i(u_1) + \eta_i(u_2)\eta_k(a_2) + \cdots + \eta_i(u_m)\eta_k(a_m) = 0, \qquad i = 1, \ldots, n.$

If we take the differences of the relations in (23.2.2) and those in (23.2.3), we have

$$\eta_i(u_2)(a_2 - \eta_k(a_2)) + \cdots + \eta_i(u_m)(a_m - \eta_k(a_m)) = 0, \qquad i = 1, \ldots, n.$$

Hence $(0, a_2 - \eta_k(a_2), a_3 - \eta_k(a_3), \ldots, a_m - \eta_k(a_m))$ is a nontrivial solution of (23.2.1) with more zero terms than $(a_1, \ldots, a_m)$, a contradiction. Hence we have that all the a_i's are in F and $u_1, \ldots, u_m$ are linearly dependent over F. $\qquad\square$

Let E/F be a field extension. We say that E is a **normal** (algebraic) extension of F if every irreducible polynomial over F with a zero in E splits over E. Hence any irreducible polynomial over F either has no zero in E or splits over E. A normal extension E/F contains a splitting field for the minimal polynomial of every element in E.

The following theorem connects three seemingly unrelated concepts together.

Theorem 23.2.4. *Let E/F be a field extension. The following conditions on E/F are equivalent:*

 (i) *E is a splitting field of a separable polynomial $f(x)$ over F;*

 (ii) *$F = E^G$ for some finite group G of automorphisms of E;*

(iii) *E is a finite normal separable extension over F.*

Moreover, if E and F are as in (i), then $E^{\mathrm{Gal}(E/F)} = F$, and if F and G are as in (ii), then $G = \mathrm{Gal}\,(E/F)$.

Proof. (i) $\Rightarrow$ (ii): Let $G = \mathrm{Gal}\,(E/F)$ and let $F' = E^G$. Then F' is a subfield of E containing F. It follows that $G = \mathrm{Gal}\,(E/F')$ since any automorphism of E fixing F also fixes F'. It is also clear that $f(x)$ remains separable over F' and E is a splitting field of $f(x)$ over F'. Hence, by Theorem 23.1.7, $|G| = [E : F] = [E : F']$. Thus, $F = F' = E^G$.

Note that we have also proved the first half of the last part of this theorem.

(ii) $\Rightarrow$ (iii): Artin's Lemma (Theorem 23.2.3) shows that E is a finite extension of F. Let $f(x)$ be a monic irreducible polynomial over F having a zero $r \in E$. Let $\{r_1 = r, r_2, \ldots, r_m\}$ be the orbit of r under the action of G. Then

$$\{\,\eta(r_1),\ \eta(r_2),\ \ldots,\ \eta(r_m)\,\} = \{\,r_1, \ldots, r_m\,\}$$

for any $\eta \in G$ (see Exercise 5, §10.1). Now define

$$g(x) = \prod_{i=1}^{m}(x - r_i).$$

Since an automorphism fixing F can only send r to a conjugate of r over F, all the r_i's are zeros of $f(x)$. Hence we have that $g(x)$ is a divisor of $f(x)$ over E. For all $\eta \in G$, we have

$$g_\eta(x) = \prod_{i=1}^{m}(x - \eta(r_i)) = \prod_{i=1}^{m}(x - r_i) = g(x).$$

Hence $g(x) \in F[x]$ since $E^G = F$. It follows that $f(x) = g(x)$ splits and is separable over F since both $f(x)$ and $g(x)$ are monic and $f(x)$ is irreducible over F. We now have that E is a finite normal extension of F. We also just showed that the minimal polynomial of any element in E over F is separable. Hence E is also a separable extension.

(iii) $\Rightarrow$ (i): Since E is a finite extension, we can find $\alpha_1, \ldots, \alpha_n$ in E which generates E over F. For each i, let $f_i(x)$ be the minimal polynomial of α_i over F and let $f(x) = \prod_i f_i(x)$. By (iii), all the f_i's split and are separable over F. Hence $f(x)$ splits and is separable over F. Note that E is also the splitting field of $f(x)$ over F.

For the second half of the last part of this theorem, note that since G fixes F by (ii), G is a subgroup of $\mathrm{Gal}(E/F)$. By Theorem 23.1.7 and Artin's Lemma, we have that

$$|\mathrm{Gal}(E/F)| = [E : F] \leq |G| = |\mathrm{Gal}(E/F)|.$$

It follows that $|G| = |\mathrm{Gal}(E/F)|$ and $G = \mathrm{Gal}(E/F)$. $\qquad \square$

This theorem characterizes whether a field extension is a finite normal separable extension. For example, $E = \mathbb{Q}[\sqrt[4]{2}]$ is not a normal extension of $\mathbb{Q}$ since E does not contain all the zeros of $x^4 - 2$. In Example 22.2.4, we have also obtained that $\mathrm{Gal}(E/\mathbb{Q})$ is a group of order $2 < [E : \mathbb{Q}] = 4$, which again confirms this fact. In Example 23.2.1 we showed that the fixed field of $\mathrm{Gal}(E/\mathbb{Q})$ is $\mathbb{Q}[\sqrt{2}]$. Theorem 23.2.4 guarantees that E is a normal extension over $\mathbb{Q}[\sqrt{2}]$. Indeed E_2 is the splitting field of $x^2 - \sqrt{2}$ over $\mathbb{Q}[\sqrt{2}]$.

Let p be a positive prime integer. Let's consider the field $E = \mathbb{F}_p[\sqrt[p]{t}\,]$ in Example 23.1.4. In this case, E is the splitting field of $x^p - t$ over $F = \mathbb{F}_p[t]$, but it is still not a separable extension over F. In this case, $\mathrm{Gal}\,(E/F)$ is the trivial group, and its fixed field is definitely not F!

Exercises 23.2

1. Prove Lemma 23.2.2.

2. Let $E = \mathbb{Q}[\sqrt[4]{2}, i\,]$ and $G = \mathrm{Gal}\,(E/\mathbb{Q})$. Find E^G.

3. Let $E = \mathbb{Q}(r)$ be the field satisfying $r^3 + r^2 - 2r - 1 = 0$. Verify that $r' = r^2 - 2$ is also a root of $x^3 + x^2 - 2x - 1 = 0$. Determine $G = \mathrm{Gal}\,(\mathbb{Q}(r)/\mathbb{Q})$. Find E^G.

4. Let $E = \mathbb{Z}_p(t)$ where t is transcendental over $\mathbb{Z}_p$.

   (a) Show that the map $\eta\colon E \to E$ sending t to $t + 1$ is a field automorphism.

   (b) Let G be the group of automorphisms generated by η. Determine $F = E^G$ and $[E : F]$.

5. Let E be a finite extension of a field F. Use the method in the proof of Artin's Lemma to show that $\mathrm{Gal}\,(E/F)$ is a finite group and $|\mathrm{Gal}\,(E/F)| \le [E : F]$. (Hint: Let $\{u_1, u_2, \ldots, u_n\}$ be a basis for E over F and assume $\mathrm{Gal}\,(E/F)$ contains at least $n + 1$ distinct automorphisms $\sigma_1 = 1_E, \sigma_2, \ldots, \sigma_{n+1}$. Consider the linear system

$$\sigma_1(u_i)x_1 + \sigma_2(u_i)x_2 + \cdots + \sigma_{n+1}(u_i)x_{n+1} = 0, \qquad i = 1, \ldots, n.$$

of n equations in $n + 1$ unknowns.)

6. Use the method in the proof of Artin's lemma to prove the following result on differential equations. Let $y_1, y_2, \ldots, y_{n+1}$ be real analytic functions which satisfy a linear differential equation

$$y^{(n)} + a_1 y^{(n-1)} + \cdots + a_{n-1} y' + a_n y = 0$$

with constant coefficients ($a_i \in \mathbb{R}$). Then the y_i's are linearly dependent over $\mathbb{R}$.

23.3   Fundamental Theorem of Galois Pairing

We are now ready to state the main theorem of this chapter.

Theorem 23.3.1 (Fundamental Theorem of Galois Pairing). *Let E be a finite normal separable extension over a field F and $G = \mathrm{Gal}\,(E/F)$ be its Galois group. Let*

$$\mathfrak{F} = \{\, K : K \text{ is an intermediate field between } E \text{ and } F \,\},$$

and let

$$\mathfrak{G} = \{H : H \text{ a subgroup of } G\}.$$

There is an order-reversing one-to-one correspondence (pairing) between $\mathfrak{F}$ and $\mathfrak{G}$:

$$
\begin{array}{ccc}
\mathfrak{F} & \longleftrightarrow & \mathfrak{G} \\
K & \longmapsto & \mathrm{Gal}\,(E/K) \\
E^H & \longleftarrow & H.
\end{array}
$$

In other words, $E^{\mathrm{Gal}(E/K)} = K$ and $\mathrm{Gal}\,(E/E^H) = H$.

Moreover, the following further properties are satisfied by the pairing.

(a) *We have the relations $|H| = [E : E^H]$ and $[G : H] = [E^H : F]$.*

(b) *The following three conditions are equivalent:*

  (i) *$H \lhd G$;*

  (ii) *$\eta(E^H) \subseteq E^H$ for all $\eta \in G$;*

  (iii) *E^H is normal over F.*

In this case, we have that $\mathrm{Gal}\,(E^H/F) \simeq G/H$.

At this point, all the groundwork has been laid, and we are just one step away from completing the proof of this incredibly long theorem. However, instead of plunging into the proof right away, let's first look at a few examples to get a feeling on this theorem.

Example 23.3.2. Let $E = \mathbb{Q}[\sqrt{2}]$. Then E is the splitting field of $x^2 - 2$ and hence a finite normal separable extension over $\mathbb{Q}$. In Example 23.1.2, we have seen that $G = \mathrm{Gal}\,(E/\mathbb{Q})$ is the group of order 2. There are no proper nontrivial subgroup in G. Thus there are no intermediate fields between E and $\mathbb{Q}$. We could have proved this result directly (*Cf.* Exercise 5, §21.2), but we use it now to demonstrate Theorem 23.3.1.

Example 23.3.3. Let E be the splitting field of $p(x) = x^3 - 2$ over $\mathbb{Q}$ and let $\omega = e^{2\pi i/3}$. Then $E = \mathbb{Q}[\sqrt[3]{2}, \sqrt[3]{2}\omega] = \mathbb{Q}[\sqrt[3]{2}, \omega]$ is a finite normal separable extension of $\mathbb{Q}$. Since $\omega \notin \mathbb{Q}[\sqrt[3]{2}]$, we have that

$$[E : \mathbb{Q}] = [E : \mathbb{Q}[\sqrt[3]{2}]][\mathbb{Q}[\sqrt[3]{2}] : \mathbb{Q}] = 2 \times 3 = 6$$

(see also Exercise 2, P. 350). By Theorem 23.1.7, $G = \mathrm{Gal}\,(E/\mathbb{Q})$ is a group of order 6. Since $\sqrt[3]{2}$ can only be mapped to $\sqrt[3]{2}$, $\sqrt[3]{2}\omega$ or $\sqrt[3]{2}\omega^2$, and ω can only be mapped to ω or ω^2, we have our 6 automorphisms over $\mathbb{Q}$. Proposition 23.1.5 tells us that G can be embedded into S_3. Thus, G is isomorphic to S_3. The proper nontrivial subgroups of G include one subgroup of order 3 and three subgroups of order 2, each generated by an element of order 2. In the following table, we give an automorphism σ of order 3 and three automorphisms τ_1, τ_2, τ_3 of order 2 in G.

	σ	τ_1	τ_2	τ_3
$\sqrt[3]{2} \mapsto$	$\sqrt[3]{2}\omega$	$\sqrt[3]{2}$	$\sqrt[3]{2}\omega$	$\sqrt[3]{2}\omega^2$
$\omega \mapsto$	ω	ω^2	ω^2	ω^2

With all this information, we can construct the subgroup lattice of G. Labeled beside the line connected two groups is the subgroup index. For example, the subgroup $\langle \sigma \rangle$ is of index 2 in G.

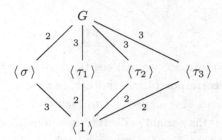

We now demonstrate in detail how to find $E^{\langle \sigma \rangle}$, the subfield fixed by σ. Note that $1, \sqrt[3]{2}, \sqrt[3]{4}, \omega, \sqrt[3]{2}\omega, \sqrt[3]{4}\omega$ form a basis for E over $\mathbb{Q}$. Let

$$z = a + b\sqrt[3]{2} + c\sqrt[3]{4} + d\omega + e\sqrt[3]{2}\omega + f\sqrt[3]{4}\omega, \qquad a, b, c, d, e, f \in \mathbb{Q}.$$

Since $\omega^2 = -\omega - 1$, we have that

$$\sigma(z) = a + b\sqrt[3]{2}\omega + c\sqrt[3]{4}\omega^2 + d\omega + e\sqrt[3]{2}\omega^2 + f\sqrt[3]{4}$$
$$= a - e\sqrt[3]{2} + (-c + f)\sqrt[3]{4} + d\omega + (b - e)\sqrt[3]{2}\omega - c\sqrt[3]{4}\omega.$$

Note that $z \in E^{\langle \sigma \rangle}$ if and only if $\sigma(z) = z$ if and only if

$$b = -e, \quad c = -c + f, \quad e = b - e \quad \text{and} \quad f = -c.$$

This implies that $b = c = e = f = 0$. Hence

$$E^{\langle \sigma \rangle} = \mathbb{Q} \oplus \mathbb{Q}\omega = \mathbb{Q}[\omega].$$

We leave it to the reader to calculate the other fixed subfields. We now use the Galois pairing to construct the (inverted) subfield lattice of E over $\mathbb{Q}$. In this diagram, the subfields are placed above the fields they are contained in. Labeled beside the line connecting two fields is the field extension degree. For example $[\mathbb{Q}[\omega] : \mathbb{Q}] = 2$.

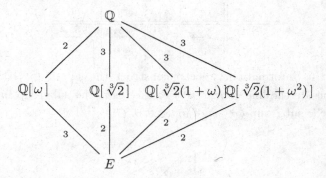

Comparing the two lattices, we will see that not only the subgroups and the fixed subfields correspond to each other, but the subgroup indices and field extension degrees also correspond to each other.

The subgroup $\langle \sigma \rangle$ is normal in G. Its corresponding fixed subfield $E^{\langle \sigma \rangle}$ is $K = \mathbb{Q}[\omega]$. Note that K is the splitting field of $x^2 + x + 1$ over $\mathbb{Q}$. Thus, K is a normal extension of $\mathbb{Q}$.

Example 23.3.4. Let E be the splitting field of the polynomial $x^7 - 2$ over $\mathbb{Q}$. Then $E = \mathbb{Q}(\sqrt[7]{2}, \zeta)$, where $\zeta = e^{2\pi i/7}$. Let $K = \mathbb{Q}[\zeta]$ and $L = \mathbb{Q}[\sqrt[7]{2}]$.

Since $[K : \mathbb{Q}] = 6$ and $[L : \mathbb{Q}] = 7$ are both divisors of $[E : \mathbb{Q}]$, we have that $42 \mid [E : \mathbb{Q}]$. Moreover, we have that $[E : \mathbb{Q}] \leq 42$ since $[E : L] \leq 6$ and $[L : \mathbb{Q}] = 7$. We conclude that $[E : \mathbb{Q}] = 42$, $[E : K] = 7$ and $[E : L] = 6$. The automorphisms of E fixing $\mathbb{Q}$ sends $\sqrt[7]{7}$ to $\sqrt[7]{7}\zeta^i$, $i = 0, 1, \ldots, 7$ and ζ to ζ^j, $j = 1, \ldots, 6$. There are exactly 42 of these.

Let $G = \mathrm{Gal}\,(E/\mathbb{Q})$. The field K is the splitting field of $\Phi_7(x)$ over $\mathbb{Q}$. Thus $\mathrm{Gal}\,(K/\mathbb{Q})$ is isomorphic to $\mathbb{Z}_7^*$ by Example 23.1.1. Hence $\mathrm{Gal}\,(K/\mathbb{Q})$ is a cyclic group of order 6 by Corollary 22.5.4. Since K is a normal separable extension of $\mathbb{Q}$, by the Galois pairing, $N = \mathrm{Gal}\,(E/K)$ is a normal subgroup of order 7 in G and $G/N \cong \mathrm{Gal}\,(K/\mathbb{Q})$ is cyclic of order 6.

Recall that $\mathbb{Z}_7^* = \langle 3 \rangle$. Let

		σ	τ
$\sqrt[7]{2}$	$\mapsto$	$\sqrt[7]{2}\zeta$	$\sqrt[7]{2}$
ζ	$\mapsto$	ζ	ζ^3

Then σ is an automorphism of order 7 fixing K. Thus, $N = \langle \sigma \rangle$. The automorphism τ is of order 6 such that $\tau^i \notin N$ for $1 \leq i \leq 5$. Hence $G/N = \langle \overline{\tau} \rangle$. It follows that G is generated by σ and τ. Since N is normal in G, we have the additional relation

$$\tau \sigma \tau^{-1} = \sigma^3$$

since $\tau^{-1}(\zeta) = \zeta^5$. We leave it to the reader to check that $\mathrm{Gal}\,(E/\mathbb{Q})$ can be presented as

$$\langle x, y : x^7 = y^6 = 1, \; yxy^{-1} = x^3 \rangle.$$

Note that $H = \langle \tau \rangle$ has two proper nontrivial subgroups $H_1 = \{ e, \tau^3 \}$ and $H_2 = \{ e, \tau^2, \tau^4 \}$. The fixed subfield of H_1 in E is $L_1 = \mathbb{Q}[\sqrt[7]{2}, \zeta + \zeta^{-1}] = \mathbb{Q}(\cos(2\pi/7))$ since $\tau^3(\zeta) = \zeta^6 = \zeta^{-1}$ and $\tau^3(\zeta + \zeta^{-1}) = \zeta + \zeta^{-1}$. The Galois pairing tells us that $[E : L_1] = 2$. This can be directly verified by observing that ζ is a zero of

$$f(x) = (x - \zeta)(x - \zeta^{-1}) = x - (\zeta + \zeta^{-1})x + 1$$

over L_1. Similarly, the fixed subfield of H_2 in E is $L_2 = \mathbb{Q}[\zeta + \zeta^2 + \zeta^4]$.

We now prove the main theorem.

Proof of Theorem 23.3.1. It is clear from Lemma 23.2.2(a) and (b) that the mappings from $\mathfrak{F}$ to $\mathfrak{G}$ and from $\mathfrak{G}$ to $\mathfrak{F}$ are both order-reversing. We now show that these two mappings are inverse to each other. To do so, we need to show that $E^{\mathrm{Gal}(E/K)} = K$ and $\mathrm{Gal}\left(E/E^H\right) = H$ for all H and K. This is simply the last part of Theorem 23.2.4 if we choose the F in the theorem to be K and the G in the theorem to be H respectively.

To prove (a), note that E remains a normal separable extension over E_H for any subgroup H of G. We then have

$$\begin{cases} |H| = |\,\mathrm{Gal}\left(E/E^H\right)| = [E : E^H]; \\ [G : H] = \dfrac{|G|}{|H|} = \dfrac{[E : F]}{[E : E^H]} = [E^H : F]. \end{cases}$$

To prove (b), we do so by proving (i) $\Rightarrow$ (ii) $\Rightarrow$ (iii) $\Rightarrow$ (i).

(i) $\Rightarrow$ (ii): Let $a \in E^H$. Take any $\eta \in G$. For any $\sigma \in H$,

$$\sigma(\eta(a)) = \eta(\eta^{-1}\sigma\eta(a)) = \eta(a)$$

since $\eta^{-1}\sigma\eta \in H$. Hence $\eta(a) \in E^H$. We have shown that $\eta(E^H) \subseteq E^H$.

(ii) $\Rightarrow$ (iii): Let $f(x) \in F[x]$ have a zero α in $K = E^H$. Since E is normal over F, $f(x)$ splits over E. Let β be any zero of $f(x)$ in E. Then there is an automorphism η mapping α to β. However, since $\eta(K) \subseteq K$, we have that $\beta \in K$. Thus, K contains all zeros of $f(x)$ and $f(x)$ splits over K. Thus, K is normal over F.

(iii) $\Rightarrow$ (i): Let $\sigma \in H$ and $\eta \in G$. We want to show that $\eta^{-1}\sigma\eta \in H$. Since $H = \mathrm{Gal}\left(E/E^H\right)$, it suffices to show that $\eta^{-1}\sigma\eta$ fixes E^H. Let $a \in K = E^H$. Since K is normal over F, K contains all conjugates of a over F, including $\eta(a)$. Hence

$$\eta^{-1}\sigma\eta(a) = \eta^{-1}\sigma(\eta(a)) = \eta^{-1}(\eta(a)) = a.$$

Thus, $\eta^{-1}\sigma\eta$ fixes K.

For the last part of the theorem, we define a map φ from $G = \mathrm{Gal}\left(E/F\right)$ to $\mathrm{Gal}\left(E^H/F\right)$ by sending η to $\eta|_{E^H}$ with both the domain and the codomain restricted to E^H. This map is clearly onto since any automorphism of E^H can be extended to E. It is also clear that $\eta \in \ker\varphi$ if and only if $\eta|_{E^H} = \mathbf{1}_E^H$ if and only if η fixes E^H if and only if $\eta \in \mathrm{Gal}\left(E/E^H\right) = H$. $\qquad\square$

Exercises 23.3

1. Let E and L_2 be as in Example 23.3.4. By the Galois pairing we know that $[E : L_2] = 3$. Find the minimal polynomial of ζ over L_2.

2. Let $E = \mathbb{Q}[\sqrt{2}, \sqrt{3}]$.

   (a) Find all the proper subfields of E.

   (b) Is $\alpha = 9\sqrt{2} - 17\sqrt{3}$ a primitive element for E over $\mathbb{Q}$? In other words, it is true that $E = \mathbb{Q}[\alpha]$?

3. Let E be the splitting field of $x^{12} - 1$ over $\mathbb{Q}$. Exhibit the correspondence between the subgroups of $G = \mathrm{Gal}\,(K/\mathbb{Q})$ and the subfields of E by constructing the subgroup lattice of G and the subfield lattice of E.

4. Construct the subfield lattice of $E = \mathbb{Q}(\sqrt[4]{2}, i)$. Find all subfields of E which are normal extensions of $\mathbb{Q}$.

5. Find all subfields of $\mathbb{Q}[\sqrt[4]{2}]$.

6. Let E be the splitting field of $p(x) = x^6 + x^3 + 1$ over $\mathbb{Q}$ and let G be $\mathrm{Gal}\,(E/\mathbb{Q})$. Set up the one-to-one correspondence between subfields of E and subgroups of G.

Review Exercises for Chapter 23

1. Is $\mathbb{Q}[\sqrt[3]{2}, i]$ a normal extension over $\mathbb{Q}$?

2. Show that $K = \mathbb{Q}(\sqrt{2}, \sqrt{3}, u)$ where $u^2 = (9 - 5\sqrt{3})(2 - \sqrt{2})$ is a normal extension of $\mathbb{Q}$. Determine $\mathrm{Gal}\,(K/\mathbb{Q})$.

3. Let E be the splitting field of $f(x) = x^n - 1$ over $\mathbb{Q}$. Show that $\mathrm{Gal}\,(E/\mathbb{Q})$ is a commutative group.

4. Let E be the splitting field of $(x^2 - 2)(x^3 - 3)$ over $\mathbb{Q}$. Let K and L be the splitting fields of $x^2 - 2$ and $x^3 - 3$ over $\mathbb{Q}$ respectively inside E.

   (a) Find $[E : \mathbb{Q}]$, $[E : K]$ and $[E : L]$.

   (b) Show that $\mathrm{Gal}\,(E/\mathbb{Q}) \cong S_3 \times \mathbb{Z}_2$.

   (c) Find all subfields of E.

   (d) Find a primitive element for E over $\mathbb{Q}$.

5. Let x be an indeterminate over $\mathbb{Q}$. For a, b, c, $d \in \mathbb{Q}$ with $ad - bc \neq 0$, show that the map

$$\varphi \colon x \longmapsto \frac{ax + b}{cx + d}$$

   is an automorphism of $\mathbb{Q}(x)$.

CHAPTER 24

Applications of
the Galois Pairing

In 1918, Emmy Noether (1882–1935) posed the *Inverse Galois problem*, asking if given a field F and a group G, whether it is possible to find a field extension E of F such that $G = \mathrm{Gal}\,(E/F)$. This problem remains unsolved until today. Noether later reduced this problem to *Noether's problem*, whether the fixed subfield of a subgroup of S_n acting on $F(x_1, \ldots, x_n)$ is a purely transcendental extension of F. We will look at a special case of this problem in this chapter.

We are familiar with the formula that the solutions to the quadratic equation $f(x) = ax^2 + bx + c \in \mathbb{R}[\,x\,]$, $a \neq 0$, are

$$\frac{-b \pm \sqrt{b^2 - 4ac}}{2a}.$$

Formulas for the zeros of cubic and quartic equations over reals have also existed for a long time. After years of futile search for the "radical formula" for zeros of an arbitrary quintic function, it was Niels Henrik Abel (1802–1829) who first proved that a quintic equation may not be "solvable by radicals". We will also present a proof in this chapter.

24.1   Fields of Invariants

Let F be a field. Consider the quotient field $F(x_1, \ldots, x_n)$ of the polynomial ring $F[x_1, \ldots, x_n]$ of n indeterminates over F. Let π be a permutation in S_n. Then there is an associated automorphism $\zeta(\pi)$ on $F[x_1, \ldots, x_n]$, which fixes elements of F and sends x_i to $x_{\pi(i)}$, $1 \leq i \leq n$. Moreover, this map can be extended in one and only one way to an automorphism on $E = F(x_1, \ldots, x_n)$, which, for the sake of simplicity, will be denoted by $\zeta(\pi)$ again. Let

$$(24.1.1) \qquad\qquad G = \{\, \zeta(\pi) : \pi \in S_n \,\}.$$

Then G is isomorphic to S_n (see Exercise 1). This is a group of automorphisms of the field $E = F(x_1, \ldots, x_n)$. The fixed elements of G are called **symmetric rational expressions**, and E^G is called the **field of symmetric rational expressions**.

We proceed to compute E^G using the Galois pairing. We introduce a new indeterminate t over E and consider the polynomial

$$g(t) = (t - x_1)(t - x_2) \cdots (t - x_n) = t^n - p_1 x t^{n-1} + p_2 t^{n-2} - + \cdots + (-1)^n p_n$$

where

$$(24.1.2) \quad p_1 = \sum_{1}^{n} x_i, \qquad p_2 = \sum_{i<j} x_i x_j, \qquad \ldots, \qquad p_n = x_1 x_2 \cdots x_n.$$

The p_i's are called the **elementary symmetric polynomials**.

The automorphism $\zeta(\pi)$ can be extended to an automorphism $\overline{\zeta}(\pi)$ of $E[t]$ fixing t. Clearly,

$$\overline{\zeta}(\pi)(g(t)) = (t - x_{\pi(1)})(t - x_{\pi(2)}) \cdots (t - x_{\pi(n)}) = g(t).$$

Thus $\overline{\zeta}(\pi)$, and hence $\zeta(\pi)$, fixes all the coefficients of $g(t)$. We have the following result.

Lemma 24.1.1. *For all $\pi \in S_n$, we have that $\zeta(\pi)(p_i) = p_i$ for all i. Thus, $p_i \in E^G$ for all i.*

From the previous lemma we see that $K = F(p_1, \ldots, p_n) \subseteq E^G$. Also note that E is the splitting field of $g(t)$ over K and that $g(t)$ is a separable polynomial over K. Hence, E is a finite normal separable extension of degree at most $n!$ over K by Proposition 22.1.6. We can use Galois pairing to determine E^G.

Lemma 24.1.2. *Let $x_1, \ldots, x_n$ be indeterminates over F and $p_1, \ldots, p_n$ be the elementary symmetric polynomials of the x_i's. Then*

$$\text{Gal}\,(F(x_1, \ldots, x_n)/F(p_1, \ldots, p_n)) \simeq S_n.$$

Proof. Let $E = F(x_1, \ldots, x_n)$ and $K = F(p_1, \ldots, p_n)$. Then from our previous discussion we know that $|\text{Gal}\,(E/K)| \leq n!$. However, we already have a group of automorphisms of order $n!$ on E. This forces that $\text{Gal}\,(E/K) \cong S_n$. $\qquad\square$

Theorem 24.1.3. *Let $x_1, \ldots, x_n$ be indeterminates over the field F. The field of symmetric rational expressions over F is $F(p_1, \ldots, p_n)$ where the p_i's are the elementary symmetric polynomials of the x_j's.*

Proof. This follows from the fact that $E = F(x_1, \ldots, x_n)$ is a finite normal separable extension of $K = F(p_1, \ldots, p_n)$. Hence by Lemma 24.1.2 and the Galois pairing we have that $K = E^{S_n}$. $\qquad\square$

We conclude that any symmetric rational expression in the indeterminates x_i can be expressed as a quotient of polynomials in terms of the p_i's. Thanks to the Galois pairing, we have a simple yet elegant result in *Invariant Theory*.

Next, we are going to use our knowledge on invariants to derive other interesting results.

Let E be a finite normal separable field extension over F and let $G = \text{Gal}\,(E/F)$. Define the **trace**, denoted $\text{Tr}_{E/F}$, and the **norm**, denoted by $\text{N}_{E/F}$, by

$$\text{Tr}_{E/F}(a) = \sum_{\tau \in G} \tau(a) \quad \text{and} \quad \text{N}_{E/F}(a) = \prod_{\tau \in G} \tau(a)$$

for $a \in E$. Since each τ is an F-linear transformation from E to itself, so is $\text{Tr}_{E/F}$. Note that for all $\sigma \in G$, we have that

$$\sigma(\text{Tr}_{E/F}(a)) = \sigma\left(\sum_{\tau \in G} \tau(a)\right) = \sum_{\tau \in G} \sigma(\tau(a)) = \text{Tr}_{E/F}(a).$$

Hence $\text{Tr}_{E/F}(a) \in E^G = F$. In fact, $\text{Tr}_{E/F}$ is an F-linear transformation from E to F. We leave it to the reader to check that $N_{E/F}$ is a group homomorphism from the multiplicative group E^* to F^* (see Exercise 2).

For the following results, let p be a positive prime integer and $q = p^n$ where n is a positive integer. In this case, $\text{Gal}\,(\mathbb{F}_q/\mathbb{F}_p) = \langle \sigma \rangle$ where σ is the Frobenius map (see Exercise 11, §22.5).

Proposition 24.1.4. *The trace map* $\text{Tr}_{\mathbb{F}_q/\mathbb{F}_p}$ *from* $\mathbb{F}_q$ *to* $\mathbb{F}_p$ *is onto with*

$$\ker \text{Tr}_{\mathbb{F}_q/\mathbb{F}_p} = \{\, a - \sigma(a) \in \mathbb{F}_q : a \in \mathbb{F}_q \,\}.$$

Proof. Note that

$$\text{Tr}_{\mathbb{F}_q/\mathbb{F}_p}(a) = \sum_{i=0}^{n-1} \sigma^i(a) = a + a^p + a^{p^2} + \cdots + a^{p^{n-1}}.$$

If $\text{Tr}_{\mathbb{F}_q/\mathbb{F}_p}$ is the trivial map, then the polynomial of degree p^{n-1}

$$f(x) = x + x^p + x^{p^2} + \cdots + x^{p^{n-1}}$$

would have p^n solutions in $\mathbb{F}_q$, a contradiction. Since $\text{Tr}_{\mathbb{F}_q/\mathbb{F}_p}$ is an F-linear transformation, this means that the image of $\text{Tr}_{\mathbb{F}_q/\mathbb{F}_p}$ is at least one-dimensional over $\mathbb{F}_p$. Hence $\text{Tr}_{\mathbb{F}_q/\mathbb{F}_p}$ is onto. This also implies that $\ker \text{Tr}_{\mathbb{F}_q/\mathbb{F}_p}$ is an $n - 1$-dimensional subspace of $\mathbb{F}_q$, and must contain exactly p^{n-1} elements. Clearly, we have that

$$\text{Tr}_{\mathbb{F}_q/\mathbb{F}_p}(\sigma(a)) = \sum_{i=0}^{n-1} \sigma^{i+1}(a) = \sum_{i=1}^{n} \sigma^i(a) = \text{Tr}_{\mathbb{F}_q/\mathbb{F}_p}(a).$$

The kernel contains $a - \sigma(a)$ for all $a \in F$. Let S be the set of all the $(a - \sigma(a))$'s. Observe that $a - \sigma(a) = b - \sigma(b)$ if and only if $a - b = \sigma(a) - \sigma(b) = \sigma(a - b)$. This means that $a - b$ is fixed by σ and thus fixed by $\langle \sigma \rangle$. We conclude that $a - \sigma(a) = b - \sigma(b)$ if and only if $a - b \in \mathbb{F}_p$. It follows that $|S| = |\mathbb{F}_q/\mathbb{F}_p| = q/p = p^{n-1}$. Hence $S = \ker \text{Tr}_{\mathbb{F}_q/\mathbb{F}_p}$. $\qquad\square$

Proposition 24.1.5. *The norm map* $N_{\mathbb{F}_q/\mathbb{F}_p}$ *from* $\mathbb{F}_q^*$ *to* $\mathbb{F}_p^*$ *is onto with*

$$\ker N_{\mathbb{F}_q/\mathbb{F}_p} = \{\, a/\sigma(a) \in \mathbb{F}_q^* : a \in \mathbb{F}_q^* \,\}.$$

Proof. Note that

$$N_{\mathbb{F}_q/\mathbb{F}_p}(a) = \sum_{i=0}^{n-1} \sigma^i(a) = a a^p a^{p^2} \cdots a^{p^{n-1}} = a^{(p^n-1)/(p-1)}.$$

Each of the $p^n - 1$ elements in $\mathbb{F}_q^*$ is a root of

$$g_b(x) = x^{(p^n-1)/(p-1)} - b = 0$$

for some $b \in \mathbb{F}_p^*$. There are $p - 1$ such equations, each of which has at most $(p^n - 1)/(p - 1)$ roots. This forces that each equation is with exactly $(p^n - 1)/(p - 1)$ roots. Hence $N_{\mathbb{F}_q/\mathbb{F}_p}$ is onto and

$$|\ker N_{\mathbb{F}_q/\mathbb{F}_p}| = \frac{p^n - 1}{p - 1}$$

since the kernel consists of roots of $g_1(x) = 0$. Since

$$N_{\mathbb{F}_q/\mathbb{F}_p}(\sigma(a)) = \prod_{i=0}^{n-1} \sigma^{i+1}(a) = \prod_{i=1}^{n} \sigma^i(a) = N_{\mathbb{F}_q/\mathbb{F}_p}(a),$$

we have that $a/\sigma(a)$ belongs in the kernel of $N_{\mathbb{F}_q/\mathbb{F}_p}$ for all $a \in \mathbb{F}_q^*$. Let T be the collection of all the $(a/\sigma(a))$'s. Note that for a and $b \neq 0$, $a/\sigma(a) = b/\sigma(b)$ if and only if $a/b = \sigma(a/b)$ if and only if $a/b \in \mathbb{F}_q^{\langle \sigma \rangle} = \mathbb{F}_p$. Hence

$$|T| = |\mathbb{F}_q^*/\mathbb{F}_p^*| = \frac{p^n - 1}{p - 1}.$$

It implies that $T = \ker N_{\mathbb{F}_q/\mathbb{F}_p}$. $\qquad\qquad\qquad\qquad\qquad\qquad$ □

This theorem is a special case of Hilbert's Theorem 90 and it can be proved by using the first cohomology of the Galois group $\mathrm{Gal}\,(\mathbb{F}_q/\mathbb{F}_p)$.

Exercises 24.1

1. Let $G = \{\zeta(\pi) : \pi \in S_n\}$ be as in (24.1.1). Show that $\varphi : S_n \to G$ sending $\pi \mapsto \zeta(\pi)$ is an isomorphism.

2. Let E be a finite normal separable extension over F. Show that $N_{E/F}$ is a group homomorphism from the multiplicative group E^* to F^*.

3. Compute $N_{\mathbb{F}_9/\mathbb{F}_3}(a)$ for all $a \in \mathbb{F}_9^*$.

4. Compute $\mathrm{Tr}_{\mathbb{F}_9/\mathbb{F}_3}(a)$ for all $a \in \mathbb{F}_9$.

5. Let $q_i = \sum_{k=1}^n x_k^i \in K[x_1, \ldots, x_n]$ for $1 \le i \le n$.

   (a) Show that $q_1 = p_1$, $q_2 = p_1^2 - 2p_2$ for $n \ge 2$, $q_3 = p_1^3 - 3p_1p_2 + 3p_3$ for $n \ge 3$ and $q_4 = p_1^4 - 4p_1^2p_2 + p_2^2 + 6p_1p_3 - 6p_4$ for $n \ge 4$.

   (b) Let $\mathrm{Char}\, K = 0$. Show that $p_1 \in K[q_1]$, $p_2 \in K[q_1, q_2]$ for $n \ge 2$, $p_3 \in K[q_1, q_2, q_3]$ for $n \ge 3$ and $p_4 \in K[q_1, q_2, q_3, q_4]$ for $n \ge 4$.

24.2   Solvable Groups

We next proceed to demonstrate the insolvability of the quintic (polynomial equations of degree 5). However, we need to know a little more about groups before we can go on.

Definition 24.2.1. A **subnormal series** of a group G is a finite sequence of subgroups $H_0, H_1, \ldots, H_n$ in G such that $H_n = G$ and $H_i \lhd H_{i+1}$ for all i. We usually use

$$H_0 \lhd H_1 \lhd \cdots \lhd H_{n-1} \lhd H_n = G$$

to denote a subnormal series of **length** n. If in addition, in a subnormal series, all the subgroups are normal in G, we call it a **normal series**.

A subnormal (normal) series $(K_j)_j$ is called a **refinement** of a subnormal (normal) series $(H_i)_i$ of a group G if $\{H_i\}_i \subseteq \{K_j\}_j$.

A normal series is always a subnormal series, but not vice versa. However, every subnormal series of an abelian group is a normal series.

Example 24.2.2. Both series

$$\{0\} \lhd 12\mathbb{Z} \lhd 6\mathbb{Z} \lhd \mathbb{Z} \qquad \text{and} \qquad \{0\} \lhd 9\mathbb{Z} \lhd \mathbb{Z}$$

are normal series of $(\mathbb{Z}, +)$. The series

$$0\mathbb{Z} \lhd 72\mathbb{Z} \lhd 36\mathbb{Z} \lhd 9\mathbb{Z} \lhd 3\mathbb{Z} \lhd \mathbb{Z}$$

is a refinement of $\{0\} \lhd 9\mathbb{Z} \lhd \mathbb{Z}$.

Example 24.2.3. Let $\rho = (1\ 2\ 3\ 4)$, $\gamma_1 = (1\ 3)$ and $\gamma_2 = (2\ 4)$. The series

$$\{1\} \lhd \{1,\ \gamma_1\} \lhd \{1,\ \rho^2,\ \gamma_1,\ \gamma_2\} \lhd D_4$$

is subnormal but not normal. This series has no further refinements.

A subnormal series of a group G is called a **composition series** if it has no redundant components and no further meaningful refinements.

Lemma 24.2.4. *A subnormal series*

$$H_0 \lhd H_1 \lhd \cdots \lhd H_{n-1} \lhd H_n = G$$

is a composition series if and only if all the factor groups H_{i+1}/H_i are nontrivial and simple.

Proof. It is possible to insert a normal subgroup H of H_{i+1} with $H_i \subsetneq H \subsetneq H_{i+1}$ if and only if H/H_i is a proper nontrivial normal subgroup of H_{i+1}/H_i. And this is true if and only if H_{i+1}/H_i is not simple. □

Note that composition series may not exist for some groups. For example, the additive group $\mathbb{Z}$ has no composition series. However, any finite group has a composition series. We start with finding a maximal proper normal subgroup N in G. If it is the trivial subgroup then we are done. Otherwise, we repeat the process with N. Since G is finite, this process must stop eventually.

Definition 24.2.5. Two subnormal series (H_i) and (K_j) are called **isomorphic** if the two series are of the same length, say n, and there is a $\sigma \in S_n$ such that $H_i/H_{i-1} \simeq K_{\sigma(i)}/K_{\sigma(i)-1}$ for $i = 1, \ldots, n$.

Example 24.2.6. The two series

$$0\mathbb{Z} \lhd 6\mathbb{Z} \lhd 2\mathbb{Z} \lhd \mathbb{Z} \qquad \text{and} \qquad 0\mathbb{Z} \lhd 6\mathbb{Z} \lhd 3\mathbb{Z} \lhd \mathbb{Z}$$

are isomorphic. Both series are of length 3. The factor groups of the first series are $6\mathbb{Z}, \mathbb{Z}_3, \mathbb{Z}_2$, while the factor groups of the second series are $6\mathbb{Z}, \mathbb{Z}_2, \mathbb{Z}_3$.

Theorem 24.2.7 (Schreier). *Any two subnormal series of a group G have isomorphic refinements.*

Proof. Let G be a group and let

$$(24.2.1) \qquad\qquad H_0 \lhd H_1 \lhd \cdots \lhd H_{n-1} \lhd H_n = G$$

and

$$(24.2.2) \qquad\qquad K_0 \lhd K_1 \lhd \cdots \lhd K_{n-1} \lhd K_m = G$$

be two subnormal series of G. Let

$$H_{ij} = H_i(H_{i+1} \cap K_j) \qquad \text{and} \qquad K_{ji} = K_j(K_{j+1} \cap H_i).$$

The subgroup H_{ij} contains both H_i and $H_{i+1} \cap K_j$. Hence, we have obtained a new series

$$H_{i0} \subseteq H_{i1} \subseteq \cdots H_{i2} \subseteq \cdots \subseteq H_{im}.$$

Note that $H_{i,0} = H_i(H_{i+1} \cap K_0) = H_i$ while $H_{im} = H_i(H_{i+1} \cap G) = H_{i+1}$. Thus, we obtain a series

$$\cdots \subseteq H_{i-1,m-1} \subseteq H_{i-1,m}$$
$$= H_{i0} \subseteq H_{i1} \subseteq \cdots H_{i2} \subseteq \cdots \subseteq H_{im}$$
$$= H_{i+1,0} \subseteq H_{i+1,1} \subseteq \cdots$$

which is a refinement of (24.2.1). Similarly, we have the series

$$\cdots \subseteq K_{n-1,j-1} \subseteq K_{n,j-1}$$
$$= K_{0j} \subseteq K_{1j} \subseteq \cdots K_{2j} \subseteq \cdots \subseteq K_{nj}$$
$$= K_{0,j+1} \subseteq K_{1,j+1} \subseteq \cdots$$

which is a refinement of (24.2.2). These two refinements are of the same length. It follows from the following Zassenhaus Lemma that the following properties hold for all i and j:

  (i) $H_{ij} \lhd H_{i,j+1}$ and $K_{ij} \lhd K_{i+1,j}$;

  (ii) $\dfrac{H_{i,j+1}}{H_{ij}} \simeq \dfrac{K_{i+1,j}}{K_{ij}}$.

Thus, the two refinements are isomorphic. $\qquad\qquad\qquad\qquad\qquad\qquad$ $\square$

Lemma 24.2.8 (Zassenhaus Lemma). *Let* $H \lhd H'$ *and* $K \lhd K'$ *be subgroups of the group* G. *Then the following relations hold:*

(a) $H' \cap K \lhd H' \cap K'$ *and* $H \cap K' \lhd H' \cap K'$;

(b) $H(H' \cap K) \lhd H(H' \cap K')$ *and* $K(H \cap K') \lhd K(H' \cap K')$;

(c) $\dfrac{H(H' \cap K')}{H(H' \cap K)} \simeq \dfrac{H' \cap K'}{(H \cap K')(H' \cap K)} \simeq \dfrac{K(H' \cap K')}{K(H \cap K')}$.

Proof. To prove (a), let $a \in H' \cap K$ and $g \in H' \cap K'$. Then $gag^{-1} \in H'$ and $gag^{-1} \in K$ since $K \lhd K'$. Hence $H' \cap K \lhd H' \cap K'$. Similarly, $H \cap K' \lhd H' \cap K'$.

We prove (b) and (c) together. Since $H \lhd H'$ and $H' \cap K$, $H' \cap K'$ are subgroups of H', $H(H' \cap K)$ and $H(H' \cap K')$ are both subgroups of H' by the second isomorphism theorem. By (a), $L = (H \cap K')(H' \cap K)$ is a normal subgroup of $H' \cap K'$ (see Exercise 4(c), §8.1). We now define a map φ from $H(H' \cap K')$ to $(H' \cap K')/L$. Let $h \in H$ and $x \in H' \cap K'$. Define $\varphi(hx) = xL$. We check the following properties.

(1) *The map φ is well-defined.* Suppose $h_1, h_2 \in H$ and $x_1, x_2 \in H' \cap K'$ are such that $h_1 x_1 = h_2 x_2$. Then $x_1 x_2^{-1} = h_2 h_1^{-1}$ lies in $H \cap H' \cap K' = H \cap K' \subseteq L$. Hence $x_1 L = x_2 L$.

(2) *The map φ is a group homomorphism.* Let $h_1, h_2 \in H$ and x_1, $x_2 \in H' \cap K'$. Since $H \lhd H'$, $x_1 h_2 x_1^{-1} = h_3 \in H$. Thus

$$\varphi((h_1 x_1)(h_2 x_2)) = \varphi((h_1 h_3)(x_2 x_2)) = x_1 x_2 L$$
$$= (x_1 L)(x_2 L) = \varphi(h_1 x_1)\varphi(h_2 x_2).$$

(3) *The group homomorphism φ is onto.* Since $H' \cap K' \subseteq H(H' \cap K')$, $\varphi(x) = xL$ for all $x \in H' \cap K'$.

(4) *The kernel of φ is $H(H' \cap K)$.* Let $h \in H$ and $x \in H' \cap K'$. Then $hk \in \ker \varphi$ if and only if $xL = L$ if and only if $hx \in HL = H(H \cap K')(H' \cap K) = H(H' \cap K)$.

It now follows from the first isomorphism theorem that

$$H(H' \cap K) \lhd H(H' \cap K') \quad \text{and} \quad \frac{H(H' \cap K')}{H(H' \cap K)} \simeq \frac{H' \cap K'}{(H \cap K')(H' \cap K)}.$$

The proof of the other half of (b) and (c) is similar. $\square$

Zassenhaus Lemma is also called the *Butterfly Lemma*. It derives this name because the subgroup lattice of the groups involved in this lemma can be shaped like a butterfly. The interested reader can try drawing his or her own version of butterfly.

Using Theorem 24.2.7, we have the following immediate result.

Corollary 24.2.9 (Jordan-Hölder Theorem). *Any two composition series of a group (if they exist) are isomorphic.*

Example 24.2.10. Both

$$0\mathbb{Z}_{12} \lhd 6\mathbb{Z}_{12} \lhd 3\mathbb{Z}_{12} \lhd \mathbb{Z}_{12} \quad \text{and} \quad 0\mathbb{Z}_{12} \lhd 4\mathbb{Z}_{12} \lhd 2\mathbb{Z}_{12} \lhd \mathbb{Z}_{12}$$

are composition series of $\mathbb{Z}_{12}$. Even though these two series are different, they are isomorphic. Both are of the same length and both have factor groups $\mathbb{Z}_3$, $\mathbb{Z}_2$ and $\mathbb{Z}_2$.

A group G is called a **solvable** group if it has a composition series such that all of its factor groups are cyclic. Thanks to Jordan-Hölder Theorem, to test if a group is solvable, we only need to check one of its composition series.

Obviously finite cyclic groups are solvable since the factor groups in any of its subnormal series are homomorphic images of subgroups of cyclic groups and thus are cyclic as well (see Proposition 4.2.1 and Exercise 4, §8.2).

Next we provide more criteria for checking whether a group is solvable or not.

Lemma 24.2.11. *Let $H \lhd G$ be groups. Then G is solvable if and only if H and G/H are solvable.*

Proof. The subnormal series $\{e\} \lhd H \lhd G$ of G can be refined to a composition series

$$\{e\} \lhd H_1 \lhd \cdots \lhd H_m = H \lhd K_1 \lhd \cdots \lhd K_n = G.$$

This series induces a composition series

$$\{e\} \lhd H_1 \lhd \cdots \lhd H_m = H$$

for H and a composition series

$$\frac{H}{H} \lhd \frac{K_1}{H} \lhd \cdots \lhd \frac{K_n}{H} = \frac{G}{H}$$

for G/H. The factor groups of H and the factor groups of G/H together make up the factor groups of G. Hence the result. $\qquad \square$

Proposition 24.2.12. *If a group G has a subnormal series whose factor groups are solvable, then G is solvable.*

Proof. Suppose given a subnormal series

$$H_0 \lhd H_1 \lhd \cdots \lhd H_{n-1} \lhd H_n = G$$

which satisfies the stated requirement. Since H_0 and H_1/H_0 are solvable, we have that H_1 is solvable. We have a shorter subnormal series

$$H_1 \lhd H_2 \lhd \cdots \lhd H_{n-1} \lhd H_n = G$$

which satisfies the stated requirement. Now the result follows from induction on the length of the subnormal series. $\qquad \square$

Corollary 24.2.13. *Any finite abelian group is solvable.*

Proof. Clearly cyclic groups are solvable. Let G be a finite abelian group. By the Structure Theorem of Finite Abelian Groups, $G \cong G_1 \times \cdots \times G_n$ where G_i is cyclic for each i. Let

$$H_i = G_1 \times \cdots \times G_i \times \{e\} \times \cdots \times \{e\}$$

for $1 \leq i \leq n$. Then

$$\{1\} \lhd H_1 \lhd H_2 \lhd \cdots \lhd H_{n-1} \lhd H_n = G$$

is a subnormal series of G. Note that the factor groups are $H_i/H_{i-1} \cong G_i$ is cyclic and hence solvable for $1 \leq i \leq n$. Thus G is solvable by Proposition 24.2.12. $\qquad \square$

Example 24.2.14. In Example 24.2.3 we can see that the group D_4 is solvable. In fact, any group of order p^n, p a prime, is solvable. By Sylow's first theorem, there exists a subnormal series

$$\{e\} \lhd H_1 \lhd H_2 \lhd \cdots \lhd H_{n-1} \lhd H_n = G$$

in which $|H_i| = p^i$. Thus the factor groups are all isomorphic to $\mathbb{Z}_p$.

It is easy to see that S_n is solvable for $1 \leq n \leq 4$. Note that

$$\{1\} \lhd \langle (12)(34) \rangle \lhd \langle (1\ 2), (3\ 4) \rangle \lhd A_4 \lhd S_4$$

is a composition series for S_4 with cyclic factor groups. Hence S_4 is solvable.

Proposition 24.2.15. *The group A_5 is simple and the group S_5 is not solvable.*

Proof. A normal subgroup of A_5 must be the union of $\{1\}$ and some other conjugacy classes. The class equation of A_5 is

$$60 = 1 + 12 + 12 + 20 + 15$$

(*Cf.* Exercise 7, §6.2 and Exercise 5, §10.2). No combination of 1 and 12, 12, 20, 15 can make up a divisor of 60 other than 1 and 60. Thus, A_5 is simple. This implies that

$$\{1\} \rhd A_5 \lhd S_5$$

is a composition series for S_5. We conclude that S_5 is not solvable. $\square$

In general, the group S_n is not solvable for $n \geq 5$ since A_n is simple for $n \geq 5$ (we will not prove this fact in this book). The composition series

$$\{1\} \lhd A_n \lhd S_n$$

of S_n has a non-cyclic factor group A_n for $n \geq 5$.

Exercises 24.2

1. Determine if D_{10} is solvable by finding a composition series of D_{10}.

2. Find all composition series of $\mathbb{Z}_2 \times \mathbb{Z}_2$.

3. Find all composition series of $S_3 \times \mathbb{Z}_2$.

4. Show that an abelian group has a composition series if and only if it is finite.

5. If G is a group of order pq where p and q are distinct prime integers, show that G is solvable.

6. Show that the direct product of a finite number of solvable groups is solvable.

24.3 Insolvability of the Quintic

We all are familiar with the extremely useful formula for the solutions of the general quadratic polynomial equation. Let

$$f(x) = x^2 + ax + b = 0$$

where a and b are variables over $\mathbb{Q}$. Then

$$x = \frac{-a \pm \sqrt{a^2 - 4b}}{2}.$$

In this section, our goal is to show that it is impossible to find a formula for the solutions of the general quintic (degree 5) polynomial equation

$$(24.3.1) \qquad f(x) = x^5 + a_1 x^4 + a_2 x^3 + a_3 x^2 + a_4 x + a_5 \in \mathbb{Q}[x],$$

where $a_1, \ldots, a_5$ are indeterminates over $\mathbb{Q}$, so that the solutions are obtained by taking a finite sequence of sums, differences, product, quotients and kth roots (radicals) of rational numbers and $a_1, \ldots, a_5$. We make a formal definition as follows.

Definition 24.3.1. A field extension E/F is a **field extension by radicals** if there are $\alpha_1, \ldots, \alpha_r \in E$ and positive integers $n_1, \ldots, n_r$ such that

$$E = F[\alpha_1, \ldots, \alpha_r],$$

where $\alpha_1^{n_1} \in F$ and $\alpha_i^{n_i} \in F[\alpha_1, \ldots, \alpha_{i-1}]$ for $2 \leq i \leq r$. The finite chain of fields

$$F \subseteq F[\alpha_1] \subseteq F[\alpha_1, \alpha_2] \subseteq \cdots \subseteq F[\alpha_1, \ldots, \alpha_r]$$

is called a **root tower** over F.

A polynomial $f(x) \in F[x]$ is **solvable by radicals over** F if the splitting field of $f(x)$ over F is contained in a field extension by radicals over F. If we simply say that $f(x)$ is **solvable by radicals**, we mean that $f(x)$ is solvable by radicals over the field generated by the coefficients of $f(x)$ over $\mathbb{Q}$.

Thus, to find a formula for the solutions of (24.3.1) in the fashion described is to show that $f(x)$ is solvable by radicals over $F = \mathbb{Q}(a_1, \ldots, a_5)$. However, to say that the quintic is not solvable in general is not to say that no quintic is solvable, as the following example shows.

Example 24.3.2. The polynomial $f(x) = x^5 - 1$ is solvable by radicals over $\mathbb{Q}$. The splitting field of $f(x)$ over $\mathbb{Q}$ is $\mathbb{Q}[\zeta]$, where $\zeta = e^{2\pi i/5}$. Clearly, $\mathbb{Q}[\zeta]$ is an extension by radicals over $\mathbb{Q}$ since $\zeta^5 \in \mathbb{Q}$.

The polynomial $g(x) = x^5 - 2$ is solvable by radicals. The splitting field of $g(x)$ over $\mathbb{Q}$ is $L = \mathbb{Q}[\sqrt[5]{2}, \zeta]$. Since $(\sqrt[5]{2})^5$ and $\zeta^5 \in \mathbb{Q}$, L is an extension by radicals over $\mathbb{Q}$.

Let $f(x) \in F[x]$ and let E be the splitting field of $f(x)$ over F. We also call $\operatorname{Gal}(E/F)$ the **Galois group** of $f(x)$ over F.

Lemma 24.3.3. *Let F be a field of characteristic 0 and n be a positive integer. The Galois group of $x^n - 1$ over F is a finite abelian group and hence is solvable.*

Proof. Remember that $\mathbb{Q}[\zeta]$, where $\zeta = e^{2\pi i/n}$, is the splitting field of $x^n - 1$ over $\mathbb{Q}$. Let E be the splitting field of $x^n - 1$ over F. Since $\operatorname{Char} F = 0$, $\mathbb{Q}$ is the prime field of F. Thus, E contains a copy of splitting field of $x^n - 1$ over $\mathbb{Q}$. Hence, by Theorem 22.2.7, there is a monomorphism from $\mathbb{Q}[\zeta]$ into E. We will still use ζ to denote the image of ζ in E. Then $E = F[\zeta]$.

Any automorphism in $\operatorname{Gal}(E/F)$ is completely determined by its image of ζ. Let $\eta \in \operatorname{Gal}(E/F)$. Then $\eta(\zeta)$ is a conjugate of ζ over F. Hence $\eta(\zeta) = \zeta^i$ for some i. Let σ be another automorphism in $\operatorname{Gal}(E/F)$ with $\sigma(\zeta) = \zeta^j$. Then

$$(\eta\sigma)(\zeta) = \eta(\sigma(\zeta)) = (\zeta^j)^i = (\zeta^i)^j = \sigma(\eta(\zeta)) = (\sigma\eta)(\zeta).$$

This implies that $\eta\sigma = \sigma\eta$. Thus, $\operatorname{Gal}(E/F)$ is abelian. Remember that $[E : F] \leq n!$. Thus, $\operatorname{Gal}(E/F)$ is a finite group. It follows from Corollary 24.2.13 that $\operatorname{Gal}(E/F)$ is solvable. $\square$

Lemma 24.3.4. *Let F be a field of characteristic 0, n a positive integer and $a \in F$. The Galois group of $x^n - a$ over F is a solvable group.*

Proof. Let E be the splitting field of $x^n - a$ over F. Enlarge E to L to include $\zeta = e^{2\pi i/n}$ as in the proof of Lemma 24.3.3. Let α be any zero of $x^n - a$. Then

$$x^n - a = (x - \alpha)(x - \alpha\zeta) \cdots (x - \alpha\zeta^{n-1})$$

in $E[\zeta]$. This implies that $\alpha\zeta^i \in E$ for $0 \le i \le n - 1$. It follows that $\zeta = \alpha\zeta/\zeta$ is in E. Let $K = F[\zeta]$. Then K is the splitting field of $x^n - 1$ over F and K is a normal extension over F. Let $G = \mathrm{Gal}\,(E/F)$ and $H = \mathrm{Gal}\,(E/K)$. Then by Theorem 23.3.1, H is normal in G and $\mathrm{Gal}\,(K/F) \cong G/H$. By Lemma 24.2.11, it suffices to show that H and $\mathrm{Gal}\,(K/F)$ are solvable. The fact that $\mathrm{Gal}\,(K/F)$ is solvable follows from Lemma 24.3.3. It remains to show that $\mathrm{Gal}\,(E/K)$ is solvable.

Observe that $E = K[\alpha]$. Let η and $\sigma \in \mathrm{Gal}\,(E/K)$. Then $\eta(\alpha) = \alpha\zeta^i$ for some i and $\sigma(\alpha) = \alpha\zeta^j$ for some j. It then follows that

$$(\eta\sigma)(\alpha) = \eta(\alpha\zeta^j) = \alpha\zeta^i\zeta^j = \alpha\zeta^j\zeta^i = \sigma(\alpha\zeta^i) = (\sigma\eta)(\alpha).$$

Thus, $\eta\sigma = \sigma\eta$ and $\mathrm{Gal}\,(E/K)$ is finite abelian and hence solvable. $\square$

Lemma 24.3.5. *Let F be a field of characteristic 0. Let $n_1, \ldots, n_r$ be positive integers. Let $F \subset K_1 \subset K_2 \subset \cdots \subseteq K_r$ be a sequence of field extensions such that K_1 is a splitting field of $x^{n_1} - a_1$ for some $a_1 \in F$, K_2 is a splitting field of $x^{n_2} - a_2$ for some $a_2 \in K_1$, and so on. Then $\mathrm{Gal}\,(K_r/F)$ is solvable.*

Proof. Let $K_0 = F$. For each i, K_i is the splitting field of $x^{n_i} - a_i$ over K_{i-1}. Thus, K_i is a finite separable normal extension over K_{i-1}. By Theorem 23.3.1, $\mathrm{Gal}\,(K_r/K_i) \lhd \mathrm{Gal}\,(K_r/K_{i-1})$ for $0 \le i \le r - 1$. We obtain a subnormal series

$$\{1\} = \mathrm{Gal}\,(K_r/K_r) \lhd \mathrm{Gal}\,(K_r/K_{r-1}) \lhd \cdots \lhd \mathrm{Gal}\,(K_r/K_1) \lhd \mathrm{Gal}\,(K_r/F).$$

of $\mathrm{Gal}\,(K_r/F)$. The factor groups are

$$\frac{\mathrm{Gal}\,(K_r/K_{i-1})}{\mathrm{Gal}\,(K_r/K_i)} \cong \mathrm{Gal}\,(K_i/K_{i-1})$$

for $0 \leq i \leq r - 1$. They are all solvable by Lemma 24.3.4. It follows that $\mathrm{Gal}\,(K_r/F)$ is solvable by Proposition 24.2.12. $\qquad\square$

Theorem 24.3.6. *Let F be a field of characteristic 0. If $f(x) \in F[x]$ is solvable by radicals over F, then its Galois group over F is also solvable.*

Proof. Let E be the splitting field of $f(x)$ over F. Then E is contained in an extension K by radicals over F. We assume that

$$K = F[\alpha_1, \ldots, \alpha_r],$$

where $\alpha_1^{n_1} \in F$ and $\alpha_i^{n_i} \in F[\alpha_1, \ldots, \alpha_{i-1}]$ for $1 \leq i \leq r$. Let $n = n_1 n_2 \cdots n_{r-1}$. We may enlarge K to contain $\zeta = e^{2\pi i/n}$. Let $K_1 = F[\zeta]$ and $K_{i+1} = F[\zeta, \alpha_1, \ldots, \alpha_i]$ for $1 \leq i \leq r$. Then K_1 is the splitting field of $x^n - 1$ over F and K_{i+1} is the splitting field of $x^{n_i} - a_i$ over K_i for $1 \leq i \leq r$. By Lemma 24.3.5, $\mathrm{Gal}\,(K_{r+1}/F)$ is solvable. Since E is a finite normal separable extension over F, $\mathrm{Gal}\,(E/F) \cong \mathrm{Gal}\,(K_{r+1}/F)\,/\,\mathrm{Gal}\,(K_{r+1}/E)$ is solvable by Lemma 24.2.11. $\qquad\square$

Theorem 24.3.7. *Let $a_1, \ldots, a_n$ be indeterminates over $\mathbb{Q}$. The general quintic polynomial $f(x)$ in (24.3.1) is not solvable by radicals.*

Proof. If $f(x)$ is solvable by radicals, then it would give us a formula which works for all quintic polynomials. To show this is impossible, it suffices to find one counterexample.

Let $y_1, \ldots, y_5$ be indeterminates over $\mathbb{Q}$ and consider the quintic polynomial

$$g(x) = (x - y_1)(x - y_2)(x - y_3)(x - y_4)(x - y_5)$$
$$= x^5 - p_1 x^4 + p_2 x^3 - p_3 x^2 + p_4 x - p_5,$$

where the p_i's are elementary symmetric polynomials of $y_1, \ldots, y_5$ (see (24.1.2)). Let $F = \mathbb{Q}(p_1, \ldots, p_5)$ and $E = \mathbb{Q}(y_1, \ldots, y_5)$. Then E is the splitting field of $g(x)$ over F. By Lemma 24.1.2, $\mathrm{Gal}\,(E/F) \cong S_5$. Thus, by Proposition 24.2.15 and Theorem 24.3.6, $g(x)$ is not solvable by radicals. $\qquad\square$

If the proof of this theorem still does not convince you, we provide an example of a quintic polynomial over $\mathbb{Q}$ which is not solvable in Exercise 2.

Exercises 24.3

1. Let F be a field of characteristic $\neq 2$. Let a and b be indeterminates over F. Show that $x^2 + ax + b$ is solvable by radicals over $F(a, b)$.

2. This exercise gives a polynomial of degree 5 in $\mathbb{Q}[x]$ which is not solvable by radicals.

   (a) Show that the subgroup of S_5 which contains a 5-cycle and a transposition is S_5 itself.

   (b) Show that if $f(x)$ is an irreducible polynomial in $\mathbb{Q}[x]$ of degree 5 having two complex and three real zeros, then the Galois group of $f(x)$ over $\mathbb{Q}$ is S_5.

   (c) Show the polynomial $f(x) = 2x^5 - 5x^4 + 5$ is irreducible over $\mathbb{Q}$. Use Calculus to show that $f(x)$ has exactly 3 real zeros. Conclude from part (b) that $f(x)$ is not solvable by radicals.

3. Is the polynomial $3x^5 - 15x + 5$ solvable by radicals over $\mathbb{Q}$?

The following problems demonstrate how to solve polynomial equations of degree 3 in the characteristic 0 case.

4. Let F be an algebraically closed field of characteristic 0, and let

$$f(x) = x^n + ax^{n-1} + bx^{n-2} + \text{ lower degree terms}$$

be a monic polynomial of degree $n \geq 1$ over F.

   (a) If $a \neq 0$, we may make the transformation $y = x + (a/n)$ so that $f(x) = g(y)$ where the coefficient of degree $n - 1$ is 0. Thus we conclude any quadratic polynomial equation is solvable by radicals.

   (b) Let $x_1, \ldots, x_n$ be the roots of $f(x)$ (counted with multiplicity). Show that $a = b = 0$ if and only if

$$\begin{cases} x_1 + \cdots + x_n = 0, \\ x_1^2 + \cdots + x_n^2 = 0. \end{cases}$$

(c) (Tschinhaus transformations) Let $n \geq 3$. Assume $a = 0$ but $b \neq 0$. Suppose $x_1, \ldots, x_n$ are the n roots (counted with multiplicity) of $f(x) = 0$. Let

$$y_i = x_i^2 + rx_i + s, \qquad \text{for } i = 1, \ldots, n.$$

Let $g(y) = \prod_{i=1}^{n}(y - y_i)$. Show that it is possible to find r, s in F so that the coefficients of $g(y)$ in degree $n - 1$ and $n - 2$ are both 0. Conclude that any cubic polynomial equation is solvable by radicals.

Review Exercises for Chapter 24

1. Let F be a field of characteristic $\neq 2$ and let $E = F(x_1, \ldots, x_n)$ the rational field of n indeterminate over F. Let G be as in (24.1.1) and let H be the subgroup $G = \{\zeta(\pi) \in G : \pi \in A_n\}$. Show that $E^H = K(p_1, \ldots, p_n, \Delta)$ where $\Delta = \Pi_{i<j}(x_i - x_j)$.

2. Let G be a group of order pqr where p, q and r are positive prime integers with $p > q > r$. Show that G is solvable.

3. Let E/F be a field extension with Char $F \neq 2$ and let $f(x) = ax^6 + bx^3 + c$ where $a, b, c \in E$ and $a \neq 0$. Show that $f(x)$ is solvable by radicals over $F(a, b, c)$.

4. Construct a quintic polynomial of your own which is not solvable by radicals over $\mathbb{Q}$.

Index

Printed in the United States
By Bookmasters